METHANE
AND ITS
DERIVATIVES

CHEMICAL INDUSTRIES

A Series of Reference Books and Textbooks

Consulting Editor

HEINZ HEINEMANN

ADDITIONAL VOLUMES IN PREPARATION

METHANE AND ITS DERIVATIVES

Sunggyu Lee

The University of Akron
Akron, Ohio

CRC Press
Taylor & Francis Group
Boca Raton London New York

CRC Press is an imprint of the
Taylor & Francis Group, an **informa** business

CRC Press
Taylor & Francis Group
6000 Broken Sound Parkway NW, Suite 300
Boca Raton, FL 33487-2742

First issued in paperback 2019

© 1997 by Taylor & Francis Group, LLC
CRC Press is an imprint of Taylor & Francis Group, an Informa business

No claim to original U.S. Government works

ISBN-13: 978-0-8247-9754-6 (hbk)
ISBN-13: 978-0-367-40118-4 (pbk)

**Visit the Taylor & Francis Web site at
http://www.taylorandfrancis.com**

**and the CRC Press Web site at
http://www.crcpress.com**

Preface

In recent years there has been a considerable increase in research and development in the area of environmentally acceptable alternative fuels as well as the use of alternative feedstocks as building blocks for applied chemicals and petrochemicals. Conventional process industries are also faced with stricter emission control requirements. This trend is going to intensify for the next several decades because of the dwindling supply of petroleum reserves, emphasis on environmentally safe chemical technologies, fierce competitiveness in the global petrochemical market, and new discoveries of high-reserve gas wells and gas hydrates. Among all fossil fuels, natural gas generates the least carbon dioxide, a major greenhouse chemical, upon combustion. The utilization of natural gas could therefore help solve the greenhouse-effect problems, even though methane itself is also considered a greenhouse gas.

More and more books, monographs, and research articles in this area are currently being published, and most of them are narrowly focused, often emphasizing fine details (e.g., those of kinetic mechanisms, bench-scale experimental results, analytical procedures, and process hardware). Furthermore, many of the books are edited, thus lacking consistency and completeness. Researchers, students, professors, and field engineers need a comprehensive textbook covering the entire topic and presenting an overview of the subject, from the viewpoints of process chemistry and engineering. This single-authored book is designed to fill this need.

This book aims at presenting all the scientific and technological information regarding methane and its derivatives, and at providing an overview of the petrochemical industries that are (or could be) based on methane or natural gas. Special attention has been given to the methane derivatives and their commercial exploitation. Therefore, the coverage follows the tree structure of chemicals, not industries. Some distinctions have been made among the key building blocks, other building blocks, first-level derivatives, secondary derivatives, etc. Process chemistry, synthesis technology, historical developments, material properties, and market trends for each major chemical are discussed in detail and are compared with other alternatives. Because of their growing significance and close relationship with energy conversion, environmental issues related to process technology are addressed whenever applicable. Essential aspects of thermodynamics and process data as well as recent advances in process technology are also presented.

Engineers and scientists working in fuel science and engineering and specialty chemicals, particularly in methane and its derivatives, will find this to be useful as a reference. This book also compiles a great deal of information and data, yet condensed into a concise volume that will be very convenient for researchers as well as students. It is also a good reference text for professors.

This book can be used as a textbook for a three credit-hour course on methane chemistry, methane and ammonia technology, or natural gas engineering. It may be used for a more general course in fuel engineering, gas and coal technology, or alternative energy and feedstocks, or as a reference text for engineering design courses or chemistry electives. The background required of a reader is an understanding of organic chemistry at the college level. For scientists and engineers in the energy and chemical industries, this book will serve as a one-volume comprehensive reference providing necessary information on process chemistry, manufacturing technology, alternative routes, market trends, and environmental issues.

Even though quite a bit of effort was called for to develop a book of this nature, the encouragement and assistance that I received from various people made the work enjoyable. I would like first of all to thank my wife, Kyung, for her constant support and love while writing this book. I would also like to thank my children for the countless little things that mean a lot.

The comments and suggestions made by my colleagues at the university and at the American Institute of Chemical Engineers, Fuels and Petrochemicals Division, have been a great help in putting together the various parts of this book. The support and encouragement provided by the late Robert Iredell's family made this work possible and are greatly appreciated.

Last but not least, I wish to acknowledge the direct and indirect help in the compiling of the scientific and technical information by Mr. Brian S. Kocher, Mr. Timothy I. Tartamella, Mrs. Medha Joshi, Mr. Mark Brundage, Mr. Ben Lopez, Mr. Cher-Dip Tan, Mr. Ramesh Rachapudi, Mr. Sanjay Shah, Mr. Abhay Sardesai, Mr. J. David Tucker, Mr. H. Bryan Lanterman, Dr. Kathy

L. Fullerton, and Dr. Teresa J. Cutright. I would also like to thank Ms. Jeanine L. Gray and Mrs. Maria Peters for editing and assembling the manuscript. Without their contribution, this book would have not been possible.

Sunggyu Lee

Contents

METHANE
AND ITS
DERIVATIVES

1

Sources and Availability of Methane

I. ENERGY OVERVIEW

A. United States Energy Supplies

The U.S. energy production during March 1995 totaled 5.9 quadrillion Btu, which is a 1.0% increase from the level during March 1994. Crude oil and natural gas plant liquids decreased 0.9%, coal production decreased 0.3%, and natural gas production remained about the same. However, all other forms of energy production combined were up 10.4% from the level of production during March 1994 [1]. Such statistical information is available from the Monthly Energy Review by the Energy Information Administration (EIA). Table 1.1 shows the energy summary for March 1995.

Energy consumption during March 1995 totaled 7.5 quadrillion Btu, 1.3% above the level of consumption during March 1994. For the month of March 1995, the total energy production for the month is substantially lower than the total energy consumption for the same month, indicating that the consumption/production pattern is highly seasonal, and the storage of energy and fuel is a very important issue. Consumption of natural gas increased 2.7%, petroleum products consumption increased 0.5%, and coal consumption decreased 2.5%. Consumption of all other forms of energy combined increased 8.1% from the level of March 1994.

Net energy imports by the United States during March 1995 totaled' 1.6 quadrillion Btu, which is 5.0% above the level of March 1994. Net imports of natural gas were up by 12.2% and net imports of petroleum increased 6.3%. Net exports of coal rose 17.2% from the level of the previous year.

Table 1.1 Energy Summary for March 1995. (Quadrillion Btu).

	March			Cumulative January through March				
	1995	1994	Percent change[a]	1995	1995 daily rate	1995	1995 daily rate	Percent change[a]
Production[b]	5.947	5.886	1.0	17.394	0.193	16.698	0.186	4.2
Coal	2.045	2.052	−0.3	5.835	0.065	5.4332	0.060	7.4
Natural gas (dry)	1.659	1.658	0.0	4.905	0.054	4.826	0.054	1.6
Crude oil[c] and natural gas plant liquids	1.396	1.409	−0.9	4.068	0.045	4.098	0.046	−0.7
Other[d]	0.847	0.768	10.4	2.586	0.029	2.342	0.026	10.4
Consumption[b]	7.481	7.384	1.3	22.909	0.255	23.147	0.257	−1.0
Coal	1.557	1.596	−2.5	4.807	0.053	4.992	0.055	−3.7
Natural gas[e]	2.146	2.091	2.7	6.905	0.077	7.041	0.078	−1.9
Petroleum products[f]	2.898	2.883	0.5	8.516	0.095	8.649	0.096	−1.5
Other[g]	0.881	0.815	8.1	2.681	0.030	2.464	0.027	8.8
Net imports	1.565	1.491	5.0	4.273	0.047	4.313	0.048	−0.9
Coal[h]	−0.166	−0.141	17.2	−0.455	−0.005	−0.346	−0.004	31.7
Natural gas	0.223	0.199	12.2	0.686	0.008	0.614	0.007	11.8
Petroleum[i]	1.474	1.386	6.3	3.947	0.044	3.922	0.044	0.6
Other[j]	0.033	0.047	−28.8	0.094	0.001	0.122	0.001	−23.0

a Based on daily rates prior to rounding.

b Due to a lack of consistent historical data, some renewable energy sources are not included. For example in 1992, 3.0 quadrillion Btu of renewable energy consumed by U.S. electric utilities to generate electricity for distribution is included, but an estimated 3.0 quadrillion Btu of renewable energy used by other sectors is not included.

c Includes lease condensate.

d "Other" is hydroelectric and nuclear electric power, and electricity generated for distribution from wood, waste, geothermal, wind, photovoltaic, and solar thermal energy.

e Includes supplemental gaseous fuels.

f Products obtained from the processing of crude oil (including lease condensate), natural gas, and other hydrocarbon compounds.

g "Other" is hydroelectric and nuclear electric power, electricity generated for distribution from wood, waste, geothermal, wind, photovoltaic, and solar thermal energy; and net imports of electricity and coal coke.

h Minus sign indicates exports are greater than imports.

i Crude oil, lease condensate, petroleum products, pentanes plus, unfinished oils, gasoline blending components, and imports of crude oil for the Strategic Petroleum Reserve.

j "Other" is net imports of electricity and coal coke.

Notes: Totals may not equal sum of components due to independent rounding. Geographic coverage is the fifty States and the District of Columbia.
Source: [1]

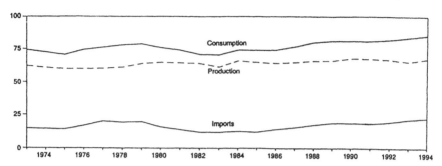

Figure 1.1   Energy overview: consumption, production, and imports, 1973–1994 (units: quadrillion Btu). *Source*: [1].

Figure 1.1 shows the U.S. energy consumption, production, and imports for 1973–1994. During this period, the lowest production was recorded in 1975 at 59.86 quadrillion Btu, whereas the highest production was realized in 1990 at 67.85 quadrillion Btu. For the same period, the U.S. consumption ranged from 70.52 (1983) to 85.57 (1994) quadrillion Btu, whereas the lowest and highest energy imports were recorded at 12.03 (1983) and 22.58 (1994) quadrillion Btu, respectively. From the statistics, one can also observe that

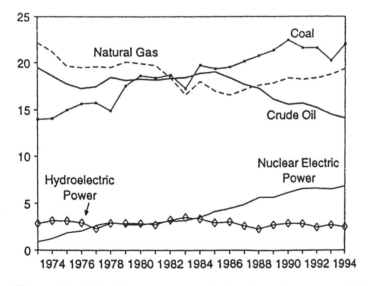

Figure 1.2   United States energy production by major sources, 1973–1994 (units: quadrillion Btu). *Source*: [1].

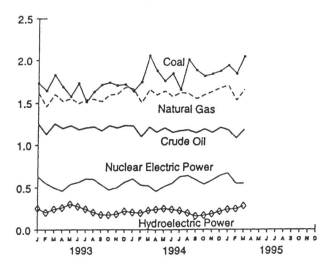

Figure 1.3 Monthly production by major sources, 1973–1994 (units: quadrillion Btu). *Source*: [1].

the energy consumption is very strongly related with the economic strength of the country for a given year.

Figure 1.2 shows the U.S. energy production by major source for the period of 1973–1994. As can be clearly seen, the coal production is the most important, followed by the production of natural gas, crude oil, nuclear electric power, and hydroelectric power, in decreasing order of importance. The U.S. production of crude oil continues to decrease, while all other major sources are increasing.

The monthly production of energy by major sources is given in Figure 1.3. Coal production is the most fluctuating source, and nuclear electric power is the next. The fluctuation in monthly production is due to the effect of weather on production and manufacturing operations.

The yearly energy consumption pattern of the United States for the period of 1973–1994 is shown in Figure 1.4. The monthly consumption pattern by major sources for the period of 1992–1995 is shown in Figure 1.5. Interestingly, the consumption of natural gas was most fluctuating from month to month (i.e., high in the winter season and low in the summer season).

Figures 1.6 and 1.7 show the yearly and monthly energy net imports by the United States, respectively. The net imports of petroleum products followed an up-and-down pattern, first increasing (1974–1977), then decreasing (1979–1982), and then increasing again (1985–1994). This fluctuating change

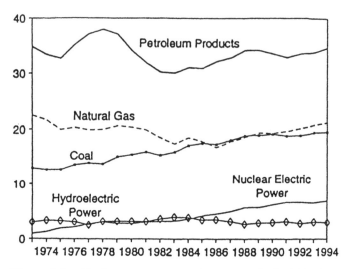

Figure 1.4   Yearly energy consumption patterns in the United States (units: quadrillion Btu). *Source*: [1].

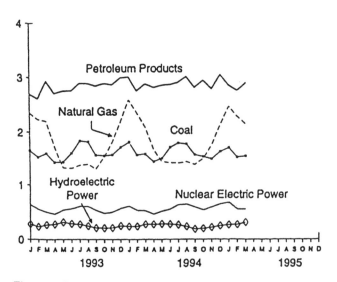

Figure 1.5   Monthly energy consumption pattern in the United States (units: quadrillion Btu). *Source*: [1].

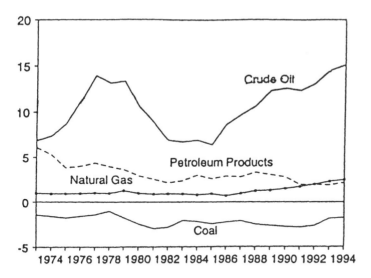

Figure 1.6   Energy net imports, 1973–1994 (units: quadrillion Btu). *Source*: [1].

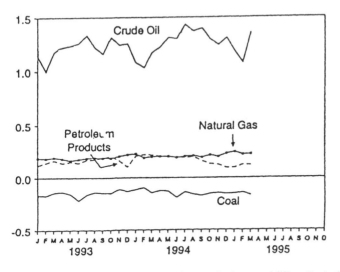

Figure 1.7   Monthly energy net import (units: quadrillion Btu). *Source*: [1].

may be attributed to the energy policy changes, domestic crude oil price, international market price of crude oil, and international politics.

B. World Primary Energy Production

In every phase of human life, consumption of energy in various forms is involved. Macroscopically speaking, the energy consumption of a country is directly related to the country's total production activities as well as the living standards of the country. This is why the energy resources are the most important resources for any country or region. The world has already experienced a major energy crisis and several minicrises.

According to the U.S. Energy Information Administration, *International Energy Annual (1994)*, the total world primary energy production was 278.78 quadrillion Btu in 1983 and 343.10 quadrillion Btu in 1992. The U.S. energy production was the highest at 63.79 quadrillion Btu in 1992, followed by the former U.S.S.R. with 57.88 quadrillion Btu in 1992. Table 1.2 shows the world primary energy production by region and country [2].

II. CRUDE OIL RESERVES

Since the world energy crisis ignited by the oil embargo in 1973, we have been constantly hearing about the world's rapidly depleting oil reserves. We have also heard about the estimated number of years supply of petroleum from various sources. All these talks and statistics have been closely linked to the governmental policy making, programs of alternative fuels, long range planning of industries, etc. Table 1.3 shows the estimated proved world crude oil reserves in billions of barrels.

From Table 1.3, it is very interesting that the estimated proven world crude oil reserves have been in an increasing trend (i.e., 338.67 billion barrels in 1965 versus 999.76 billion barrels in 1995). This is due to the advances in recovery processes as well as the increased activities in geological exploration. Table 1.4 shows the statistical data for the world crude oil production by regions.

Table 1.4 shows that the worldwide crude oil production has remained fairly constant at the level of 21–22 billion barrels per year. Considering the current proven crude oil reserves of the world and the current rate of worldwide crude oil production, the current proven reserve will last 45 years, if 100% recovery is assumed. This number gives much of the optimism for the future petroleum supply. In previous years, the same calculation would have given 30 (1965), 33 (1970), 37 (1975), 19 (1980), 35 (1985), 47 (1990), 45 years (1995), respectively.

III. SUPPLIES OF NATURAL GAS IN THE UNITED STATES

The total dry natural gas production in the United States during April 1995 was an estimated 1.6 trillion cubic feet, 1% higher than the production during the previous April. However, consumption of natural and supplemental gas in April 1995 was 1.7 trillion cubic feet, 5% above the level in April 1994. Imports of natural gas in April 1995 were 225 billion cubic feet, 10% higher than imports in April 1994. Stocks of working gas (gas available for withdrawal) in underground natural gas storage reservoirs at the end of April 1995 totaled 1.4 trillion cubit feet, 17% above the previous year's level.

Tables 1.5 and 1.6 show the U.S. natural gas production data for the period of 1973 through 1994. Several important facts can be pointed out:

1. For the 20-yr period, the yearly gross withdrawals changed between 18,659 (1983) and 24,067 (1973) billion cubic feet.
2. By viewing the 1994 statistics, the monthly gross drawings of natural gas stayed at a relatively constant level (i.e., about 2,000 billion ft^3 per month).

In Tables 1.5 and 1.6, the following terms are used:

Total dry gas production = Total wet gas production −

Extraction loss

Total wet gas production = Total marketed production

= Gross withdrawals − Repressuring − Nonhydro−

carbon gases removed − Vented and flared

In general, the total dry gas production is approximately 80% of the gross withdrawals of natural gas.

Figure 1.8 shows the U.S. consumption, dry production, and imports of natural gas for the period of 1973–1994. Figure 1.9 shows the monthly data for consumption, dry production, and imports for 1993–1995. It is interesting to note that the dry production and imports are not fluctuating monthly, while consumption varies seasonally. Figures 1.10 and 1.11 show the breakdown of natural gas consumption by different sectors. While the industrial consumption fluctuates substantially from year to year, the residential consumption stayed nearly constant. Figures 1.12 and 1.13 show the yearly and monthly data for underground storage. The total underground storage amount by the end of 1994 reached a level of 7 trillion cubic feet.

Table 1.7 shows the data for the U.S. natural gas trade by country. The data indicates that the U.S. import of natural gas mainly depends on Canada.

Table 1.2 World Primary Energy Production[a] (Quadrillion Btu)

Region and country	1973	1974	1975	1976	1977	1978	1979
Canada	9.50	9.38	8.89	8.68	9.03	9.14	9.99
Mexico	1.87	2.15	2.44	2.74	3.14	3.76	4.51
United States	61.95	60.72	59.73	59.81	60.14	61.03	63.71
Total: North America	73.32	72.25	71.06	71.22	72.31	73.93	78.22
Venezuela	8.13	7.29	5.90	5.79	5.69	5.52	6.04
Total: Central & South America	12.85	12.09	10.62	10.67	10.88	11.19	12.23
France	1.78	1.82	1.88	1.71	2.00	2.02	2.03
Germany[b]	4.92	4.95	4.87	5.14	5.12	5.12	5.44
Italy	1.06	1.07	1.08	1.11	1.15	1.13	1.10
Netherlands	2.52	2.86	3.16	3.44	2.90	2.49	2.69
Norway	0.83	0.88	1.22	1.47	1.47	2.07	2.68
United Kingdom	4.66	4.47	4.97	5.39	6.58	7.19	8.24
Total: Western Europe	19.58	20.01	21.36	22.34	23.93	24.81	27.40
Iran	13.27	13.68	12.23	13.43	12.88	11.93	7.46
Iraq	4.35	4.25	4.89	5.25	5.08	5.53	7.49
Kuwait	6.75	5.67	4.68	4.84	4.46	4.90	5.78
Saudi Arabia	16.63	18.68	15.68	18.88	20.51	18.45	21.24
United Arab Emirates	3.27	3.56	3.56	4.20	4.45	4.09	4.12
Total: Middle East	46.61	48.33	43.58	49.13	49.77	47.30	48.65
Algeria	2.47	2.35	2.34	2.65	2.63	3.28	3.16
Libya	4.72	3.32	3.27	4.26	4.53	4.40	4.63
Nigeria	4.45	4.89	3.88	4.51	4.50	4.06	4.95
South Africa	1.48	1.57	1.65	1.84	2.03	2.15	2.46
Total: Africa	14.81	13.93	13.19	15.49	16.44	16.70	18.30
Australia	3.25	3.44	3.64	3.35	3.43	3.51	3.66
China	13.14	14.25	15.11	15.35	16.16	18.05	18.53
India	2.22	2.36	2.68	2.88	2.95	3.14	3.30
Indonesia	2.90	2.99	2.90	3.38	3.85	3.71	3.83
Total: Far East & Oceania	26.17	28.04	29.71	30.59	32.09	34.53	35.99
Poland	4.94	5.09	5.37	5.00	5.18	5.36	5.51
Former USSR	39.28	41.77	43.62	47.46	49.79	52.11	53.88
Total: Eastern Europe/Former USSR	51.44	54.17	56.51	60.23	62.95	65.40	67.52
Total: World	244.78	248.81	246.03	259.67	268.37	273.87	288.31

[a] Includes only crude oil, lease condensate, natural gas plant liquids, dry natural gas, coal, net hydroelectric power and net nuclear power.
[b] Beginning with 1982, Germany includes both the former East Germany and the former West Germany.
Source: U.S. Energy Information Administration, International Energy Annual, 1994.

1980	1981	1982	1983	1984	1985	1986	1987	1988	1989	1990	1991	1992
10.04	9.75	9.66	10.14	11.01	11.80	11.71	12.32	13.17	13.10	13.07	13.96	14.36
5.80	6.78	7.82	7.70	7.88	7.74	7.07	7.28	7.33	7.37	7.56	7.78	7.76
64.64	64.30	63.79	61.08	65.71	64.58	64.05	64.62	65.79	65.85	67.66	67.36	66.68
80.48	80.83	81.27	78.92	84.60	84.12	82.82	84.22	86.30	86.31	88.29	89.10	88.80
5.71	5.58	5.22	5.00	5.02	4.78	5.17	5.14	5.59	5.74	6.38	6.86	6.80
12.15	12.13	12.01	12.20	13.14	13.44	14.18	14.08	15.04	15.52	16.32	17.35	17.60
2.26	2.65	2.61	2.96	3.37	3.54	3.81	3.97	4.09	4.09	4.20	4.49	4.66
5.44	5.57	7.47	7.41	7.78	8.27	8.02	7.87	8.11	7.90	7.29	6.31	6.14
1.07	1.11	1.16	1.11	1.17	1.18	1.26	1.19	1.27	1.20	1.19	1.28	1.32
3.32	3.10	2.67	2.62	2.71	2.82	2.71	2.78	2.55	2.63	2.63	2.94	2.91
3.02	3.06	3.12	3.42	3.66	3.83	3.98	4.48	4.74	5.77	5.96	6.22	6.80
8.40	8.71	9.44	9.85	8.78	10.11	10.55	10.24	9.96	8.95	8.78	9.29	9.23
28.65	29.73	32.42	33.75	34.43	37.19	38.02	38.59	39.10	38.58	38.30	38.96	39.31
3.91	3.28	5.12	5.67	5.29	5.57	5.06	5.66	5.70	7.01	7.67	8.28	8.53
5.45	2.16	2.19	2.17	2.61	3.09	3.66	4.58	5.97	6.47	4.54	0.69	1.01
3.88	2.70	1.98	2.51	2.76	2.44	3.36	3.77	6.34	4.32	2.83	0.43	2.48
22.48	22.57	14.86	11.69	11.29	8.55	11.91	10.73	12.73	12.68	15.92	19.75	20.63
3.89	3.45	3.00	2.91	3.00	3.29	3.68	4.21	4.25	4.99	5.51	6.27	6.23
42.13	36.73	29.54	27.29	27.65	25.66	30.62	32.09	36.04	39.61	41.02	40.23	43.94
2.75	2.95	3.11	3.46	3.71	3.77	3.55	4.01	4.02	4.28	4.52	4.81	4.83
4.03	2.57	2.61	2.52	2.53	2.46	2.43	2.29	2.73	2.69	3.18	3.44	3.46
4.50	3.18	2.86	2.77	3.12	3.35	3.30	3.04	3.29	3.88	4.05	4.26	4.49
2.74	3.09	3.24	3.45	3.87	4.17	4.26	4.23	4.39	4.28	4.19	4.19	5.06
17.35	15.09	15.43	16.12	17.52	18.43	18.14	18.53	19.54	20.47	21.46	23.39	23.70
3.53	3.89	4.04	4.24	4.41	5.33	5.49	6.02	5.81	6.15	6.71	6.30	6.66
18.32	18.10	19.14	20.47	22.39	21.60	25.34	26.25	27.50	29.15	29.29	29.89	30.18
3.35	4.09	4.43	4.88	5.22	5.53	5.97	5.71	5.97	6.53	6.96	6.71	6.94
4.16	4.27	3.68	3.84	4.27	4.24	4.33	4.36	4.49	4.86	5.14	5.70	6.09
36.53	37.47	38.88	41.57	45.18	49.50	51.56	53.24	54.94	58.26	59.93	60.94	62.75
5.28	4.54	5.16	5.25	5.37	5.54	5.72	5.79	5.87	5.49	4.72	3.78	3.74
55.67	56.71	57.88	59.27	61.30	63.17	65.73	67.61	69.82	69.94	68.86	62.51	NA
69.20	69.67	69.23	70.80	73.02	75.04	77.92	79.86	81.99	81.76	79.08	70.80	67.01
286.50	281.66	278.78	280.65	295.54	303.37	313.26	320.61	332.97	340.51	344.41	340.76	343.10

Table 1.3 Estimated Proved World Crude Oil Reserves Annually as of January 1
(Billions of Barrels)

	1965	1970	1975	1980	1985	1990	1995
United States	30.99	29.63	34.25	29.81	28.45	26.50	22.96
Canada	6.18	8.62	7.17	6.80	7.08	6.13	5.04
Latin America	25.53	29.18	40.58	56.47	83.32	125.03	129.07
Middle East	212.18	333.51	403.86	361.95	398.38	660.25	660.29
Africa	19.40	54.68	68.30	57.07	55.54	58.84	62.18
Australia and Asia	11.61	13.14	21.05	19.36	18.53	22.55	44.45
Western Europe	2.04	1.78	25.81	23.48	24.43	18.82	16.57
Communist Nations	30.75	60.00	111.40	90.00	84.10	84.10	NA
USSR and Eastern Europe	NA	NA	NA	NA	NA	NA	59.20
Total	333.67	530.53	712.42	644.93	699.81	1002.21	999.76

Sources: 1965-1975 U.S. API; 1980-1990 U.S. EIA; 1995 *Oil and Gas Journal*.

Table 1.4 World Crude Oil Production by Area (Billion Barrels)

	1965	1970	1975	1980	1985	1990	1993
United States	2.85	3.52	3.05	3.15	3.27	2.68	2.52
Canada	0.30	0.46	0.52	0.52	0.53	0.57	0.61
Latin America	1.68	1.92	1.60	2.04	2.28	2.49	2.67
Western Europe	0.15	0.14	0.20	0.89	1.37	1.53	1.70
Middle East	3.05	5.09	7.16	6.74	3.76	6.00	6.69
Africa	0.81	2.22	1.83	2.23	1.77	2.17	2.25
Asia	0.25	0.50	0.81	1.00	1.14	2.26	2.37
Communist Nations	1.98	2.87	4.33	5.20	5.35	NA	NA
USSR	NA	NA	NA	NA	NA	3.96	2.86
Total	11.06	16.72	19.50	21.76	19.49	21.66	21.67

Sources: 1965-1970: International Petroleum Annual; 1975: USEIA, World Crude Oil Production
Annual; 1980-1990: USEIA, Petroleum Supply Annual; 1993: *Oil and Gas Journal*, "Worldwide
Production Report" Issue.

Table 1.5 United States Natural Gas Production (Billion Cubic Feet)

	Gross withdrawals	Total dry gas production
1973	24,067	21,731
1975	21,104	19,236
1977	21,097	19,163
1979	21,883	19,663
1981	21,587	19,181
1983	18,659	16,094
1985	19,607	16,454
1987	20,140	16,621
1989	21,074	17,311
1991	21,750	17,698
1992	22,132	17,840
1993	22,729	18,244
1994	23,679	18,244

Sources: Energy Information Administration; *Natural Gas Monthly*, June, 1995.

Table 1.6 1994 United States Natural Gas Production Monthly (Billion Cubic Feet)

Month	Gross withdrawal	Total dry gas production
January	2,045	1,623
February	1,843	1,462
March	2,037	1,614
April	1,943	1,552
May	2,003	1,597
June	1,906	1,533
July	1,965	1,579
August	1,951	1,568
September	1,890	1,516
October	1,987	1,562
November	2,014	1,599
December	2,096	1,640
Total	23,679	18,845

Sources: Energy Information Administration; *Natural Gas Monthly*, June, 1995

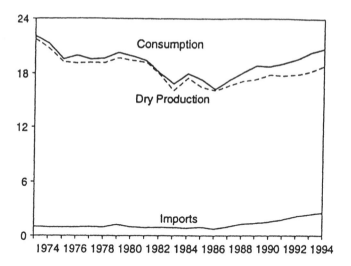

Figure 1.8 Overview of natural gas consumption, dry production, and import, 1973–
1994 (units: trillion cubic feet). *Source*: [1].

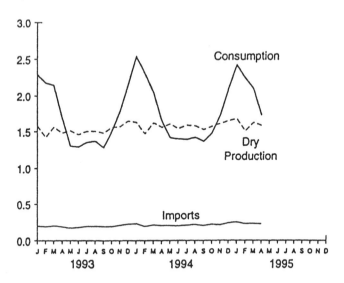

Figure 1.9 Monthly overview of natural gas consumption, dry production and imports, 1973–1994 (units: trillion cubic feet). *Source*: [1].

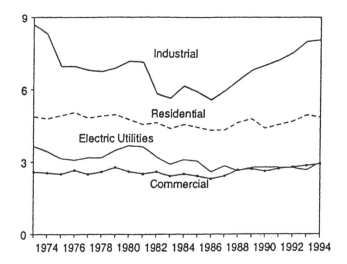

Figure 1.10 Natural gas consumption by major sectors, 1973–1994 (units: trillion cubic feet). *Source*: [1].

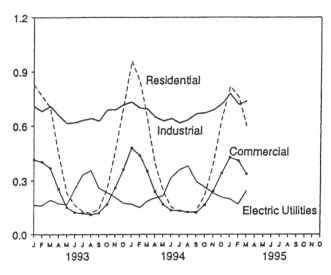

Figure 1.11 Monthly consumption of natural gas by major sectors (units: trillion cubic feet). *Source*: [1].

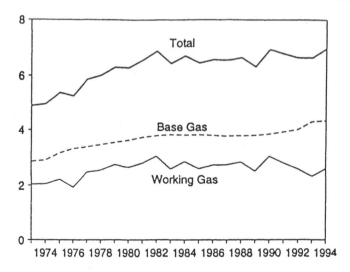

Figure 1.12   Year-end underground storage of natural gas, 1973–1994 (units: trillion cubic feet). *Source*: [1].

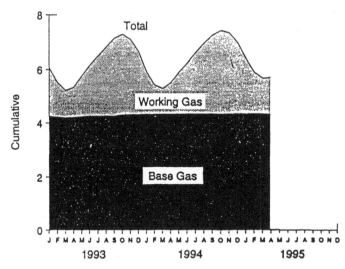

Figure 1.13   Month-end underground storage of natural gas, 1973–1994 (units: trillion cubic feet). *Source*: [1].

Table 1.7 United States Natural Gas Trade (Billion Cubic Feet)

Country	Imports			Exports		
	1974	1984	1994	1974	1984	1994
Canada	959	755	2,500			
Algeria	0	36	51			
Other	0	52	7			
Total	959	843	2,558			
Canada				13		48
Mexico				13	2	47
Japan				50	53	63
Total				76	55	157

IV. METHANE FROM NATURAL GAS

Natural gas is a mixture of hydrocarbons ranging from methane to C_7 or higher. The lighter hydrocarbons are paraffinic; those corresponding to a naphtha cut also consist mostly of paraffins but include small amounts of naphthenes and occasionally aromatics. In addition to hydrocarbons, natural gas may contain as much as 45% acid gases (i.e., H_2S and CO_2). A typical composition of natural gas is given in Table 1.8 [3].

There are various types of natural gas which are classified according to their composition as [3]:

1. Dry or lean gas: mostly methane
2. Wet gas: considerable amounts of higher hydrocarbons
3. Sour gas: high concentration of H_2S
4. Sweet gas: little H_2S
5. Residue gas: higher paraffins having been extracted
6. Casing head gas: derived from an oil well by extraction at the surface.
Note: Some natural gas contains substantial amounts of helium, nitrogen, and carbon dioxide.

The principal purpose of processing natural gas is to produce a stream containing only methane and ethane. In the cold regions of the world or in winter, natural gas is used mainly for space heating in large urban centers. In countries where natural gas is plentiful, methane is also the raw material for producing synthesis gas. Synthesis gas is very important as a building block in petrochemical industry, and as a raw material for alternative synthetic fuels. Purified methane containing no heavier hydrocarbons is the raw material

Table 1.8 Approximate Composition Data for
"Wet" and "Dry" Natural Gas (Calorific Value:
900–1100 Btu/ft^3)

Constituents	Composition (vol %)	
	"wet"	"dry"
Hydrocarbons		
Methane	84.6	96.0
Ethane	6.4	2.0
Propane	5.3	0.6
Isobutane	1.2 .	0.18
n-Butane	1.4	0.12
Isopentane	0.4	0.14
n-Pentane	0.2	0.06
Hexanes	0.4	0.10
Heptanes	0.1	0.08
Nonhydrocarbons		
Carbon dioxide	0–5	
Helium	0–0.5	
Hydrogen sulfide	0–5	
Nitrogen	0–10	
Argon	0–0.05	
Radon, krypton, xenon	Traces	

Source: Speight, J., *Gas Processing: Environmental Aspects
and Methods*, Butterworth-Heinemann, 1993. Oxford, U.K.

for making hydrogen cyanide, chloromethanes, carbon disulfide, and hydrogen, etc. Efforts are being made to explore diverse avenues of utilizing methane directly for synthesis of various essential petrochemicals. This is even more important considering that the dwindling reserve of petroleum in the world will prompt the world to switch the raw material for petrochemicals (including polymers) to other sources.

Natural gas constituents heavier than methane are also excellent petrochemical feedstocks [1]. Ethane is the most desirable starting material for producing ethylene whenever a minimum amount of byproducts is desired. Propane and butane can also be dehydrogenated to olefins, propylene, and butene, and butene can be further dehydrogenated to butadiene. The naphtha fraction, which is also known as natural gasoline, has a low octane number, but is an excellent feedstock for cracking to olefins and/or steam reforming.

V. VERTICAL AND HORIZONTAL INTEGRATION OF PETROCHEMICAL INDUSTRIES

Petrochemical economics depend strongly upon the feedstock valuation. This is especially true since the petrochemical industry is, in general, capital intensive and the cost of raw material is extremely important in the final manufacturing cost.

Petroleum and natural gas derivatives can be attributed values that may vastly differ depending upon the local market distribution of various fractions, transportation or pipeline costs, other feedstock availability, other infrastructure for manufacturing, and the effects of taxation and import duties.

The subject of economics in the petrochemical industry can be approached in several different ways. The first approach is building an organizational infrastructure along family-tree lines that originate from one or more building block chemicals. The second approach is building a network or an organizational structure according to chemical function. The third approach is based on a combination of the above two.

The first approach is normally referred to as "vertical integration," while the second approach is often called "horizontal integration." Both vertical and horizontal integration approaches have been vital factors in restructuring and stream-lining modern chemical industries.

As an example of horizontal integration, consider some companies producing several chlorine derivatives, hypochlorites, and chlorites. They may be often the result of a particular raw material position. This is the case of firms specializing in the production of compounds having a common chemical function or that can be made by a given unit process. However, a more powerful motivation of horizontal integration comes from a desire to provide edges against changes in market structure or to complement a line of products. For example, consider a company producing a variety of polymeric resins and synthetic fibers, or a firm producing trimethylolpropane (TMP) deciding to produce isocyanates.

In the second half of the twentieth century, the main influence in the chemical business has been the need to integrate vertically. For example, firms that until recently were content to produce a limited number of intermediates or end-products, have been under constant pressure either to integrate backward by acquiring their own sources of raw materials, or to integrate forward by gaining control of their clients or final customers. As a crucial evidence of such efforts, the percentage of captive utilization of most major chemical intermediates has been growing steadily. This is because the unit profit is normally maximized at the finished product end of the production chain. Most gigantic oil and chemical companies in the world have rapidly expanded and enlarged the scope of their activities both by acquiring other targeted firms and by gaining stronger market shares thus consolidating their positions. In the United States, there are limitations to forward vertical integration by acquisition, imposed by antitrust legislation.

From the standpoints of vertical and horizontal integration of chemical manufacturing industries, it is of crucial importance to understand that there are certain chemicals which are used in preparation of a large family of other chemicals possessing diverse functions and properties. In petrochemical or organic chemical industries, the following seven building blocks are frequently used:

methane
ethylene
propylene
acetylene
C_4 hydrocarbons
aromatic hydrocarbons (or, BTX)
synthesis gas

Many chemicals can be made from two or more building blocks. For example, styrene can be made from two building blocks of ethylene and aromatic hydrocarbons.

Methane is very important as a building block. Both the vertical and horizontal integration with methane as a starting building block is a key topic of this book.

VI. BIOMASS CONVERSION TO METHANE

Until the middle of the 19th century, the main source of energy for heating and industrial processes in the United States was wood. The trend has sharply changed toward the use of fossil fuels, such as coal and petroleum. Heavy dependence on the fossil fuels as well as dwindling reserves of resources have prompted the question: Is it possible to use this renewable source of energy for our basic energy needs? The process of using biomass as fuel is basically an alternative form of solar energy use. In other words, the energy storage process is handled through the production of plant matters via photosynthesis.

One pound of dry plant material will generate approximately 7,500 Btu when burned directly. The amount of energy available from biomass is close to the heating value of average, uncleaned coal.

The most publicized utilization of biomass is gasohol production, an excellent example of biomass conversion to liquid fuel. Although other grains can be used, the principal crop utilized in the United States has been corn. After harvesting, the entire stalk and cobs are chopped up, ground, and mixed with water. Then, the resulting mixture is cooked to help convert the starches

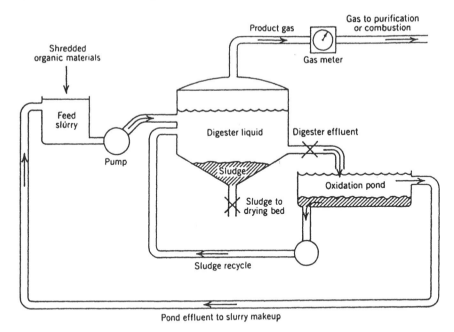

Figure 1.14 A unit for the continuous conversion of biomass by anaerobic fermentation into methane and other combustible gases. *Source*: [4].

into sugars by enzymatic reactions. The sugars then go through a so-called alcohol fermentation process to produce ethyl alcohol. Distillative purification separates ethyl alcohol from the rest of the material. The product ethanol is then blended with unleaded gasoline and directly usable in automobile engines. The added ethanol provides necessary oxygen contents for the gasoline, thus helping to alleviate CO emission problems. This is why such a fuel is called an oxygenated fuel.

In addition to producing a liquid fuel such as ethanol, biomass can also be converted into methane. In areas where natural gas resources are scarce, biomass conversion into methane serves as an attractive process, since methane is a very conveniently usable gaseous fuel for domestic energy needs. There have been a number of processes developed for this conversion.

The process that is most direct and simple is the fermentation of organic matter by bacteria in the absence of oxygen (i.e., anaerobic fermentation). The organic material that can be used for this process include; crops, vegetable wastes, animal wastes or refuses, lumber mill wastes, brewery wastes, sludge from sewage treatment plants, and municipal wastes. In the absence of oxygen, but in the presence of water, the organic matters will ferment naturally, and 60–80% of the carbon is converted into carbon dioxide, methane, and small amounts of hydrogen sulfide and nitrogen. After purification, the resultant gas is almost pure methane with a heating value of about 1,000 Btu/ft^3. The typical yield of such a process is 4.5–6.5 ft^3 of methane per pound of dry organic material, or 9–13 × 10^6 Btu/ton [4]. This biomass conversion process is very widely used on small scales in under-developed countries and rural areas of fossil energy deficient countries. A schematic of this process is given in Figure 1.14.

REFERENCES

1. Energy Information Administration (EIA) of U.S. Department of Energy, *Monthly Energy Review*, June: 1–105 (1995).
2. *Basic Petroleum Data Book*, American Petroleum Institute, Vol. 15, No. 1, January 1995, New York.
3. Speight, J. G., *Fuel Science & Technology Handbook*, Marcel Dekker, Inc., New York, 1990, p. 31.
4. Kraushaar, J. J. and Ristinen, R. A., *Energy and Problems of a Technical Society, (revised edition)*, John Wiley and Sons, New York, 1988.

2

Synthesis Gas

I. SYNTHESIS GAS: PAST AND FUTURE

Crucially important building blocks of the chemical industry are carbon monoxide (CO) and hydrogen (H_2) which are used in a variety of processes. The production of H_2 and CO, synthesis gas (syngas), is carried out by the steam reforming of methane, partial oxidation of fuel oil, and coal gasification. Synthesis gas is widely used and can be converted into petrochemicals, higher alcohols, and synthetic fuels. Hydrogen is used in ammonia synthesis and petroleum refining industries, while carbon monoxide is widely used in the production of plastics, paints, foams, pesticides, and insecticides.

Common names for mixtures of CO/H_2 are derived from their origin such as: "water gas" ($CO + H_2O$) from steam gasification of coal and "crack gas" ($CO + 3H_2$) from the steam reforming of methane, or from their application such as "methanol synthesis gas" ($CO + 2H_2$) for the manufacture of methanol and "oxo gas" ($CO + H_2$) for hydroformylation reactions. Another synthesis gas is ammonia synthesis gas ($N_2 + 3H_2$) which does not contain CO as a higher constituent.

Initially, the manufacture of synthesis gas was carried out by gasifying coke from coal at low temperatures with air and steam. Coal gasification dates back to and continued throughout the 1800s when it was used for heating and lighting in both industry and private sectors. By 1930, there were over 11,000 coal gasifiers in the United States. However, the inefficient processes were replaced in the 1940s by oil and gas. Oil and natural gas became commonly employed as a synthesis gas feedstock due to their ease of handling and high hydrogen content (H:C for coal 1:1, 2:1 for oil, and 4:1 for methane-rich natural gas).

In the 1990s, methane reforming is still the predominant method of producing syngas, supplying more than 80% of the world's synthesis gas [1]. It is currently the most cost effective method of synthesis gas production and is expected to remain that way well into the 21st century [2]. Other synthesis gas production processes include naphtha reforming, fuel oil partial oxidation, and coal gasification.

II. METHANE REFORMING

The production of syngas by methane reforming is carried out by one or more of the following processes: steam reforming, CO_2 reforming, and/or partial oxidation (oxygen-enhanced reforming). The typical feedstock for methane reforming is natural gas, which has a typical composition of 85% methane, 9% ethane, 3% propane, 1% butane, and 1% nitrogen. However, the composition varies from site to site (Table 2.1) [3].

Steam reforming of methane is the most widely used process for the production of syngas from methane. The endothermic reaction between H_2O and CH_4 typically takes place over a nickel catalyst to produce H_2/CO in a 3:1 ratio. Carbon dioxide reforming is used when an abundant source exists to replace water in the endothermic reactions with CH_4; this produces a synthesis gas with a 1:1 ratio of H_2/CO. Another methane reforming process, partial oxidation (POX) of methane, utilizes substoichiometric amounts of oxygen in a highly exothermic reaction to produce syngas in a H_2/CO ratio of 1:2 without the use of a catalyst. However, the steam reforming produces the most hydrogen-rich syngas product.

The production of syngas may involve the use of prereforming, primary reformers, secondary reformers, and oxygen-enhanced reforming, depending on the composition requirements. The syngas loop configuration schemes will vary depending on the composition of the natural gas feedstock and end use

Table 2.1 Natural Gas Compositions [3]

	Gas #1	Gas #2	Gas #3	Gas #4	Gas #5
CO_2	—	—	6.5	0.8	—
N_2	5.0	0.8	—	8.4	—
CH_4	90.0	83.4	77.5	84.1	36.7
C_2H_6	5.0	15.8	16.0	6.7	14.8
C_3H_8	—	—	—	—	23.5
C_4H_{10}	—	—	—	—	14.7
C_5H_{12}	—	—	—	—	10.4

of the syngas produced. In many instances, methanol and ammonia plants are coupled to the same synthesis gas plant to utilize more efficiently the energy produced from the reforming of natural gas.

A. Thermodynamics

Steam reforming (allothermal process) is the most widely used process for the production of syngas from methane. The important steam reforming reactions are shown in Reactions 2.1–2.3. Reaction 2.1 represents the endothermic steam reforming reaction where synthesis gas is produced in a 3:1 H_2/CO ratio.

$$CH_4 + H_2O = CO + 3H_2 \qquad \Delta H_{298}^o = 206 \text{ kJ/mol} \tag{2.1}$$

Reaction 2.2 represents the exothermic water–gas shift (WGS) reaction, which produces CO_2 and H_2.

$$CO + H_2O = CO_2 + H_2 \qquad \Delta H_{298}^o = -41 \text{ kJ/mol} \tag{2.2}$$

From these two reactions it can be gathered that at higher temperatures less CH_4 and more CO will be present in the equilibrium gas. However, by LeChatelier principle, increasing the pressure will increase the methane equilibrium content. Reaction 2.3 represents the steam reforming of higher hydrocarbons, which are present in small quantities in natural gas.

$$C_nH_m + nH_2O = nCO + \left(n + \frac{m}{2}\right)H_2 \tag{2.3}$$

The presence of higher molecular weight hydrocarbons will enhance the carbon deposition effect, which can lead to reduced catalyst and reformer life. The carbon deposition involves the conversion of hydrocarbons by three different reaction mechanisms, the Boudouard reaction (Reaction 2.4) and direct deposition reaction of hydrocarbons by Reactions 2.5 and 2.6. The Boudouard reaction involves two CO molecules reacting to form $C_{(s)}$ and CO_2.

$$2CO = C\ (s) + CO_2 \qquad \Delta H_{298}^o = -172 \text{ kJ/mol} \tag{2.4}$$

The two other direct carbon deposition reactions are shown in Reactions 2.5 and 2.6, which involve the release of H_2 as well as the formation of $C_{(s)}$.

$$CH_4 = C\ (s) + 2H_2 \qquad \Delta H_{298}^o = 75 \text{ kJ/mol} \tag{2.5}$$

$$C_nH_m \dashrightarrow nC + \frac{m}{2}H_2 \tag{2.6}$$

Unfortunately, the nickel catalyst used to catalyze Reactions 2.1–2.3 also catalyzes Reactions 2.4–2.6. Eventually, carbon formation can lead to the deactivation of catalyst and the build-up of carbon deposits, which may lead to blockage of reformer tubes or the formation of hot spots.

Steam cracking (pyrolysis) will occur if process temperatures are in excess of 923 K on the higher molecular weight hydrocarbons; this forms olefins, which easily polymerize and then degrade into coke as shown in Reaction 2.7 [4].

$$C_n \longrightarrow \text{Olefins} \longrightarrow \text{Polymers} \longrightarrow \text{Coke} \tag{2.7}$$

The coke usually forms on the tube wall or as deposits encapsulating the catalyst, thus resulting in areas of increased pressure drop and hot spots.

The steam reforming reaction (Reaction 2.1) and the water–gas shift reaction (Reaction 2.2), are reversible reactions at steam reforming temperatures (723–1223 K), meaning that thermodynamics govern the equilibrium conversion (the maximum attainable conversion) [1,5]. A reaction that is limited by chemical equilibrium has a low net rate and limited conversion potential. The reaction equilibrium constants versus temperature are shown for Reactions 2.1, 2.2, 2.4, and 2.5 in Figure 2.1 [1]. The water–gas shift reaction is least dependent on temperature as is evident from the equilibrium constant. Because of this, the water–gas shift reaction affects many commercial chemical processes including coal gasification and methanol synthesis [1]. It is evident from Figure 2.1 that the steam reforming reaction (Reaction 2.1) and the decomposition of methane reaction (Reaction 2.5) are not thermodynamically favored until the temperature exceeds 950 K. However, Reaction 2.4, the formation of $C_{(s)}$ and CO_2 from CO, is favored at lower temperatures (< 950 K). As a result, the reforming operating temperature must be in excess of 950 K for appreciable conversion to occur and make the process most efficient.

In areas where CO_2 is abundantly available, it can be used in place of steam and will give an enhanced H_2/CO ratio as shown in Reaction 2.8, although carbon deposition is enhanced in this reaction system. Typically, CO_2 is recycled from downstream in methanol- and oxo-synthesis gas operations.

$$CH_4 + CO_2 = 2CO + 2H_2 \qquad \Delta H^o_{298} = 247 \text{ kJ/mol} \tag{2.8}$$

Another method commonly used for the conversion of methane is partial oxidation, shown in Reaction 2.9, which may or may not be catalytic. The reaction will be accompanied by complete gasification.

$$CH_4 + 0.5O_2 = CO + 2H_2 \qquad \Delta H^o_{298} = -38 \text{ kJ/mol} \tag{2.9}$$

Partial oxidation (oxygen enhanced reforming, autothermal reforming) of methane utilizes substoichiometric amounts of oxygen for the incomplete combus-

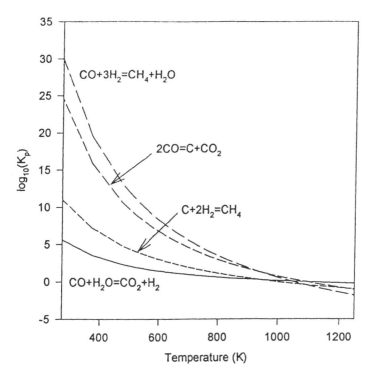

Figure 2.1 Equilibrium constants versus temperature. *Source*: [1,5].

tion of methane. The partial oxidation reaction is highly exothermic and can result in soot formation if reactor temperatures are not homogeneous [4,6,7].

B. Catalyst

The metal catalysts active for steam reforming of methane are the group VIII metals, usually nickel. Although other group VIII metals are active, they have drawbacks; for example, iron rapidly oxidizes, cobalt cannot withstand the partial pressures of steam, and the precious metals (rhodium, ruthenium, platinum, and palladium) are too expensive for commercial operation. Rhodium and ruthenium are ten times more active than nickel, platinum, and palladium. However, the selectivity of platinum and palladium are better than rhodium [1]. The supports for most industrial catalysts are based on ceramic oxides or oxides stabilized by hydraulic cement. The commonly-used ceramic supports include α-alumina, magnesia, calcium-aluminate, or magnesium-aluminate [4,8]. Supports used for low temperature reforming (< 770 K) are γ-alumina and chromia, since they suffer from substantial sintering and weakening at temperatures above 770 K [1].

The support predominately used in North America is α-alumina, which has a surface area of 2–5 m^2/g but can be fired to a high surface area (15–20 m^2/g). Although high surface area α-alumina tends to sinter rapidly in steam, it can be stabilized by zirconia. Unfortunately, the acidic support increases the potential for carbon deposition reactions, meaning that the catalyst can only be used with clean natural gas or stabilized by an alkalized species. Magnesia aluminate supports have a high surface area of 12–20 m^2/g and are stable if they have been calcined at high temperatures. However, care must be taken to ensure that the catalyst temperature does not fall below 300°C during exposure to steam, since the magnesium oxide will be hydrated to magnesium hydroxide, destroying the crystal structure. Calcium aluminate is the predominant catalyst used outside of North America. It has good thermal stability and is naturally alkaline [8].

Steam reforming catalysts must meet stringent requirements such as: high activity, reasonable life, good heat transfer, low pressure drop, high thermal stability, excellent mechanical strength (withstanding startups and shutdowns), in-situ regeneration from poisoning, operation at low steam-to-carbon ratios, and heavier molecular weight hydrocarbons. Although present catalysts cannot meet all these requirements, some trade-offs must occur for the optimal selection for each particular process [2].

A typical nickel catalyst is made from nickel(15–25 wt%) that is finely dispersed onto support material. Unfortunately, the nickel catalyst is highly susceptible to poisoning by sulfur compounds. The catalyst life is dependent upon several factors such as sulfur poisoning, sintering, and carbon deposition. Therefore, sulfur compounds must first be removed from the natural gas feedstock to ensure catalyst life.

Sulfur poisoning occurs through two competing reaction systems. The first is dissociative chemisorption of hydrogen sulfide according to Reaction 2.10 [1,9,10]. The second system begins in Reaction 2.11 defined as the adsorption of steam, which further reacts with hydrogen sulfide, as in Reaction 2.12, to form inactive sulfur poisoned nickel catalyst and water.

$$Ni + H_2S = NiS + H_2 \tag{2.10}$$

$$Ni + H_2O = NiO + H_2 \tag{2.11}$$

$$NiO + H_2S = NiS + H_2O \tag{2.12}$$

The exact mechanism of sulfur poisoning is not well understood, but is believed to be a combination of these two reaction systems [2,9].

Table 2.2 Routes for Carbon Formation [1]

	Whisker carbon	Encapsulating polymers	Pyrolytic carbon
Formation	Diffusion of C through Ni-crystal Nucleation and whisker growth with Ni-crystal at the top	Slow polymerization of C_nH_m radicals on Ni-surface into encapsulating film	Thermal cracking of hydrocarbon Deposition of C-precursors on catalyst
Effects	No deactivation of Ni-surface Breakdown of catalyst and increasing ΔP	Progressive deactivation	Encapsulation of catalyst particle Deactivation and increasing ΔP
Temperature (K)	> 720	< 770	> 870
Variables	High temperature Low H_2O/C_nH_m No enhanced H_2O adsorption Low activity Aromatic feed	Low temperature Low H_2O/C_nH_m Low H_2/C_nH_m Aromatic feed	High temperature High void fraction Low H_2O/C_nH_m High pressure Acidity of catalyst

Sintering involves the loss of surface nickel from exposure to temperatures exceeding the Tammann temperature $0.5T_m$ which for nickel is 864 K. Recrystallization from sintering may change the nickel crystals available and cause a decrease in activity. Another reason for loss of activity is surface diffusion which occurs above the Huttig temperature $1/3T_m$, 576 K for nickel, resulting in the reorganization of nickel crystals. The nickel crystal growth is impeded by the support pore size. This means that the metal particles cannot grow larger than the pore size, thus preventing any or further sintering or surface diffusion of the nickel catalyst [1,9].

Carbon formation on steam reforming catalysts takes place in three different forms: whisker-like carbon, encapsulated carbon, and pyrolytic carbon as described in Table 2.2 [1]. Whisker-like carbon grows as a fiber from the catalyst surface with a pear-shaped nickel crystal on the end. Strong fibers can even break down catalyst particles increasing the pressure drop across the reformer tubes [4]. The carbon for whisker formation is formed by the reaction of hydrocarbons as well as CO over transition metal catalysts [1]. The whisker growth is a result of diffusion through the catalyst and nucleation to form a long carbonaceous fiber.

Encapsulated carbon is formed from the reaction of adsorbed hydrocarbons onto nonreactive areas (deposits), which polymerize on the catalyst. This forms graphitic carbon that encapsulates into the film and progressively de-activates the catalyst [1]. This degradation process occurs at mild temperatures and low steam to hydrocarbon ratios.

Pyrolytic carbon is formed mainly by three different reactions, namely, the reversible decomposition of methane (Reaction 2.5), the irreversible cracking of higher hydrocarbons (Reaction 2.6), and/or coke formation (Reaction 2.7). The formation of these carbon deposits leads to the breakdown of the catalyst and hot spots in the reactor. Pyrolytic carbon is usually found as dense shales on the reformer wall or encapsulating the catalyst particles. The process leads to the deactivation of the catalyst and increase of pressure drop across the reformer tubes. The thermal cracking of hydrocarbon occurs at high temper-atures and at low steam to hydrocarbon ratios.

C. Processes

The catalytic steam reforming of methane was developed by BASF in the first quarter of this century and was first used in 1931 in Baton Rouge by Standard Oil of New Jersey (Exxon). The first steam reforming plant used refinery off-gas (CH_4, H_2, and other light hydrocarbons) as a feedstock. Although the process was improved over the years, North America was the main user of the technology due to the abundance of natural gas. It was not until the 1960s when natural gas fields were developed in Europe did the technology become wide spread [8]. Since the early days, the range of feed-stocks (natural gas to naphtha), catalyst formulation, furnace design, desulfur-ization processes, and process efficiency have increased considerably, making steam reforming of methane the most economical choice for synthesis gas production.

The steam reforming of methane has many different configurations depend-ing on the end-use of the gas and the plant manufacturer. However, reforming is generally performed in four stages: desulfurization of natural gas, prere-forming, primary reforming, and secondary reforming (Figure 2.2). First, the natural gas is purified of H_2S and organic sulfur compounds such as mercap-tans and carbon disulfide. Organic sulfur compounds are hydrogenated to H_2S and then adsorbed onto packed beds, reducing the sulfur content to less than 0.1 ppm. Second, prereforming is utilized in gasification plants for protection against carbon formation when feedstock composition varies or when highly aromatic (> 30% aromatic). Prereforming limits the amount of steam neces-sary for reforming. The third step, primary reforming, uses the addition of steam and heat (allothermal) to support the highly endothermic reactions at temperature of 825°C over a nickel catalyst. The fourth step, the secondary

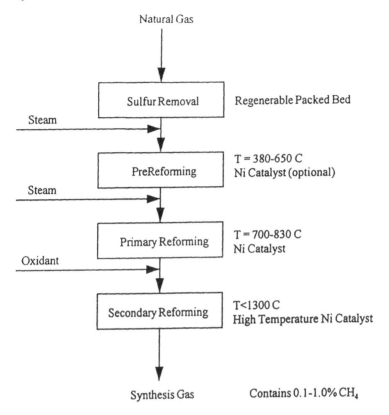

Figure 2.2   Block diagram of a steam reforming process.

reformer, utilizes oxygen or air to produce heat (autothermal) through exo-thermic combustion reactions increasing the temperature (1300°C) and forcing the steam reforming reactions to a low methane equilibrium (0.2–1%) which reduces the load on the primary reformer [11]. Typical steam reforming con-ditions are shown in Table 2.3 [8].

The incoming natural gas stream is desulfurized, decreasing the chances of catalyst poisoning. Typically, desulfurization is performed over beds packed with high surface area granular solids. The process is exothermic and great care must be taken to remove any heat liberated in the adsorption process. Fixed-bed processes can attain high product purity with long life of adsorbent. The process is insensitive to variations in pressure, feed rate, temperature, and contaminant concentration. Currently, the industry primarily utilizes ZnO packed beds. If the incoming gas contains organic sulfur it is removed by

Table 2.3 Typical Process Steam Reforming Conditions [8]

Primary Reformer	
Inlet temperature (°C)	400–600
Inlet pressure (bar a)	20–40
Steam ratio	2.5–4
Exit temperature (°C)	750–850
Exit CH_4 (mol%, dry basis)	9–13
Catalyst loading (kg/hr/l)	2–7
Secondary reformer	
Process gas inlet temperature (°C)	700–800
Process air inlet temperature (°C)	200–600
Exit temperature (°C)	900–1000
Exit CH_4 slip (mol%, dry basis)	0.5–1.5
Approach to methane-steam equilibrium (°C)	5–10
Catalyst loading (exit wet gas space velocity, hr^{-1})	8–1000

hydrogenating to H_2S over a suitable catalyst such as $Co-MoO_3/Al_2O_3$ and then adsorbed onto ZnO [12].

The prereformer is a fixed-bed adiabatic reactor loaded with a highly active catalyst. The reactor is capable of handling feedstocks ranging from natural gas to naphtha with aromatic contents of < 30%. All higher hydrocarbons are converted to CO, CO_2, H_2, and CH_4 by steam reforming reactions. The catalysts used have high activity at low temperatures (operating temperatures range from 380 to 650°C) with a thermal stability of 650°C and high resistance to carbon formation. Process configurations for adiabatic prereforming are shown in Figure 2.3, and the configurations are as follows: Case A, adiabatic prereforming without process gas preheating; Case B, adiabatic prereforming with preheat from process gas before the preheater; and Case C, adiabatic prereforming with heat of process gas both before and after prereforming [13,14]. The feed to the prereformer is preheated to a temperature in the range of 380–520°C, depending on the feedstock and the steam to carbon ratio. The overall process reaction in the prereformer is endothermic for natural gas feeds and exothermic or thermoneutral for heavier feedstocks. For natural gas feeds it is not uncommon to have a 60–70°C temperature drop across the catalyst. Prereforming removes any of the remaining sulfur from the gas stream acting as a guard reactor for the primary reformer [15]. The preformer can reduce the duty required on the primary reformer by 20% which can lead to a possible increase in synthesis gas production [16,17].

The primary reformer heats the incoming gas to reaction temperatures (825°C) and then provides the proper environment for steam reforming to

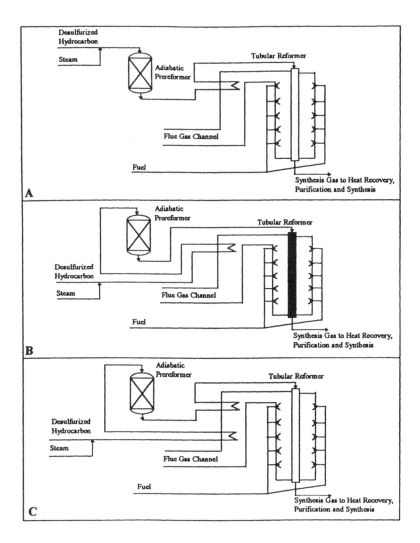

Figure 2.3 Adiabatic prereforming configuration. *Source*: [2,15,29].

occur. Reforming conditions vary from reformer configuration, catalyst activity, catalyst shape, and necessary methane slip, but temperatures generally range in the area of 825°C and pressures range from atmospheric to 44 bars. Higher pressures (< 44 bars) are used in ammonia synthesis plants to reduce compression costs before the ammonia reactor. The reformer is usually heated by natural gas fed to external burners, consuming about 1/3 of the total natural gas to the process. Other fuels can be used, but they must be nonashing and produce no carbon deposition since this can create hot spots and reduce the reformer life. Typically, the methane concentration exiting the reformer is 8–9% and is referred to as methane slip. Allowing for methane slip to the secondary reformer reduces the duty required by the primary reformer.

The primary reformer is constructed of vertically suspended chrome-nickel alloy tubes filled with a nickel catalyst. The function of the catalyst tube is to retain the catalyst and reaction gases under pressure while ensuring efficient heat transfer to promote the endothermic reaction. The reformer tube life, generally 100,000 hours, is limited by creep which is a function of the differential pressure across the tube and the maximum operating temperature [1,4,18]. The life of a reformer tube is based on the design temperature, deviation from this by as much as 20° can result in halving the tube life even if for only a short period of time (Figure 2.4). The tubes are generally 10–11 m in length, but have been known to be as long as 14 m and have a diameter of 95 mm. The length of the tube is limited by the number of welded seams allowed in the reformer heated section. Welded seam areas are more sensitive to stresses in the reformer that can lead to catastrophic failure. The tubes are supported individually by semiflexible connections, pigtails, to eliminate the stresses caused by tube expansion and contraction at start-up and shut-down. The reformer tubes typically grow by 175 mm in length and 1.3 mm in diameter when heated to operating temperatures. Modern plants have each row of tubes in a subheader which, in turn, have individual pigtails for each tube. The pigtails are either attached to the catalyst-loading flange or to the outside of the tube before it enters the firebox [18,20].

The steam-to-carbon ratio is a very important parameter in the steam reforming of methane in order to reduce or eliminate carbon deposition. The theoretical minimum ratio of steam to carbon for a natural gas feed is 1, but in practice it is a minimum of 2.2 and typically recommended to be 2.5. If the feedstock is a light naphtha the ratio is increased to 3 and even higher for naphtha containing aromatics (3.5–4.0) [8].

Severe carbon formation can easily shatter the strongest catalyst by growing inside the pore. Such physical degradation is not prevented by the strength of the support. Instead, the key is to prevent or minimize the degree of carbon formation. The single best solution is by careful incorporation of controlled amount of alkali into the catalyst formulation, such as potassium oxide to the

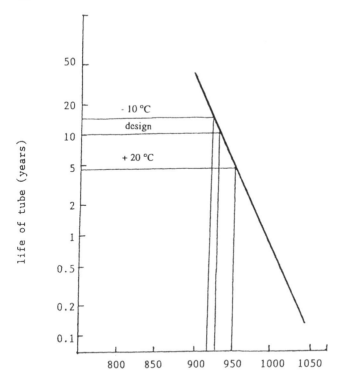

Figure 2.4   Temperature versus tube life. *Source*: [8,17].

nickel catalyst. To a certain extent, alkali can prevent carbon formation in the primary reformer. However, the addition of alkali has the disadvantage that alkali lowers the catalyst activity. Topsoe developed a low alkali-promoted catalyst, designated by RK-69–7H, which lowers the catalyst activity much more at low temperatures than at high temperatures. The lower activity of the alkali-promoted catalyst results in higher tube skin temperatures compared with the alkali-free catalyst. Therefore, alkali-promoted catalyst should only be used when necessary, and only in the part of the tube where there is a potential risk of carbon formation.

The primary reformer is a high stress environment for reforming catalysts. The catalyst must have high activity (high surface area), minimum resistance to flow (low pressure drop), good physical strength, and good heat transfer characteristics. In most catalytic systems the catalytic bed operates under

Table 2.4 BASF Primary Reforming Catalysts [35]

Catalyst type	G 1-80	G 1-50	G 1-25	G 1-25 S
Chemical composition (wt%)				
NiO	53	20	25	16
MgO	14	Balance[a]	—	—
CaO	—	Balance[a]	8	—
Al_2O_3	—	Balance[a]	Balance	Balance
SiO_2	Balance	Balance[a]	< 0.15	< 0.15
Alkali	—	5	< 0.3	< 0.2
Ignition loss (900°C)	12	10	15	1
Physical properties				
Form	Tablet	Ring	Ring	Ring
Dimensions (mm)	5 × 5	16 × 10 × 8	16 × 16 × 8	16 × 16 × 9
			12.7 × 9.5 × 4.75	16 × 10 × 9
Bulk density (kg/l)	0.9	1.1	1.2/1.3	1
Crushing strength (kg)	10	35	45/35	40/30
Surface area (m^2/g)	200	10	10	5

[a]: balance between MgO, CaO, Al_2O_3, and SiO_2

adiabatic conditions, but the opposite happens in the primary reformer. The heat supplied to the gas comes from the catalyst bed making heat transfer across the bed very important. A reduction of 30°C in temperature will halve the reaction rate. The heat is transferred by convection and the efficiency of this transfer is dependent on how random the gas flow is in the catalyst. Clearly the shape of the catalyst is an important factor in the heat transfer. The catalyst must pack in a manner that does not "bridge" (catalyst particles jam leaving a void space below them), stack up in columns, or have a significant pressure drop. The catalyst particles are in the form of solid cylindrical pellets or Raschig rings (Table 2.4) [35].

Loading of the catalyst into the reformer tubes is an important operation since the catalyst must be packed in such a way as not to be crushed and must be arranged in such a way that bridging does not occur. Currently the three methods commonly used to pack reformer tubes with catalyst are water, trickle, and socking. The water method requires that the catalyst tubes be filled with water and then filled by pouring the catalyst into the tubes. Once loaded, the catalyst and tubes then dry completely. The second method, trickle, uses a knotted rope or similar device to break the fall of the catalyst particles as they are poured into the reforming tube. The last method, socking, utilizes catalyst prepacked in polypropylene socks. The socks have a diameter slightly

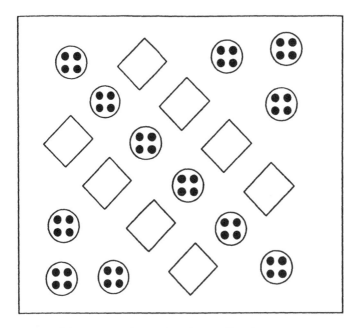

Figure 2.5   ICI 4-hole catalyst. *Source*: [8,20].

narrower than the tube. This allows for the socks to be easily lowered into the tube. In many instances this method (along with vibrational methods) are used to ensure proper catalyst distribution [22].

More recent catalysts developed by ICI and Katalco have four holes in a cylindrical pellet (Figure 2.5), which increases the surface area by 69% [35]. Haldor Topsoe introduced a tablet-shaped catalyst with 7 holes (Figure 2.6) [35]. In 1978, United Catalysts and Catalysts & Chemicals (CCE) introduced a spoked-ring catalyst. There are two variations of this catalyst: one with 5 spokes and one with 7 spokes. The 5 spokes are designed to give minimum pressure drop with high activity and heat transfer, whereas the 7 spokes are for maximum activity and heat transfer with low temperature drop. Recently, CCE developed a 9–ribbed catalyst which has higher strength and heat transfer qualities, and equally good pressure drop characteristics and activity when compared to the spoked catalysts (Table 2.5) [35].

Several basic reformer burner designs are being used today as shown in Figures 2.7 and 2.8. These basic designs are classified as top-firing, bottom-firing, side-firing, terrace-wall, and a new autothermal design. Each configuration has its own advantages and disadvantages, although the most important factors are burner to tube ratio, tubes/radiant section, and heat transfer effi-

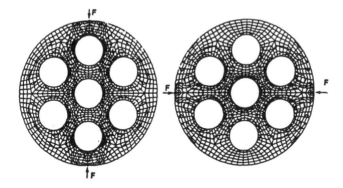

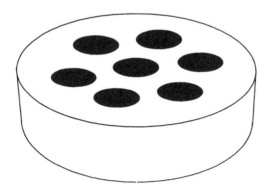

Figure 2.6   Haldor Topsoe 7-hole catalyst tablets. *Source*: [8].

Table 2.5 Comparison of UCI/CCE Primary Reforming Catalysts [35]

Type	Size	Activity	ΔP	Heat transfer
Rings	16 × 16 × 8	100	100	100
	16 × 16 × 6	100	105	105
	16 × 9 × 6	115	135	115
	12 × 9 × 6	130	155	135
5 Spokes	17 × 17	110	90	110
7 Spokes	17 × 9	135	125	125
Rib rings (9 ribs)	16 × 16 × 6	110	90	115
	16 × 9 × 6	130	125	120

ciency. The configurations can be broken down by manufacturers: Exxon manufactures bottom-fired; Foster Wheeler manufactures terraced-wall; Haldor-Topsoe manufactures side-fired; and Humphreys and Glasgow, M. W. Kellogg, Uhde Gmbh and ICI make top-fired reformers.

Industrial reformers are manufactured by a number of companies including Exxon (Figure 2.9), Foster Wheeler (Figure 2.10), Haldor Topsoe (Figure 2.11), Humphreys and Glasgow, ICI (Figure 2.12), M. W. Kellogg, and Uhde Gmbh. The design configurations differ from company to company as shown in Table 2.6 [20]. Each company offers their own unique reformer design based on the configurations discussed previously.

In the top-fired reformer, the burners are located only on the roof between the rows of reformer tubes. The heating is direct from the combustion products to the tube walls. In this configuration firing occurs only at one level, heating the natural gas as it enters the reformer. The burner to tube ratio is low and the combustion air product distribution is simplified. The unit is compact, using less steel, and has large tube capacities (600–1000) per radiant section. However, the operating environment above the box is uncomfortable and the control of heat input into the reformer is limited.

The side-fired furnace has burners located on the walls with the fire directed at one or two rows of reformer tubes or, more commonly, to the reformer walls. Heating in most of these configurations radiates from the reformer walls. This process allows for uniform heating and heating control. However, this configuration is limited to 100–150 reformer tubes per radiant section leading to poor burner to tube ratio and increased complexity of the air distribution. Since the system uses radiation from the refractory lining the thermal efficiency is low. A modification of this design is the terraced-wall design which has several levels of heating allowing for more controlled heating.

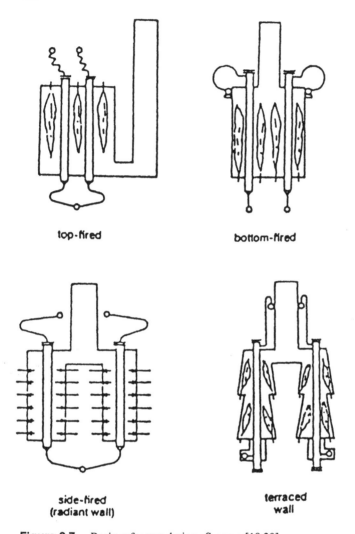

top-fired

bottom-fired

side-fired
(radiant wall)

terraced
wall

Figure 2.7 Basic reformer design. *Source*: [18,29].

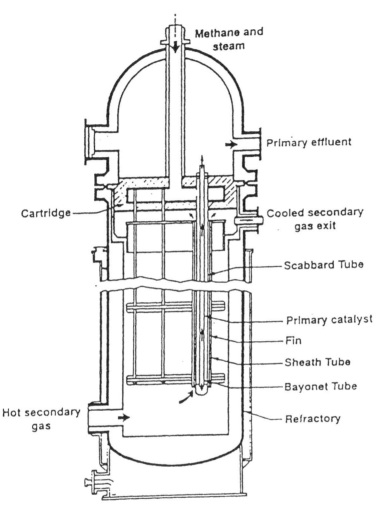

Figure 2.8 Gas-heated reformer. *Source*: [2,23,29].

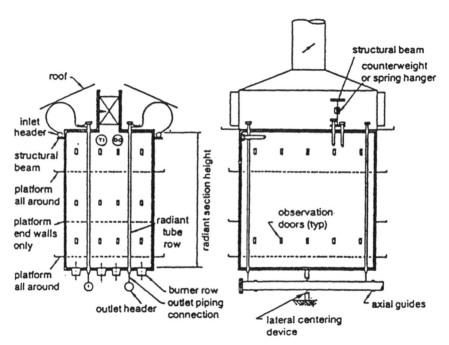

Figure 2.9   Exxon steam reformer. *Source*: [20].

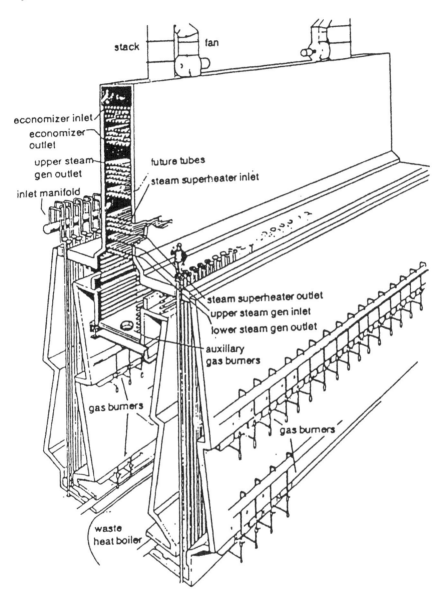

Figure 2.10   Foster Wheeler terraced-wall reformer. *Source*: [20].

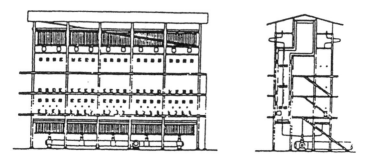

Figure 2.11   Haldor Topsoe primary reformer. *Source*: [20].

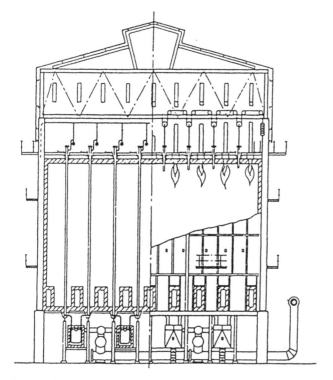

Figure 2.12   ICI top-fired reformer. *Source*: [20].

Table 2.6 Classical Reformer Designs [20]

	Reformer type	Heating	Figure
Exxon	Bottom fired	Direct	2.9
Foster Wheeler	Terraced wall	Direct	2.10
Haldor Topsoe	Side-wall fired	Radiant	2.11
Humphreys & Glasgow	Top fired	Direct	
ICI	Top fired	Direct	2.12
M. W. Kellogg	Top fired	Direct	
Uhde Gmbh	Top fired	Direct	

The bottom-fired reformer is classified into two types, one with the reforming gas flowing down the tube (as in top-fired and side-fired) and the other where the process gas flows up the reformer tubes. The burners are located on the floor on either side of two rows of reformer tubes. The flames are long and pencil thin. The system has a simplified air combustion distribution and single operating level. The system cannot handle more than 200–300 reformer tubes per radiant section and the tube metal temperatures at the process gas outlet are higher than the inlet.

Although the idea has been around since the 1950s, the new synthesis gas systems incorporate revolutionary technologies into the reforming process. The new process systems are commonly referred to as autothermal since they utilize the hot synthesis gas from the secondary reformer to heat the autothermal reformer (primary reformer). These systems have tremendous advantages over conventional processes such as isobaric conditions, smaller construction size, and more efficient heat transfer. The elimination of a pressure difference (isobaric) between the inside and the outside of the reformer tubes greatly reduces the stresses imposed on them allowing for new, more efficient tube designs. The process eliminates the waste heat that was not recoverable from the older designs except for the stream that exits the process.

The next generation of reformers—autothermal reformers—do not contain furnaces as with the classical reforming systems. The ICI LCA (Leading Concept Ammonia) plant utilizes a gas-heated reformer (GHR) (Figure 2.13) and is in operation at the 450-ton/day ammonia Severnside plant [23]. Another GHR configuration is the Tandem GHR developed by GIAP (Figure 2.14) for use in the 600 MTPD ammonia plant in Grodno-Belorussiya [20]. The M. W. Kellogg autothermal reformer is called the Kellogg Reforming Exchange System (KRES) [20,24]. Udhe Gmbh has taken the technology one step further by developing a system, Combined Autothermal Reformer (CAR), in which the primary and secondary reformers are housed in the same shell as

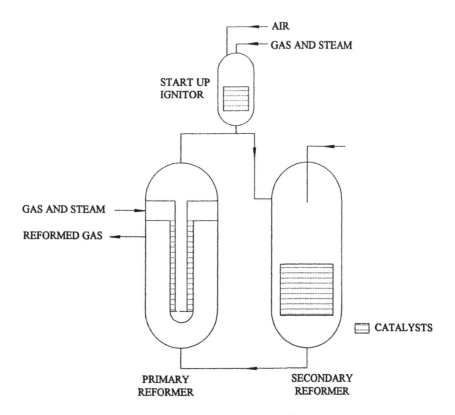

Figure 2.13 LCA gas-heated reformer. *Source*: [2,29].

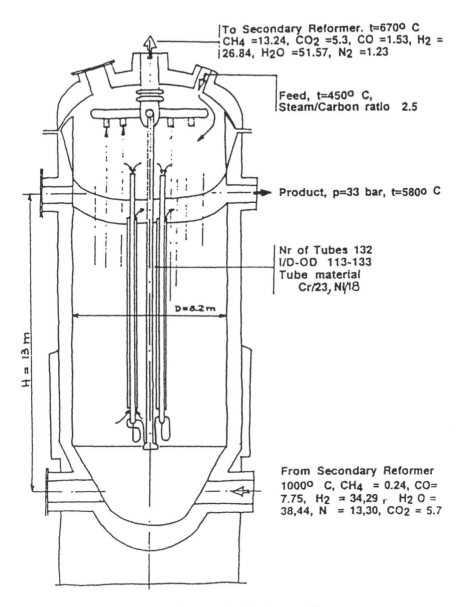

To Secondary Reformer. t=670° C
CH4 =13.24, CO2 =5.3, CO =1.53, H2 = 26.84, H2O =51.57, N2 =1.23

Feed, t=450° C,
Steam/Carbon ratio 2.5

Product, p=33 bar, t=580° C

Nr of Tubes 132
I/D-OD 113-133
Tube material
Cr/23, Ni/18

D=8.2m

H = 13 m

From Secondary Reformer
1000° C, CH4 = 0.24, CO= 7.75, H2 = 34,29 , H2 O = 38,44, N = 13,30, CO2 = 5.7

Figure 2.14 Tandem-type reformer of GIAP. *Source*: [2].

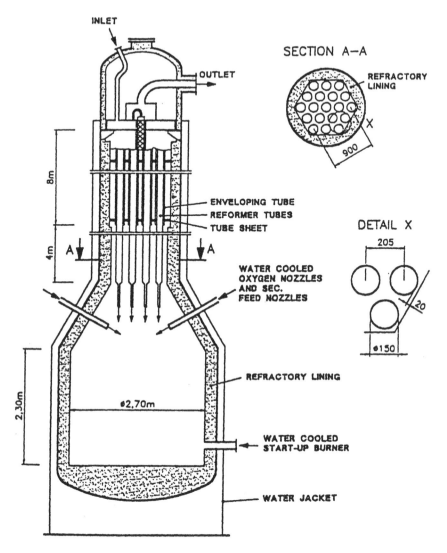

Figure 2.15   CAR demonstration reactor. *Source*: [2,29].

in Figure 2.15 [20,25,26]. The secondary zone can be operated as a noncatalytic partial oxidation reactor or as a secondary reformer.

A secondary reformer contains two distinct sections, the first for combustion (upper part) and the second a catalyst bed (lower part). A secondary reformer can alleviate 10% of the heating duty in the primary reformer by allowing methane slip. In the upper part of the reformer air, enriched air, or oxygen is mixed thoroughly with the incoming gaseous effluent from the primary reformer. The type of oxidant used depends on the final gas composition necessary; ammonia plants use air or enriched air in order to obtain the correct stoichiometric ratio of N:H, while hydrogen and methanol plants use oxygen. The oxidant mixes with the incoming gas and immediately reacts to form H_2O. The highly exothermic reaction increases the temperature inside the reformer to 1300°C. In the second stage, a catalyst bed made from nickel catalyst similar to the primary reformer is utilized to catalyze endothermic steam reforming reactions. The catalyst contains a lower amount of nickel to limit sintering at the higher temperatures. In the catalyst bed, endothermic steam reforming reactions take place that reduce the temperature and lower the methane content to 0.3–0.7%.

The stresses on the secondary reforming catalysts are not nearly as great as those on the primary reforming catalysts. From the secondary reformer design the catalyst bed is adiabatic, making the heat transfer characteristics of the catalyst less important. In general, these catalysts are much simpler than the primary reforming catalyst. The support materials are the same, but the nickel loadings are only 5–10 wt%, which results in a lower activity when compared to the primary reformer catalysts. Lower loadings are used because they inhibit sintering, which increases in rate at the higher temperatures in the secondary. The secondary reforming catalysts are made from classical shapes (i.e., solid cylinders and Raschig rings). The catalyst bed in the secondary reformer is 3–4 m deep and can exert a significant load on the catalyst which would crush many of the new fancier shapes. The typical loading involves the top 20% being solid cylinders and the rest being Raschig rings. These solid cylinders act as heat shields as the local flame temperatures will approach 1500–1700°C.

The secondary reformer is a refractory-lined cylindrical steel vessel (Figure 2.16) [14]. The walls of the vessel are protected from the extreme conditions by two layers of refractory lining. The lining near the wall of the vessel is a good insulator and mechanically moisture proof. The second lining is resistant to high temperatures and erosion by the hot gases. These linings should be low in sulfur and silica since these materials poison catalysts. In many of the new designs, the secondary reformers are water jacketed, allowing for better heat dissipation and even wall temperatures. However, the water can mask the formation of hot spots until the unit fails catastrophically. The reformer

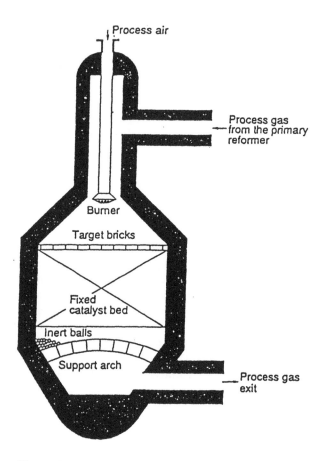

Figure 2.16   Typical secondary reformer. *Source*: [14].

walls are constructed of carbon steel; 310 or 316 stainless steels are used in places where hydrogen embrittlement can occur (thermocouples, etc.). The burners in the top of the reformer are made from alloys like HK-40 [14].

III. SYNTHESIS GAS VIA NONMETHANE ROUTES

A. Naphtha Reforming

The process of naphtha reforming is very similar to that of natural gas reforming and many reformers can handle both types of feeds. Naphtha (chemical petrol) is a low boiling point (< 200°C) liquid hydrocarbon material. Higher boiling point compounds cannot be used in steam reforming for two reasons: 1) the difficulty removing sulfur to a very low level and 2) the degradation of the H:C ratio, which makes it difficult to prevent side reactions from occurring, which, in turn, leads to carbon deposition and the potential for hot spots in the reformer.

A general reaction mechanism for steam reforming of naphtha compounds is shown in Reaction 2.13. The reactions presented previously also apply to this process.

$$C_nH_{2n+2} + nH_2O = nCO + (2n + 1)H_2 \qquad (2.13)$$

A process flow diagram is shown in Figure 2.17 for the ICI steam reforming of naphtha process [27]. Since the nickel catalyst is very sensitive to sulfur poisoning, the first stage utilizes a CoO-MoO_3/Al_2O_3 and hydrogen addition to produce H_2S at 350–450°C. Any olefins present are simultaneously hydrogenated. The resulting H_2S is adsorbed onto ZnO in an irreversible reaction. Once the sulfur is removed the feed is sent to the primary reformer. In this reformer reaction conditions vary depending on the feed distribution but they generally are in the range of 700–830°C, while the pressures range from atmospheric to 40 bars. The primary reformer, allothermal reformer, is made up of either horizontally- or vertically-mounted catalyst filled tubes that are heated externally, typically by natural gas, to reduce carbon deposition. The reformer materials and design limit the reformer tube temperature to 830°C, exceeding this temperature can reduce greatly the life of the reformer. The process utilizes a nickel (Ni-K_2O/Al_2O_3) catalyst similar to the methane reforming catalyst. The primary reformer produces a synthesis gas (H_2 and CO) containing large amounts of CH_4. In the next step, a secondary reformer is used to react unreacted CH_4. The reformer consists of a lined chamber containing a high temperature Ni catalyst. In this autothermal reformer, air or oxygen is added to produce heat raising the temperature to over 1200°C. The residual CH_4 content after exiting the reformer is 0.2–0.3%.

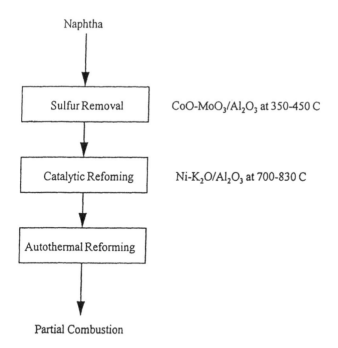

Naphtha

Sulfur Removal   CoO-MoO$_3$/Al$_2$O$_3$ at 350-450 C

Catalytic Refoming   Ni-K$_2$O/Al$_2$O$_3$ at 700-830 C

Autothermal Reforming

Partial Combustion

Figure 2.17   Flow diagram for the ICI process. *Source*: [27].

B.   Fuel Oil Partial Oxidation

The noncatalytic partial oxidation of fuel oil converts petroleum feedstock into synthesis gas, CO_2, CH_4, and H_2S if any sulfur is present. This process has the ability to handle sulfur compounds without pretreatment (sulfur removal). Eliminating a processing step from steam reforming techniques, however, requires additional equipment for oxygen addition in the reactor. The expected carbon deposition problem with hydrocarbon material containing low H:C ratios is limited by the use of carbon recovery and extinction recycle schemes [28]. The partial oxidation of fuel oil typically operates at a temperature range of 1200–1500°C and pressures of 30–80 bars in a refractory-lined reactor.

The partial oxidation reactor is broken into three distinct reaction zones. Initially, the fuel oil is injected into the reactor along with oxygen and steam. The heating vaporizes and cracks the hydrocarbon material. The steam is used to control the temperature of the reactor. The hydrocarbons, oxygen, and steam are well mixed at this stage.

In the second zone, substoichiometric amounts of oxygen partially oxidize the fuel oil into CO and H_2 as shown in Reaction 2.14. This highly exothermic reaction consumes all of the available oxygen. Other reactions occurring in this zone include hydrocarbons reacting endothermically with CO_2 and H_2O as in Reactions 2.15 and 2.16 to form synthesis gas.

$$2C_nH_{2n+2} + nO_2 = 2nCO + 2(n+1)H_2 \tag{2.14}$$

$$C_nH_m + nCO = 2nCO + \frac{m}{2}H_2 \tag{2.15}$$

$$C_nH_m + nH_2O = nCO + \left(\frac{m}{2} + n\right)H_2 \tag{2.16}$$

In the third zone, slow secondary reactions take place. The carbon particulates formed react with CO_2 and steam to form synthesis gas. However, these slow reactions do not reach equilibrium due to low residence time in the reactor, leaving some carbon formation in the reactor. The final composition of the synthesis gas is determined by the water-gas-shift reaction (Reaction 2.2). The key variables to control outlet gas composition are the oxygen:fuel ratio and steam:fuel ratio.

The current producers of partial oxidation processes are BASF, Texaco, Shell, and Hydrocarbon Research. A process block diagram showing the three process steps can be seen in Figure 2.18 [30]. The first step is the gasification process with the addition of oxygen and water. A detailed schematic of the Shell Gasification plant (SGP) is shown in Figure 2.19 [28]. The most important features of the process are the reactor, waste-heat-boiler, carbon catcher,

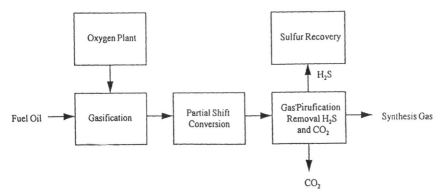

Figure 2.18   Process block diagram for partial fuel oil oxidation. *Source*: [30].

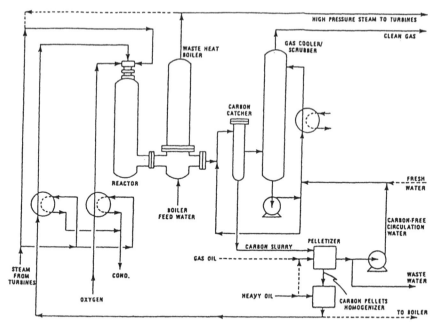

Figure 2.19   Shell gasification plant. *Source*: [30].

and the gas cooler/scrubber. The second step is the partial shift conversion to obtain the correct ratio of CO:H$_2$ and the last step is the gas purification (removal of CO$_2$ and H$_2$O).

C.   Reformer Off-Gas Purification

Synthesis gas created from the gasification of fossil fuels is contaminated by CO$_2$ and perhaps sulfur compounds (H$_2$S and COS). CO$_2$ competes and/or interferes with many chemical reactions that are utilized in the chemical industry. Sulfur gases poison catalysts, inhibiting their activity and reducing their life. Reformer off-gas purification involves the removal of these acid gases. The gases are removed by a large number of processes as shown in Table 2.7 [31]. These processes are broken down into several types; physical absorption, chemical absorption, and adsorption as shown in Table 2.8 [31]. The important variables in acid gas removal are the gas flow rate, the initial concentration of acid gases, and the final concentration required of CO$_2$ and H$_2$S.

Absorption is the most common technique used to remove acid and is broken down into three basic categories: physical, chemical, and hybrid (a combination of physical and chemical). These processes separate the acid

Table 2.7 Gas Cleaning Processes for H_2S and CO_2 Removal [31]

Process	Sorbent	Removes
Amine	Monoethanolamine, 15% in water	CO_2, H_2S
Econamine	Diglycolamine, 50-70% in water	CO_2, H_2S
Alkazid	Solution M or DIK (potassium salt of dimethyl acetic acid), 25% in water	H_2S, small amount of CO_2
Benfield, Catacarb	Hot potassium carbonate, 20-30% in water (also contains catalyst)	CO_2, H_2S, selective to H_2S
Purisol (Lurgi)	N-methyl-2-pyrrolidone	H_2S, CO_2
Fluor	Propylene carbonate	H_2S, CO_2
Selexol (Union Carbide)	Dimethyl ether polyethylene glycol	H_2S, CO_2
Rectisol (Lurgi)	Methanol	H_2S, CO_2
Sulfinol (Shell)	Tetrahydrothiophene dioxide (sulfolane) plus diisopropanol amine	H_2S, CO_2; selective to H_2S
Giammarco-Vetrocoke	K_3AsO_3 activated with arsenic	H_2S
Stretford	Water solution of Na_2CO_3 and anthraquinone disulfonic acid with activator of sodium *meta*vanadate	H_2S
Activated carbon	Carbon	H_2S
Iron sponge	Iron oxide	H_2S
Adip	Alkanolamine solution	H_2S, some COS, CO_2 and mercaptans
SNPA-DEA	Diethanolamine (DEA) solution	H_2S, CO_2
Takahax	Sodium, 1,4-napthaquinone, 2-sulfonate	H_2S

gases from the gaseous mixtures by mass transfer of solute gases into a liquid, usually organic or alkaline solutions. The processes are typically set up countercurrently to maximize mass transfer efficiency. The solvent is typically regenerated by the addition of heat. Solvents for cost-effective and efficient gas cleaning have high capacity for acid gas, a low tendency to dissolve valuable feed components, low vapor pressure, thermal stability, low tendency towards fouling and corrosion, and acceptable cost [31]. The heavy hydrocarbons must be removed from the gas stream to prevent the solvents from co-absorbing the heavy hydrocarbons. These hydrocarbons lead to foaming in the absorber which causes instability leading to higher regeneration and the hydrocarbons are lost in the regeneration process.

The equipment typically used in absorption are spray towers, tray towers, packed towers, and Venturi scrubbers. The proper equipment is selected on the basis of the acid gases to be removed, initial and final concentration, and

Table 2.8 Summary of Gas Cleaning Processes [31]

Sorbent	Nature of interaction	Regener-ation	Examples
Liquid	Absorption and chemical reaction	Yes	Many processes for the removal of CO_2 and H_2S from various gases, with solvents like water, MEA, DEA, DIPA, etc. Agents improving physical solubility may be added (Sulfinol process); H_2S may be recovered as such or oxidized to S.
Liquid + solid	Absorption + chemical reaction	Varies	Some slurry wash processes for flue gas desulfurization
Liquid	Physical adsorption	Yes	CO_2 and/or H_2S from hydrocarbon gases; solvents: N-methyl pyrroli-done, propylene carbonate, methanol
Solid	Physical adsorption	Yes	Purification of natural gas (H_2S, CO_2); with molecular sieves
		Yes	Gas drying operations (cyclic regenerative); molecular sieves
		Varies	Odor removal from waste gases (activated carbon)
Solid	Chemical reaction	No	H_2S from process gases, with ZnO
		Yes	SO_2 from flue gases, with CuO/Al_2O_3

flows. The factors affecting equipment efficiency are: gas phase nonhomogeneity, heat of reaction (if any), liquid-gas interface, and volume of gas. A simple classification list of both chemical and physical absorption processes can be found in Table 2.9 [31].

Typical physical absorption processes include the Rectisol and Selexol processes. Figure 2.20 shows a flowsheet of Selexol process [38]. The treated gas contains less than 1 ppm of the total sulfur, levels of CO_2 on the order of parts per million, and less than 7 lb/MMscf. Offgas is a highly enriched H_2S stream for Claus processing for sulfur recovery. Feed gas enters the absorber where contaminants are absorbed by Selexol solvent. Rich solvent from the bottom flows to a recycle flash drum to separate and recompress any co-absorbed product gas back to the absorber. Offgas is released from the drum by further pressure reduction. The solvent is regenerated in a stripper column. The process has a very high onstream efficiency under a wide range

Table 2.9 Simple Classification for Acid Gas
Removal Processes [31]

Chemical absorption	Physical absorption
Alkanolamines	Selexol
MEA	Rectisol
SNPA, DEA (DEA)	Sulfinol
UCAP (TEA)	
Selectamine (MDEA)	
Econamine (DGA)	
ADIP (DIPA)	
Alkaline salt solutions	
Hot potassium carbonate	
Catacarb	
Benfield	
Giammarco-Vetrocoke	
Nonregenerable	
Caustic	

of operating conditions and feed compositions, pressures to 137+ bar and temperatures from −18 to 121°C [38].

In chemical absorption, the solvent reacts with the acid gases to form weak bonds which are broken after separation by reducing the acid gas partial pressure and increasing the temperature. This process requires fewer contacting stages than physical solvent processes. Due to their nature, chemical solvents are more suited for applications at lower pressures and temperatures, although chemical absorption reaction rate is decreased at lower temperatures. The two most common types of chemical systems are amines and hot carbonates. Carbon dioxide absorption can be aided by the addition of additives such as methanol, ethanol, glycols, glycerol, sucrose, and dextrose.

The amine chemical absorption processes are broken down into primary, secondary, and tertiary amines. Table 2.10 lists some commonly used amines and their chemical names [31]. The physicochemical constraints affecting process amine losses are entrainment, vapor pressure in the gas stream, and degradation. Entrainment losses take place when foaming occurs. This can be reduced by the addition of a degasser which effectively removes the hydrocarbons between the absorption stage and the regeneration stage. The vapor pressure losses can be reduced by washing the purified gas with glycol or water at the end of the absorption process. Regenerable adsorbents such as bauxite also may be used. The amines are subject to chemical degradation typically due to oxygen attack of H_2S, thus forming free S which reacts

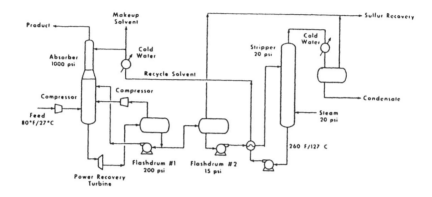

(a)

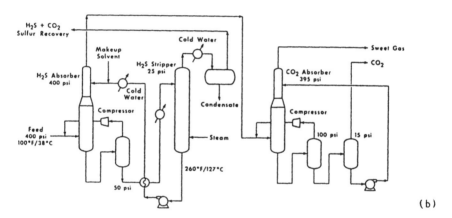

(b)

Figure 2.20 Flowsheet of Selexol process. *Source*: [38].

Table 2.10 Alkanolamines Used in Gas Treating Processes [31]

Name		Chemical formula
Ethanolamine	MEA	$HOC_2H_4NH_2$
Diethanolamine	DEA	$(HOC_2H_4)_2NH$
Triethanolamine	TEA	$(HOC_2H_4)_3N$
Diglycolamine	DGA	$H(OC_2H_4)_2NH_2$
Diisopropanolamine	DIPA	$(HOC_3H_6)_2NH$
Methyl diethanolamine	MDEA	$(HOC_2H4)_2NCH_3$

irreversibly with amines. Other compounds such as formic acid (HCO_2H), acetic acid (CH_3CO_2H), carbon disulfide (CS_2), and carbonyl sulfide (COS) react irreversibly with amines [32].

The primary amines typified by monoethanolamine (MEA) are highly reactive and capable of achieving a very low H_2S concentration. However, the process is energy intensive and monoethanolamine (MEA) degradation products are highly corrosive. Monoethanolamine processes typically use 15–20 wt% solutions. These factors make the process effective but expensive to operate and maintain. Another primary amine diglycolamine (DGA) requires less energy than MEA but has higher solvent costs since higher concentrations are used up to 60 wt% solutions [33].

A secondary amine, diethanolamine (DEA), has a lower heat of reaction and its degradation products are less corrosive than MEA. Unfortunately, the solvent is difficult to maintain at a high purity. Diisopropanolamine (DIPA), a sterically-hindered secondary amine, was the first found to be selective to H_2S in the presence of CO_2. A tertiary amine, methyl diethanolamine (MDEA), has a large share of the amine market because it is selective in removing H_2S from CO_2, resistant to degradation by organic sulfur compounds, it has less corrosive degradation compounds, and consumes less energy [33,34].

Hot carbonates are well suited for the removal of CO_2 at moderate or high levels in the presence of little or no H_2S. The process acquired its name from the use of elevated temperatures in both the absorber and the regenerator (110–115°C). Hot carbonates such as the Benfield and the Koppers Vacuum Carbonate utilize K_2CO_3 to remove H_2S, COS, and CO_2 from gas streams [35]. Their heat requirements and high solvent circulation make hot carbonates more expensive than other acid gas removal processes. Other hot carbonate processes, including the Catacarb and the Giammarco-Vetrocoke processes, use catalysts, corrosion inhibitors, and/or activators to enhance the removal of the acid gases. Hot carbonate-promoted systems are able to decrease the CO_2 level from 1% to 0.1%. Promoters include DEA, amine borates, and hindered amines [36].

Hybrid solvents attempt to combine the high gas purity offered by chemical absorption and the flash regeneration along with lower energy requirements of physical absorption. Processes such as Sulfinol are a mixture of sulfolane, DIPA or MDEA, and water. If the requirement is the total removal of the acid gases, then DIPA is used. If only H_2S is to be removed, then MDEA can be used. Another process is the Flexisorb PS which can treat gases to less than 50 ppm CO_2 and 4 ppm H_2S.

Figure 2.21 shows a schematic of Shell's Sulfinol process, which removes H_2S, COS, RSH, and CO_2 from refinery offgases or natural gases [38]. The process is also applicable to gas cleanup of synthetic and refinery gases. The total sulfur content in the treated gas can be reduced to ultra low ppm levels.

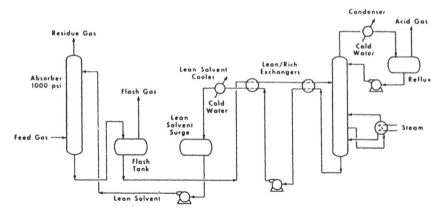

Figure 2.21 Schematic of Shell's Sulfinol process. *Source*: [38].

The mixed solvent of Sulfinol process consists of a chemical reacting alkanolamine, water and the physical solvent Sulfolane (tetrahydrothiophene dioxide). The actual formulation is customized for the species of application. The process flow scheme is similar to that of other amine processes. In most applications co-absorbed hydrocarbons from the absorber are flashed from the solvent and used as fuel gas after treating in a fuel gas absorber. The solvent is regenerated in the process. Over 180 units are either in operation or under construction as of April 1994 [38].

Figure 2.22 shows a schematic of Exxon Flexsorb solvents process. The process removes H_2S selectively or removes a group of acid impurities including H_2S, CO_2, COS, CS_2, and mercaptans from a variety of streams, depending on the solvent used. The schematic shown here is a typical amine system. Flexsorb SE solvent is an aqueous solution of a new hindered amine. Flexsorb SE plus is an enhanced aqueous solution which has improved H_2S regenerability. Flexsorb PS solvent is a hybrid solution consisting of a hindered amine, a physical solvent, and water. Flexsorb HP is a hot potassium carbonate-based system containing a hindered amine promoter. Flexsorb PS solvent yields a treated gas with: H_2S < 0.25 g/100 scf, CO_2 < 50 ppmv, COS and CS_2 < 1 ppmv, mercaptans removal > 95%. This solvent is used primarily for natural gas cleanup. Flexsorb HP is primarily for hydrogen plants or natural gas cleanup [38].

A comparison of the absorption processes is given in Tables 2.11 and 2.12. In Table 2.11 the operating pressure and selectivity of the processes are listed to show how the processes vary depending on the nature of the removal necessary [31]. In Table 2.12 the absorption processes are shown [27].

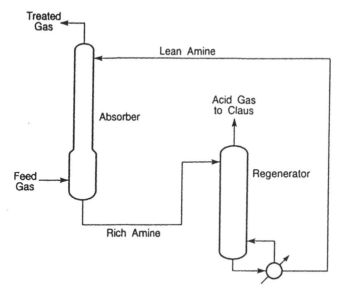

Figure 2.22   Schematic of Exxon Flexsorb solvents process. *Source*: [38].

Adsorption onto solid beds is another method of acid gas removal from reformer off-gas. Typically, adsorption beds are best suited for low to medium concentrations of H_2S and can be made selective for its removal. This type of operation is classified into two types of processes which are nonregenerable and regenerable. The nonregenerable packed beds usually operate at temperatures above 350°C and are made out of ZnO which reacts irreversibly with H_2S. These processes are capable of reducing the sulfur content to less than 0.1 ppm.

Regenerable adsorption media include molecular sieves, iron oxide, zinc titanate, and tin oxide. Molecular sieves are alkali alumino silicates that have highly localized polar charges [31,32]. These charges make it easy for polar molecules such as water, H_2S, SO_2, NH_3, carbonyl sulfide, and mercaptan to be removed. Molecular sieves can be designed to be selective in removing H_2S from gas containing appreciable amounts of CO_2 [31,32]. Total sulfur removal can be achieved with molecular sieves. Recent research has focused on zinc based materials because of their stability and ability to remove more than 99% of H_2S and COS at temperatures up to 750°C [31,32]. The best known process that removes H_2S is the Iron Sponge. The process utilizes wood chips impregnated with Fe_2O_3. The process can be regenerated by passing air over the bed. The zinc titanate is regenerated at low pressures by passing air over the bed. Systems using this process have been developed by

Table 2.11   Acid Gas Removal Processes [31]

Solvents	General operating range (CO$_2$ partial pressure)	Selectivity[a]	
		H$_2$S/CO$_2$	CO$_2$/HC
Amines			
MEA	Low	1	8
DEA	Low	1	8
DGA	Low	1	8
MDEA	Low	3	8
TEA	Low	3	8
DIPA	Low	2	8
Hot carbonate			
Catacarb	Moderate	2	9
Benfield	Moderate	6	9
Alkazid	Moderate	3	9
Giammarco-Vetrocoke	Moderate	9	9
Physical solvents			
Methanol	High	7	2
DMEPED	High	9	3
Propylene carbonate	High	4	3
N-methylpyrrolidone	High	9	3
Mixed solvents			
Sulfolane + DIPA (Sulfinol)	Moderate	2	3
Methanol + DGA (Amisol)	Moderate	1	3

[a] Arbitrary number scale 1–10: 1 = poor selectivity; 10 = excellent selectivity.

General Electric Environmental Services (GEES) and the Research Triangle Institute (RTI). Another metal oxide is tin oxide (SnO$_2$), which can remove H$_2$S and COS at temperatures ranging from 400–450°C. Regeneration takes place through steam regeneration at 1–3 MPa steam and 400–500°C. Other adsorbents that are now being investigated include manganese dioxide, regenerable dolomite, and membranes [33].

D.   Coal Conversion

Coal is defined as an organic rock composed of carbon, hydrogen, oxygen, nitrogen, sulfur, and mineral matter. Its conversion to useful products is typically carried out by reacting air, oxygen, steam, carbon dioxide, hydrogen, or a

Table 2.12 Comparative Summary of Acid Gas Removal Processes[27]

Feature	Chemical absorption		Physical absorption
	Amine process	Carbonate process	
Absorbents	MEA, DEA, DGA. MDEA	K_2CO_3; K_2CO_3 + MEA, K_2CO_3 + DEA, K_2CO_3+ arsenic trioxide	Selexol, Purisol, Rectisol
Operating pressure (psi)	Insensitive to pressure	> 200	250–1000
Operating temperature (°F)	100–400	200–250	Ambient temperatures
Recovery of absorbents	Reboiled stripping	Stripping	Flashing, reboiled, or steam stripping
Utility cost	High	Medium	Low-medium
Selectivity (H_2S, CO_2)	Selective for some amines	May be selective	Selective for H_2S
Effect of O_2 in the feed	Formation of degradation products	None	Sulfur precipitation at low temperature
COS and CS_2 removal	MEA: not removed DEA: slightly removed DGA: removed	Converted to CO_2 and H_2S and removed	Removed
Operating problems	Solution degradation, foaming, corrosion	Column instability, erosion, corrosion	Absorption of heavy hydrocarbons

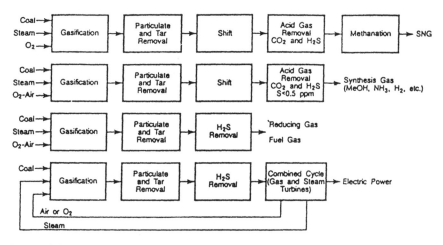

Figure 2.23 Coal gasification schemes. *Source*: [31].

mixture of these to yield gaseous, liquid, and solid products. Coal is used to produce synthesis gas, synthetic natural gas, fuel gas, and electric power (Figure 2.23) [31].

In the conversion of coal to gaseous compounds, many reactions are occurring in series and parallel. All the reactions shown in the methane reforming section (Reactions 2.1–2.10) play a role in the gasification process. The important additional coal reactions are shown in Reactions 2.17–2.20. Reaction 2.17 is the partial combustion of coal to form carbon monoxide. Reaction 2.18 is the heterogeneous steam gasification to form carbon monoxide and hydrogen. Reaction 2.19 is the complete oxidation of carbon, while Reaction 2.20 is the methanation of coal.

$$2C + O_2 = 2CO \qquad \Delta H^o_{298} = -246 \text{ kJ/mol} \qquad (2.17)$$

$$C + H_2O = CO + H_2 \qquad \Delta H^o_{298} = 119 \text{ kJ/mol} \qquad (2.18)$$

$$C + O_2 = CO_2 \qquad \Delta H^o_{298} = -394 \text{ kJ/mol} \qquad (2.19)$$

$$C + 2H_2 = CH_4 \qquad \Delta H^o_{298} = -87 \text{ kJ/mol} \qquad (2.20)$$

The important factors in coal conversion are the physical and chemical properties of the coal, heat supply (autothermal or allothermal), reactor type (fixed bed, moving bed, fluidized bed, or entrained bed), gasification agent (air, oxygen, steam, or a combination thereof), and process conditions. Typically, coal conversion is carried out at high temperatures (900–1000°C) be-

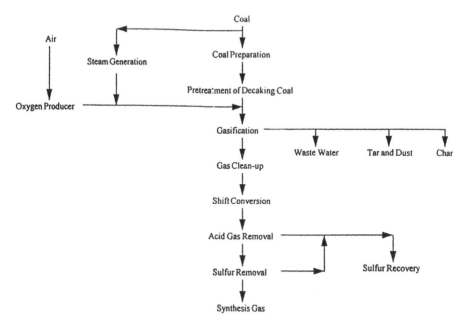

Figure 2.24   Coal gasification steps. *Source*: [5,37].

cause the gasification reactions are inherently endothermic; pressures range from atmospheric to high pressure.

Coal gasification takes place in multiple steps (Figure 2.24), but the most important processing steps are pretreatment, primary gasification, secondary gasification, and shift conversion [5,37]. Pretreatment of coal is used to reduce caking or agglomeration of coal in the gasifier. Oxygen is introduced at low temperatures to remove these compounds. The primary gasifier produces CO/ H_2 rich synthesis gas, along with other products such as CO_2, water, methane, H_2S, N_2, tars, oils, phenols, and char. The amount of char produced depends on the coal used. The gasification may be carried out in three different process systems: 1) gasifier (low Btu gas), 2) devolatizer (intermediate Btu gas), and 3) hydrogasifier (intermediate Btu gas). The difference in the systems are the reactants used (steam and H_2). The gasifier adds steam to the coal to produce CO and H_2, the hydrogasifier utilizes both steam and H_2 to produce methane, while the devolatizer uses only H_2 to produce methane and char (or coke). The secondary gasification takes the char produced from the primary gasification section and reacts it with steam to produce CO and H_2. The shift conversion is used to adjust the CO/H_2 ratio. It utilizes the water–gas shift reaction at low temperatures (< 400°C) to produce the appropriate ratio.

Table 2.13 Gasification Processes by Reactor Configuration [5,37]

Fixed bed	Fluidized bed	Entrained	Molten salt
Lurgi[a]	Winkler[a]	Koppers-Totzek[a]	Atgas
Wellman-Galusha	Hydrane	Bigas	Atomics International
Foster-Wheeler	Hygas	Combustion	Pullman Kellogg
Slagging fixed bed	U-Gas	Engineering	
Woodall-Duckham	Synthane	Texaco	
General Electric	Union Carbide	PRENFLO (Shell)	
	Westinghouse		
	Pressurized		
	CO_2 acceptor		
	Agglomerating		
	Burner		
	COED/COGAS		

[a] Conventional gasification process.

Industrially, coal gasification processes are differentiated by the type of reaction system used to gasify the coal such as: 1) fixed bed, 2) fluidized bed, 3) entrained bed, and 4) molten salt. The processes utilizing these reactor configurations are shown in Table 2.13; the typical gas distributions from some common systems shown in Table 2.14 [5,37].

The established (conventional industrial) processes are the Lurgi, the Winkler, and the Kopper-Totzek. Second generation gasification processes include Rheinbraun hydrogenative, Bertgbau-Forschung steam, Kellogg (molten Na_2CO_3), Exxon alkali carbonate catalyzed, and the Sumitomo (molten iron). In addition to these, several multistage processes have been developed such as the Westinghouse, Synthane, Bi-Gas, Hygas, U-Gas, and Hydrane processes, which are mainly for synthetic natural gas (SNG) [27,33,37].

The Winkler process was first developed in Germany (1931). The process utilizes an atmospheric fluidized bed with fine-grain (crushed) nonbaking coals. The coal is fed through a screw feeder with either O_2, air, or steam injected from the bottom of the vessel as shown in Figure 2.25, which is a schematic representation of the process [5,37]. The operating temperature varies between 800 and 1100°C, but is usually 945°C depending on the reactivity of the coal. If brown coal is used as a feed stock, the gas produced usually has a H_2/CO ratio of 1.4:1. New developments in this process include gasification under

Table 2.14 Typical Gas Compositions (vol.%) [37]

Gasifier	N_2	CO_2	CO	H_2	CH_4	C_3H_8	Btu/scf
Lurgi (O_2)	1	29	19	40	9	0	285
Winkler (O_2)	1	20	32	41	3	0	269
Koppers-Totzek (O_2)	1	9	52	34	0	0	284
Wellman (air)	46	7	26	14	3	0	160
Natural gas	0.6	0.9	0	0	91.5	1.3	1066

pressures of 10–25 bars. This decreases the reaction time while the space velocity is increased making significant economic improvements.

The Koppers-Totzek process was first commercialized in 1952 and, by 1984, was used in 19 plants in 17 different countries [27]. The process uses powdered coal at atmospheric pressures in a parallel flow of O_2 and H_2O at temperatures of 1400–2000°C. The high process temperatures eliminate the formation of hydrocarbons and produce a gas containing 85–90% syngas (CO and H_2). A process schematic is shown in Figure 2.26 [37]. Recent process developments include the Pressurized Entrained Flow Gasification (PRENFLO) with operating pressures being increased to 40 bars.

The Lurgi process dates back to the 1930s and is the most developed. Coal is fed to the gasifier from the top of the reactor. Initially, the coal is degassed with steam and product gas as it passes down through the bed at temperatures of 600–750°C. The coal is then gasified at 1000°C with steam and oxygen introduced at the bottom of the reactor. The fixed bed is equipped with a rotating blade that continually introduces lumpy coal or brown coal, and breaks up caking coal. A process schematic is shown in Figure 2.27 [5,37]. Currently, 90% of all coal is gasified by this process [27].

Catalytic gasification has gained a lot of attention in recent years due to its higher conversion at lower temperature. Salts of alkali and alkaline earth metals as well as transition metals are known to be active for gasification. The most common and effective catalysts for steam gasification are the salts, namely, the oxides and chlorides of alkali and alkaline earth metals, either separately or in combination. Batelle, Exxon, and Shell have developed their own versions of catalytic gasification processes. Exxon has reported that the impregnation of 10–20% of potassium carbonate lowers the temperature and pressure for steam gasification of bituminous coals from 1000°C to 760°C, and from 70 atm to 34 atm.

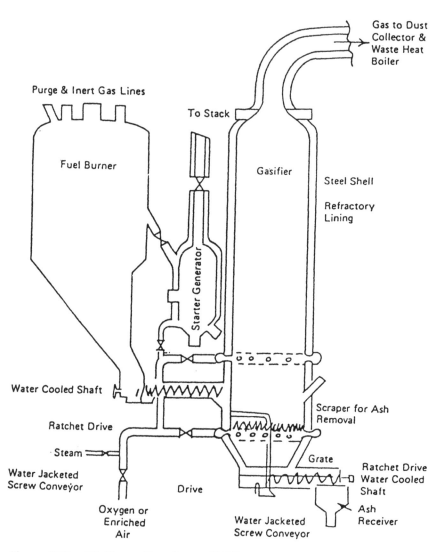

Figure 2.25 Winkler gasifier. *Source*: [5,37].

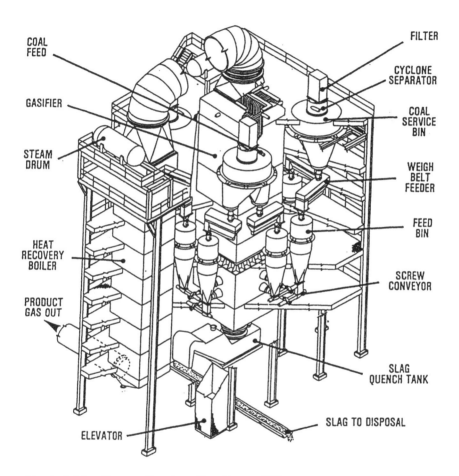

Figure 2.26   Koppers-Totzek gasifier. *Source*: [37].

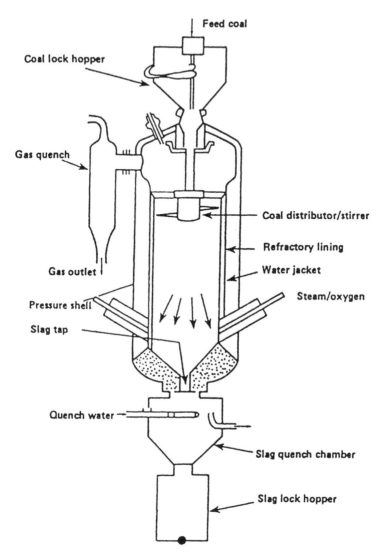

Figure 2.27   Lurgi gasifier. *Source*: [5,37].

REFERENCES

1. Rostrup-Nielson, J. R., Catalytic steam reforming, in *Catalysis Science and Technology Vol. 5* (Anderson, J. R. and Boudart, M., Eds.), Springer-Verlag, New York, 1984, pp. 1–117.

2. Orphanides, P., Developments in natural gas reforming technology of syngas, *Ammonia Plant Safety*, 34: 292–312 (1994).

3. Woodcock, K. E. and Gottlieb, M., Gas, natural, in *Kirk-Othmer Encyclopedia of Chemical Technology*, 4th edition, Vol. 12, (Kroschwitz, J. I. and Howe-Grant, M., Eds.), John Wiley & Sons, Toronto, 1994.

4. Rostrup-Nielson, J. R., Production of synthesis gas, *Catalysis Today*, 18: 305–324 (1993).

5. Speight, J. G., *The Chemistry and Technology of Coal*, Marcel Dekker, New York, 1983.

6. Teuner, S., Make CO from CO_2, *Hydrocarbon Processing*, May: 106–107 (1985).

7. Ashcroft, A. T., Cheetham, A. K., Green, M. L. H., and Vernon, P. D. F., Partial oxidation of methane to synthesis gas using carbon dioxide, *Nature*, 352(July 18): 225–226 (1991).

8. Reforming catalysts for the production of ammonia, *Nitrogen*, 174: 23–34 (1988).

9. Rostrup-Nielson, J. R., Nielson, P. E. H., Sorensen, N. K., and Carstensen, J. H., Catalyst poisoning: general rules and practical examples for ammonia plants, *Ammonia Plant Safety*, 33: 184–199, (1993).

10. Hansen, J-H. B., Storgaard, L., and Pederson, P. S., Aspects of modern reforming technology and catalysts, *Ammonia Plant Safety*, 32: 52–62 (1992).

11. Farina, G. L. and Supp, E., Produce syngas for methanol, *Hydrocarbon Processing*, March: 77–79 (1992).

12. Bixler, A. D. and Vakil, T. D., Desulfurization of oxygen containing NG to SMRs, *Ammonia Plant Safety*, 34: 89–102 (1994).

13. Vamby, R. and Madsen, S. E. L. W., Adiabatic prereforming, *Ammonia Plant Safety*, 32: 122–128 (1992).

14. Farnell, P. W., Secondary reforming: theory and application, *Ammonia Plant Safety*, 34: 24–37 (1994).

15. Verduijn, W. D., Experience with a prereformer, *Ammonia Plant Safety*, 33: 165–183, (1993).

16. Schnieder, R. V., III and LeBlanc, J. R., Jr., Choose optimal syngas route, *Hydrocarbon Processing*, March: 51–57 (1992).

17. Cromarty, B. J. and Crewdson, B. J., Preforming eases reforming furnace firing requirements, *Nitrogen*, 191: 30–34 (1991).

18. Natural gas reformer design for ammonia plants: I. primary reformer design criteria, *Nitrogen*, 166: 24–30 (1987).

19. Mohri, T., Takemura, K., and Shibasaki, T., Application of advanced material for catalyst tubes for steam reformers, *Ammonia Plant Safety*, 33: 86–100 (1993).

20. Natural gas reforming design for ammonia plants: II secondary reformer design, *Nitrogen*, 167: 31–36 (1987).

21. Cromarty, B. J., The development and application of shaped reforming catalysts, *Nitrogen*, 91: 115–125 (1991).

22. Rice, D. K., Loading of primary reformer catalyst tubes, *Ammonia Plant Safety*, 33: 216–222 (1993).
23. Elkins, K. J., Jeffrey, I. C., Kitchen, D., and Pinto, A., The ICI gas heated reformer (GHR) system, *Nitrogen*, 91: 83–95 (1991).
24. Grotz, B. J., Sosna, M. K., Bonar, I. E., and Gunko, B. M., Development and startup of tandem performing process, *Ammonia Plant Safety*, 34: 74–82 (1994).
25. Marsch, H.-D. and Thiagarajan, N., CAR: a new reformer technology, *Ammonia Plant Safety*, 29: 195–203 (1989).
26. Marsch, H.-D. and Thiagarajan, N., CAR: a new reformer technology, *Ammonia Plant Safety*, 29: 195–203 (1989).
27. Weissermel, K. and Arpe, H.-J., *Handbook of Industrial Chemistry, translated by C. R. Lindley,* 2nd, revised and extended edition, VCH, New York, 1993.
28. Pelofsky, A. H. et al., Partial oxidation, in *Heavy Oil Gasification* (Pelofsky, A. H., Ed.), Marcel Dekker, New York, 1977, pp. 39–47.
29. Reforming the front end, *Nitrogen*, 195: 22–36 (1992).
30. Pelofsky, A. H. et al., Partial oxidation, in *Heavy Oil Gasification* (Pelofsky, A. H., Ed.), Marcel Dekker, New York, 1977, pp. 48–65.
31. Speight, J. G., *Gas Processing: Environmental Aspects and Methods*, Butterworth Heinemann, Boston, 1993.
32. Yang, R. T., *Gas Separation by Adsorption Processes*, Butterwoths, Boston, 1987.
33. Kwong, V. and Meissner, R. E., Rounding up sulfur, *Chemical Engineering*, Feb.: 74–83 (1995).
34. Meibner, H., Wammes, W., and Hefner, W., Improving the performance of BASF's AMDEA-process by optimization of the activator concentration, *Nitrogen*, 91: 151–164 (1991).
35. Bartoo, R. K. and Gemborys, T. M., Recent improvements to the benfield process to extend its use, *Nitrogen*, 91: 127–139 (1991).
36. Shiyong, Z., A new promoted potassium carbonate CO_2 removal process for ammonia plants using complex activators, *Nitrogen*, 91: 141–150 (1991).
37. Funk, J. E., Coal technology, in *Riegel's Handbook of Industrial Chemistry Eighth Edition* (Kent, J. A., Ed.), Van Nostrand Reinhold Company, Cincinnati, 1983.
38. *Hydrocarbon Processing*, April 1994.

3

Methane Derivatives Via Synthesis Gas

I. AMMONIA

Ammonia is the starting block for almost all industrially produced nitrogen compounds; U.S. production in 1995 was 17.801×10^6 tons making it the 6th largest chemical manufactured [1]. The manufacture of ammonia is essentially the same as the Bosch-Haber Process developed from 1907 to 1912. The engineering fete by Bosch at BASF from 1909 to 1912 to develop a completely grassroots process while inventing many new technologies is unprecedented in engineering. The majority of the catalyst used today is still very similar to the alkali-promoted magnetite developed in Germany by Mittasch at BASF. However, new catalysts based on ruthenium have higher activities and are being developed for industrial use. The Haber-Bosch process uses a recycle loop to increase ammonia yields due to low equilibrium.

The majority of the process development occurred in western Europe and the United States. Since 1975, many plants have come on-line in other parts of the world based on this technology. A shift can be seen in the worldwide ammonia distribution as shown in Table 3.1 [2]. The shift has been to less industrial countries with large populations that require the ammonia for fertilizer production.

Ammonia compounds and its derivatives make an important part of the chemical industry and are used in almost every industry (Table 3.2) [2,3]. Ammonia is used in industries, however, the majority of the ammonia produced is either used directly or indirectly in the fertilizer industry (Table 3.3) [3].

The fertilizer industry requires a large amount of nitrogen fertilizers for food crops around the world. Modern hybrid varieties of food crops require

Table 3.1 World Ammonia Capacity Distribution [2]

Region	1975 (%)	1990 (%)
Africa	2.1	3.0
Asia	25.9	35.4
Latin America	3.9	5.3
North America	21.4	13.8
Eastern Europe	12.8	9.7
Western Europe	18.8	11.3
USSR	15.1	21.5

tremendous amounts of nitrogen compounds. The amount of ammonia required to synthesize common nonammonia nitrogen fertilizers is shown in Table 3.4 [2]. Ammonia production feeds the fertilizer industry. The other large nitrogen markets are explosives and plastics.

A. Ammonia Synthesis Technology

Historical Development

The first commercial processes used to produce ammonia were the cyanamide and electric arc processes at the beginning of the 20th century. The cyanamide

Table 3.2 Industrial Uses of Ammonia [2,3]

Uses	Chemical forms
Explosives	Nitrates, dynamite, azides
Plastics	Nitrocellulose, urea-formaldehyde, melamine, phenolic resins
Metallurgy	Bright annealing of steel, de-tinning of scrap metal, extraction of metals from ores, dry reducing gas, case hardening
Pulp and paper	Ammonium bisulfite, melamine
Rubber	Aniline, acrylonitrile, polyurethane, chemical blowing agents (for foam rubber)
Textiles	Caprolactam (monomer for nylon 6), acrylonitrile (acrylic fibers, resins, and elastomers), terephthalates
Foods	Amino acids, sodium nitrite, nitric oxides
Drugs	Vitamins, antimalarials, methionine, dentrifices, lotions, antibacterial agents (sulfanilamide, sulfathiazole, sulfapyridine), cosmetics
Miscellaneous	Refrigerant, detergents, corrosion inhibitor (petroleum refineries), stabilization of latex, insecticides, nitroparaffins, hydrazine

Table 3.3 United States Ammonia Markets in 1986 [2]

Use	Market (%)
Fertilizer	
Direct application	28.7
Urea	22.4
Ammonium nitrate	15.8
Ammonium phosphates	14.6
Ammonium sulfate	3.4
Nitrogen solutions and mixed fertilizers	0.6
Total	85.5
Industrial	
Commercial explosives	4.1
Fibers-plastics	5.1
Total	9.2
Other	5.2

process formed calcium carbide from coke and lime (Reaction 3.1). The calcium carbide then was reacted with nitrogen to form calcium cyanamide (Reaction 3.2). The calcium cyanamide then was decomposed in the presence of water to form calcium carbonate and ammonia (Reaction 3.3). The process was very energy intensive, consuming 190 GJ/ton ammonia produced.

$$CaO + 3C \xrightarrow{\;2000°C\;} CaC_2 + CO \tag{3.1}$$

Table 3.4 Ammonia Used for Production of Nitrogen Fertilizers [2]

Product	Raw material	Raw material used (ton/ton of product)
Urea	Ammonia	0.58
Nitric acid	Ammonia	0.29
Ammonium nitrate[a]	Ammonia	0.21
	Nitric acid	0.77
Nitrogen solutions	Urea	0.33
	Ammonium nitrate	0.41
Ammonium sulfate	Ammonia	0.27
Ammonium phosphate[b]	Ammonia	1.29

[a] based on UAN-30
[b] per ton of contained nitrogen

$$CaC_2 + N_2 \xrightarrow{1000°C} CaCN_2 + C \tag{3.2}$$

$$CaCN_2 + 3H_2O \dashrightarrow CaCO_3 + 2NH_3 \tag{3.3}$$

The arc process was used only in areas of very cheap electricity such as Norway and the United States. Air was passed through an electric arc raising the temperature to 3,000°C and forming NO. The process was very energy intensive using 700 GJ/ton nitrogen, which is about 17-times more than the consumption of a modern ammonia plant.

Modern synthesis technology is based on the Haber-Bosch process developed in Germany from 1909 to 1912. Although the ammonia converter used today is essentially the same as that used by BASF in 1912, many process improvements have been made. Most of the process developments in ammonia production have come from the generation of hydrogen. The first hydrogen generation was made from the chlor/alkali electrolysis process, but quickly this was found to be insufficient for the hydrogen needs. The more efficient steam reforming of coke was developed; however, this process was expensive, complex and inefficient. In 1926, the Winkler coal direct gasification process in a fluidized bed was developed (see Chapter 2). Coal gasification dominated the European synthesis gas generation until the 1950s. In the United States natural gas was cheap and steam reforming of gas was developed for generation of synthesis gas (see Chapter 2).

The development of ammonia synthesis technology may be classified into the following three historical phases: [139]

1. The *initial* phase started shortly after 1900 and lasted until the second World War.
2. The *second* phase started during World War II and culminated in the commercialization of the large single-stream plant in the 1960s.
3. The *third* phase started in early 1970s and prevails today with emphasis on improved energy efficiency and increased capacity.

In early days of Phase I, the predominant feedstock for ammonia synthesis was coke. Synthesis gas was either produced at atmospheric pressure in water–gas shift units or prepared by purification of coke oven gas. In these early plants, the process effluents from the ammonia converter were cooled without recovery of heat. Due to the lack of technology regarding the attainable size of the converter pressure shell, the physical dimensions of the converter were limiting factors for the achievable production capacity. Therefore, a particular emphasis was placed on maximization of the production capacity for a given volume [139]. During World War II, several plants were built in the United States, based on natural gas feedstock. Since then, natural gas has become the preferred feedstock in the United States and also in some other parts of the world. Other preferred feedstocks included heavy oil partial oxidation, coal gasification, and naphtha. Major developments in ammonia technology during this phase include:

1. from multitrain units to single-stream plant
2. from atmospheric pressure to high pressure (30–40 bar) in the synthesis gas preparation
3. from low capacity to higher capacity
4. improved energy efficiency from above 15 Gcal/MT (metric ton) to around 10 Gcal/MT.
5. production of steam at increased pressure
6. recovery of heat in the synthesis loop by steam production or by preheating of high pressure boiler feed water
7. change from reciprocating compressors to centrifugal compressors
8. use of quench-cooled converters instead of converters with internal cooling

The most significant technology suppliers in the second phase were Chemico, Foster Wheeler, ICI (Imperial Chemical Industries, Ltd.), Kellogg, Topsøe, and Uhde [139]. Phase III started after the energy crisis in 1973. The overall energy consumption has been reduced from about 10 Gcal/MT in the earliest large-capacity single train units to about 7 Gcal/MT in modern plants (which is less than half of the 15 Gcal/MT of pre-war coke-based plants), thus substantially reducing the production cost of ammonia. The most important modifications and enhancements introduced in this phase include: [39]

1. a major change in desulfurization by catalytic hydrogenation of sulfur compounds followed by adsorption of hydrogen sulfide on hot zinc oxide
2. enhancements in the reforming section including: an increase in operating pressure, a decrease in steam addition, the use of more sophisticated alloys for reformer tubes, an increase in exit temperatures from primary and secondary reformers, an increase in the feed preheat temperature, the use of improved catalyst with better activity, and a higher resistance to poisoning and coking
3. use of double-stage, low-temperature shift conversion, instead of single-stage, high-temperature conversion
4. enhancements in carbon dioxide removal process by physical or combined physical and chemical adsorption
5. final purification by methanation rather than copper liquor wash
6. improvements in reactor design, converter volume, and catalyst efficiency, thus dramatically increasing the single-line capacity

Table 3.5 Feedstocks for World Ammonia (%) [2]

Feedstock	1983	1987
Natural gas	66.4	69.8
Natural gas hybrid[a]	7.5	5.8
Naphtha, fuel oil, condensate	12.4	10.5
Coke oven, refinery gas, hydrogen	4.3	3.9
Coal	10.8	10.0
Total	100.0	100.0

[a] Hybrid plants are naphtha and fuel oil based plants modified to use natural gas.

Table 3.6 Total Energy Consumption and Relative Investment for 1,000 ton/day Plant [2]

Process	Total energy (GJ/ton NH_3)	Relative investment
Water electrolysis	117.3	2.2
Coal gasification (without pressure)	58.6	1.8-2.1
Coal gasification (25 bar)	48.6	1.8-2.1
Partial oxidation of heavy fuel oil	39.0	1.5
Steam reforming of naphtha	37.7	1.15
Steam reforming of natural gas	32.6	1.0

Table 3.7 Hydrogen Generation Options [2]

Raw material	Process description	Feedstock conversion reaction
Natural gas	Steam reforming[a]	$C_nH_{(2n+2)} + nH_2O = nCO + (2n + 1)H_2$
Naphtha	Steam reforming[b]	$C_nH_{(2n+2)} + nH_2O = nCO + (2n + 1)H_2$
Fuel oil	Partial oxidation	$C_nH_{(2n+2)} + (n/2)O_2 = nCO + (n + 1)H_2$
Coal	Coal gasification	$C + (1/2)O_2 = CO$
Water	Electrolysis	$H_2O = (1/2)O_2 + H_2$

[a]Nickel is used as catalyst.
[b]Promoted nickel is used as catalyst.

7. developments in compressor and turbine technology, resulting in significant energy savings

A typical, modern ammonia plant as it is designed today, has a capacity of 1,000–1,500 MT/day. The feedstock is likely to be natural gas and energy consumption is 6.7–7.2 Gcal/MT depending on the design and integration [139]. Even though the synthesis process is considered mature and conventional, significant enhancements and advances in every aspect of the process technology have been constantly realized in the past several decades.

Feedstocks, Hydrogen Sources, and Energy Consumption

Today, five different processes are used to generate hydrogen for ammonia synthesis (Table 3.5) [2]. Natural gas is the preferred feedstock and makes up 69.8% of the market due to its wide availability, low price, and high hydrogen to carbon ratio (4:1). In some areas, however, other feedstocks are more economical and make up 30.2% of the market. Coal is still used in many parts of the world, the best example being South Africa. The other comparable process of hydrogen generation is the steam reforming of naphtha feedstocks (see Chapter 2). In a very few instances ammonia plants are actually attached to other plants such as a refinery, in order to use their excess hydrogen; however, hydrogen has become more important in the refinery process, which limits the amount available.

The economical and efficient generation of hydrogen is the most important step to producing ammonia cost effectively. The energy cost per ton ammonia produced is shown in Table 3.6 along with capital cost estimates for each process [2]. In most circumstances natural gas is the most cost-effective method in both total energy requirement and initial capital costs. The nitrogen for ammonia synthesis is always taken from the air. The five options for producing hydrogen along with a description of the process and governing reaction are shown in Table 3.7 [2]. A detailed description of these processes can be found in Chapter 2.

The ammonia synthesis by coal gasification is dependent on the gasification process used (e.g., Lurgi); two processes are represented here in block diagrams (Figures 3.1 and 3.2) [2].

Various acid gas treatment processes are classified into reactive, absorptive, and hybrid systems. The processes are compared with respect to the types of solvent used, solution circulation, and acid gas concentration in the product stream (Table 3.8) [2]. The average energy requirement for ammonia production is compared in Table 3.9 [2] among different routes in the 1970s and 1980s. Even though the ammonia synthesis technology has been regarded as a very mature process, the energy savings realized over the past decades are quite substantial.

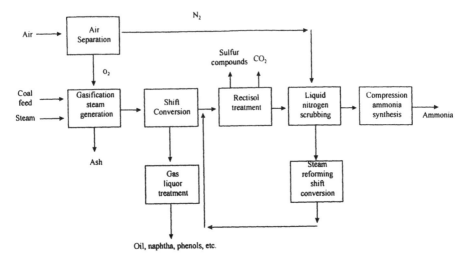

Figure 3.1   Flow sheet for ammonia production from Lurgi coal gasification. *Source*: [2].

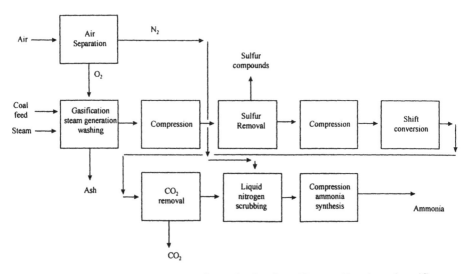

Figure 3.2   Flow sheet for ammonia production from Koppers-Totzek coal gasification. *Source*: [2].

Table 3.8 Acid Gas Treating Processes [2]

Process	Solvent	Solution circulation	Acid gas content in treated gas (ppm)
Reaction systems			
MEA	20% Monoethanolamine	Medium	< 50
Promoted MEA	25-35% Monoethanolamine plus Amine Guard	Medium	< 50
DGA	60% Diglycolamine	Medium	< 100
MDEA	40% Methyldiethanolamine plus additives	Medium	< 50
Vetrocoke	K_2CO_3 plus As_2O_3-glycine	High	500–1000
Carsol	K_2CO_3 plus additives	High	500–1000
Catacarb	25-35% K_2CO_3 plus additives	High	500–1000
Benfield	25-35% K_2CO_3 plus diethanolamine and additives	High	500–1000
Flexsorb HP	K_2CO_3 amine promoted	High	500–1000
Lurgi	25-35% K_2CO_3 plus additives	High	500–1000
Alkazid	Potassium salt of 2-(or 3-) methylaminopropionic acid	a	a
Combination reaction-physical systems			
Sulfinol	Sulfone and 1,1'-iminobis-2-propanol	Medium	< 100
TEA-MEA	Triethanolamine and monoethanolamine	High (TEA)	< 50
MDEA-sulfinol-H_2O		Low (MEA)	< 50
Physical absorption systems			
Purisol (NMP)	N-methyl-2-pyrrolidine	Medium	< 50
Rectisol	Methanol	Medium	< 10
Flour solvent	Propylene carbonate	b	b
Selexol	Dimethyl ether or propylene glycol	b	b

a dependent on service.
b dependent on pressure.

Table 3.9 Energy Required for Ammonia Production (GJ/t) [2]

	Natural gas reforming	Naphtha reforming	Fuel oil partial oxidation	Coal gasification	Water electrolysis
1970s Plant	35.9	39.6	40.6	48.5	44.3
1980s Plant	29.0	32.0		43.2	

B. Synthesis of Ammonia

Ammonia from Distillation of Coal

Coal contains about 1.5% nitrogen, and about one-eighth of this is in the form of ammonia when coal is distilled. The exact amount varies depending on the type and rank of coal. Ammonia is prepared from coke oven gas by cooling, scrubbing it with water, and later releasing the ammonia from the resulting liquor with steam. In most modern plants the ammonia is converted to $(NH_4)_2SO_4$ or $(NH_4)_3PO_4$. The production of ammonia from coal was an industrial process as early as 1810.

Synthesis of Ammonia

A process by which ammonia could be manufactured with the air was discovered by Regnault in 1840. In his discovery, a mixture of one volume of nitrogen, produced by removing the oxygen from air by heating the air over iron turnings, and three volumes of hydrogen was subjected to an electric spark while being confined over sulfuric acid.

In 1918, the British government started work on a plant at Billingham in Yorkshire; this plant was purchased in 1919 by Mond and Brunner, who developed the process and laid the foundations for the huge ICI complex at Billingham. Ammonia production began at Billingham in 1924.

Once the synthesis chemistry was established as the reaction between nitrogen and hydrogen, the remainder of the problem became finding the cheapest sources of these elements. Two simple methods of separating nitrogen from the atmospheric gas are: 1) fractionation of liquefied air, and 2) burning the oxygen in the atmosphere and thus converting into a compound that is easily separable from the residual nitrogen.

The use of fractionation of liquid air is an expensive route unless there is need fom the separated oxygen at neighboring sites. The feasibility of oxygen removal from the air by burning is demonstrated on an Ellingham diagram (Figure 3.3). The standard Gibbs free energies for several reactions are plotted against the corresponding temperatures. When the change of free energy is equal or less than zero, the reaction is feasible. For example, zinc metals and carbon function as oxygen scavengers. Methane also can work as an oxygen scavenger, since the combustion products (carbon dioxide and water) are relatively easy to remove.

Sources for hydrogen include: 1) electrolysis of a suitable electrolyte solution in water, 2) reaction of steam with a metal, 3) reaction of water with a nonmetal, and 4) synthesis gas or water gas. From the Ellingham diagram, the water–gas shift reaction is feasible below about 1100 K,

$$CO\ (g) + H_2O\ (g) = CO_2\ (g) + H_2\ (g)$$

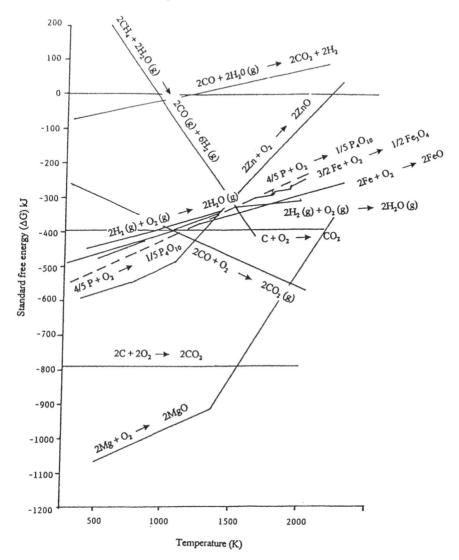

Figure 3.3 Ellingham diagram for reactions relevant to ammonia synthesis. *Source*: [5].

Table 3.10 Comparision of Various Feedstocks Used in Ammonia Production [5]

	Calorific value	Requirement per ton NH_3	
		Quantity	Energy content
Coke-oven gas			
Avaliable especially in steel-works. Typical composition 55% H_2, 25% CH_4, 8% CO, 6% N_2. If liquid nitrogen available, use to wash	19.6 MJ/m^3		
Coal	26.5 MJ/kg	1.97t	52 GJ
Naphtha			
Fraction from petroleum between 40 and 130°C, average molecular weight = 88; H/C = 2.23; Sulfur 0.1%; maximum unreacted sulfur 100 ppm; maximum Sg = 0.645–0.730 at 155°C	44 MJ/kg	0.89t	39 GJ
Fuel oil			
Fraction from petroleum between about 130 and 200°C	41 MJ/kg	1.00t	41 GJ
Natural gas			
May be associated with petroleum when CH_4 content depends on pressure. Nonassociated mostly CH_4 (H/C = 4)	33.6 MJ/m^3	1073 m^3	36 GJ
Refinery tail gases			
Organic waste decomposition			
Methane may be produced from sewage, vegetable waste/byproduct from crops, forest sawdust, and chippings.			

The steam reforming of methane reaction also is feasible above about 900 K.

$$CH_4 (g) + H_2O (g) = CO (g) + 3H_2 (g)$$

The practical routes to obtain the synthetic mixture in the molar ratio of N_2 + $3H_2$ involve:

1. Reaction with coke (Haber-Bosch process);

$$\frac{1}{2}N_2 + \frac{1}{8}O_2 + \frac{3}{2}H_2O \text{ (l)} + \frac{7}{8}C = \frac{1}{2}N_2 + \frac{3}{2}H_2 + \frac{7}{8}CO_2 - 84.6 \text{ kJ}$$

2. Reaction with methane, in natural gas;

$$\frac{1}{2}N_2 + \frac{1}{8}O_2 + \frac{5}{8}H_2O \text{ (l)} + \frac{7}{16}CH_4 = \frac{1}{2}N_2 + \frac{3}{2}H_2 + \frac{7}{16}CO_2 - 39.4 \text{ kJ}$$

This is the steam reforming process. For other hydrocarbons, the process chemistry is essentially the same. Table 3.10 compares the energy consumptions in various plants using different fuels. Interestingly, the energy consumed by the coke or coal process appears to be higher than other fuels.

Figure 3.4 shows a schematic of the TVA ammonia process, that used coke as a fuel. In this type of process, the gas producer consists of a cylinder lined with refractory brick or jacketed with water. The bottom of this cylinder is supported on a grid in a saucer so that the solid products from the cylinder may pass through the grid and out through the sides of the saucer. The particle size of coal or coke is kept within the range 5–30 nm to avoid loss through the grid mesh. To avoid particle sizing, a fluidized bed may be substituted. The remainder of the process is very similar to the hydrocarbon-based ammonia process.

Figures 3.5 and 3.6 show schematics of ammonia synthesis processes that are based on steam reforming and partial oxidation modification, respectively. The chemistry involved in that process is:

1. Primary steam-reforming:

$$CH_4 + H_2O = CO + 3H_2 - 206 \text{ kJ}$$

2. Secondary steam-reforming:

$$CH_4 + \frac{1}{2}O_2 + 2N_2 = CO + 2H_2 + 2N_2 + 35.6 \text{ kJ}$$

3. Water–gas shift reaction:

$$CO + H_2O = CO_2 + H_2 + 91.8 \text{ kJ}$$

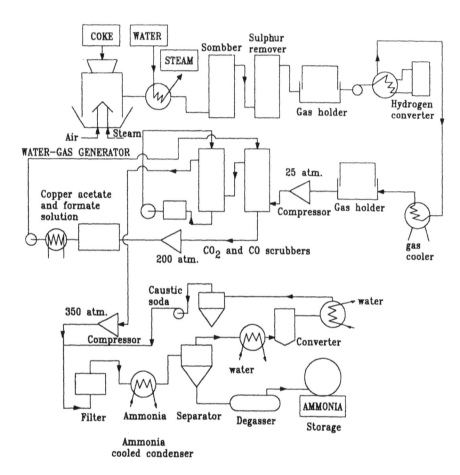

Figure 3.4 A schematic of the TVA ammonia process. *Source*: [5].

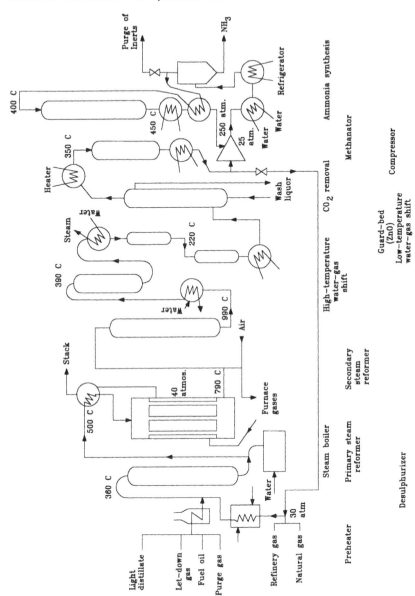

Figure 3.5 Ammonia synthesis via steam reforming of natural gas or refinery gas. *Source:* [5].

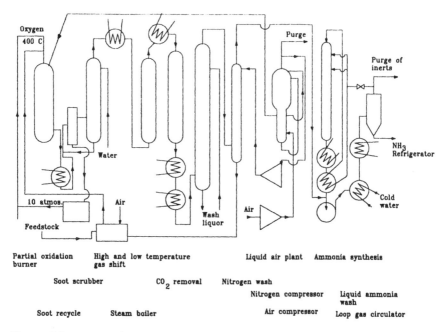

Figure 3.6   Ammonia synthesis via partial oxidation modification. *Source*: [6].

Natural gas may contain entrained dust or liquid droplets. Since the catalyst is poisoned by sulfur and chlorine, the gas must be treated and these elements have to be converted to hydrogen sulfide and hydrogen chloride, respectively. The feed is passed over a hydrotreater which has catalysts consisting of a mixture of cobalt molybdate, nickel molybdate, cobalt-nickel, and cobalt-nickel-alumina. These catalysts will cause any organic sulfur and chlorine compounds to be hydrogenated, provided the compound has a boiling point below 200°C. The gas is then passed over zinc oxide, which removes H_2S, and then over a copper-based substance which removes HCl. Filters are used to remove dust and liquid particles, and the gas may be compressed to 35–50 atm for feeding into the reforming system. The primary reformer is a huge tubular heat exchanger with furnace gases on the outside of the tubes and the reactants inside. The number of tubes can be 50–400 and the tubes are packed with catalyst (14–16% nickel on calcium aluminate) possibly with promoters such as potassium hydroxide. The reaction gas leaves the primary reformer at 800–900°C and at 25–35 atm. The secondary reformer is designed to oxidize with air the remaining methane in the gases from the primary reformer.

$$CH_4 + 2O_2 + 8N_2 = CO_2 + 2H_2O + 8N_2$$

Typically, the secondary reformer is a refractory-lined cylinder, the lower part of which is filled with a catalyst similar to that in the primary reformer. The upper part is the combustion chamber and the lower part continues the formation of carbon monoxide and hydrogen from the unburned methane. The gases leave the secondary reformer at 950–1000°C, and are passed through a heat exchanger and cooled to 375°C. The intended reaction in this stage is the water–gas shift reaction:

$$CO + H_2O = CO_2 + H_2 + 91.8 \text{ kJ}$$

For the high temperature (350–430°C) reaction the catalyst is Fe_2O_3/Cr_2O_3, whereas for the lower temperature (200–260°C) the catalyst is $Cu/ZnO/Al_2O_3$. After leaving the second shift reactor, the gases typically contain 0.3–0.4% CO. The gases are cooled and sent to remove carbon dioxide. Monoethanolamine (MEA) in 20% aqueous solution or potassium carbonate (K_2CO_3) solution are commonly used. The latter solution can act with hotter gases. For the case of K_2CO_3 absorption system, the reaction is:

$$K_2CO_3 + H_2O + CO_2 = 2KHCO_3$$

Carbon monoxide still has to be removed since it will react with ammonia to form solid ammonium carbamate. A methanation reaction is used to remove CO; this will also remove the residual CO_2.

$$CO + 3H_2 = CH_4 + H_2O$$

$$CO_2 + H_2 = CO + H_2O$$

The catalyst system used is nickel (14–16%) on calcium aluminate. The temperature of the gases to the methanator is 300–350°C.

As for compression, reciprocal compressors are used for small plants (< 500 tons/day) and centrifugal compressors are typically used for larger plants. The stoichiometric mixture of hydrogen and nitrogen is converted to ammonia:

$$\frac{1}{2}N_2 + \frac{3}{2}H_2 = NH_3 + 45.9 \text{ kJ}$$

The reaction is strongly exothermic. Since the biting temperature for catalytic reaction requires 380–540°C, the compressed gas must be preheated by heat exchange with the product gases. The conversion rate drops with a rise in temperature. Therefore, it is a good idea to keep a high temperature in the early part of the reaction to achieve a fast reaction rate, and a low temperature in the later part of the reaction to attain a good conversion. Hence the converter may consist of: 1) multiple beds of catalyst with interbed cooling by quenching with cool gases; or 2) a continuous bed containing heat exchangers. Two years is the typical period between complete overhauls of the ammonia synthesis plant, and catalysts are designed accordingly to give design capacity during this period.

C. Fertilizers

Plant life has existed for millions of years, but for a plant to be healthy at least 22 different chemical elements are necessary (Table 3.11) [4]. A deficiency in any one of these nutrients can result in limited plant growth. The addition of nutrients to the soil for enhanced plant growth has a long history dating back to early civilization. However, in modern times the fertilizer industry has become necessary to ensure plant growth.

Water may comprise up to 95 wt% of the plant. Dried plant tissue consists mainly of oxygen and carbon (Table 3.12) [4]. The plant nutrients are broken down into four categories: structural elements, primary nutrients, secondary nutrients, and micronutrients. The structural nutrients (C, H, O) are obtained from absorbing water through the root system and carbon dioxide through the leaves. These materials are the building blocks of the plant. These elements are used to produce organic compounds through photosynthesis that are necessary for life and growth. The primary nutrients (nitrogen, potassium, and

Table 3.11   Chemical Elements Essential for Plant Growth [4]

Essential macronutrients	Essential micronutrients	Beneficial
Metals		
Potassium	Iron	Aluminum
Calcium	Copper	Strontium
Magnesium	Manganese	Rubidium
	Zinc	
	Molybdenum	
	Cobalt	
	Vanadium	
	Sodium	
	Gallium	
Nonmetals		
Carbon	Boron	Selenium
Hydrogen	Silicon	
Oxygen	Chlorine	
Phosphorus	Iodine	
Nitrogen		
Sulfur		

Table 3.12 Chemical Analysis of
Dried Plant Tissue [4]

Name	Amount in tissue (wt%)
Structural elements	
Oxygen	45
Carbon	44
Hydrogen	6
Primary nutrients	
Nitrogen	2
Phosphorus	0.5
Potassium	1.0
Secondary nutrients	
Calcium	0.6
Magnesium	0.3
Sulfur	0.4
Micronutrients	
Boron	0.005
Chlorine	0.015
Copper	0.001
Iron	0.020
Manganese	0.050
Molybdenum	0.0001
Zinc	0.0100
Total	99.9011

phosphorous) are required in relative abundance compared to other elements. Plant growth quickly depletes them from the soil, since they are not principal components. Today, these elements are most often supplied by fertilizers. The secondary nutrients (calcium, magnesium, and sulfur) are typically in ample supply in the soil; however, they can be depleted over time and may need to be added. These nutrients are often contained in the compounds found in primary nutrient fertilizers. Also, they are added through limestone or dolomite control of soil pH. The micronutrients (boron, chorine, copper, iron, manganese, molybdenum, and zinc) are supplied by the soil, but can be depleted over time as well and may need to be added. However, this is an insignificant segment of the fertilizer industry.

The fertilizer industry is based on replenishing the primary nutrients (i.e., nitrogen, phosphorus, and potassium). Commercially, fertilizers are produced in three different forms which are single nutrient, binutrient, and multinutrient.

Binutrient and multinutrient fertilizers are often referred to as mixed fertilizers. The primary nutrient fertilizers have a rating convention which is designated by three weight percent numbers that identify the amount of N, P_2O_5, and K_2O available for plant fertilization.

Fertilizers are packaged in a variety of different forms including solids, liquids (solutions and suspensions), and gas (anhydrous ammonia). The important physical properties of solid fertilizers are particle size, particle strength, caking tendency, chemical stability, and hygroscopicity.

Nitrogen Fertilizers

Nitrogen fertilizer is consumed at twice the rate of either potassium or phosphorous (Table 3.13) [5]. Little nitrogen remains from one season to the next, since food crops require large amounts of nitrogen and nitrates are easily leached and volatilize as NO_x. Nitrogen utilization for crops ranges from 30% for vegetables to 60% for grain crops, but is usually not greater than 50% of the nitrogen available.

In the year ending June 30, 1991, 77.0 million metric tons of nitrogen fertilizer were used worldwide with 10.1 million metric tons in the United States alone [4]. Nitrogen can be added to soil by natural means such as crop rotation, natural organics, mineral nitrogen, or by synthetic nitrogen fertilizers.

Table 3.13 World Consumption of Fertilizer [5]

Year	N:	P_2O_5:	K_2O	Total consumption (metric tons)
1905				1.9
1913	1	3	1.4	
1921				3.86
1931				6.44
1938	1	1.6	1	9.2
1946				7.5
1951	1	1.54	1.1	14.9
1956				21.3
1961	1	1	0.8	28.5
1978	1	0.69	0.49	106
1980	1	0.51	0.39	115
1983	1	0.49	0.38	125
1987 (est.)	1	0.52	0.39	145
1992 (est.)	1	0.51	0.39	170
2000 (est.)	1	0.47	0.40	210–220

Before 1945, crop rotation was used extensively as a means of supplying nitrogen in the soil. This process took advantage of legumes (peas and clovers) symbiotic relationship with soil bacteria. The bacteria infects the roots of the plant creating nodules that fix atmospheric nitrogen. The fixed nitrogen is then absorbed into the root system. However, this system is presently ineffective because modern hybrid varieties of grains are tremendous consumers of nitrogen.

Another natural source of nitrogen is organic materials such as manure (animal and human excrement), guano (deposits of bird droppings), fish meal (dried pulverized fish and scrap fish), packing house waste, or compost. Unfortunately, this natural waste contains a low percentage of nitrogen. Although the use of these materials is still in practice today, it accounts for less than 1% of the total nitrogen fertilizer market.

Mineral nitrogen accounts for less than 0.1% of the U.S. nitrogen market (5,900 tons in 1991). This was not the case at the beginning of the 20th century as deposits such as those in Chile were commonly used. Nitrogen was rarely found in mineral form due to the high solubility of nitrates in water, although deposits of alkali metal nitrates can be found in desert areas. Chilean saltpeter (soda) contains 16% nitrogen in the form of sodium nitrate. Presently, the production of ammonia allows for more economical production of synthetic fertilizers.

The majority of commercial nitrogen fertilizers are produced from ammonia, which accounts for 95% of the production. Ammonia can be used directly in either gaseous or aqueous form or indirectly by reacting it to produce other compounds (Table 3.14). Routes for making ammonia into fertilizers are shown in Figure 3.7 [4]. The major fertilizer compounds are ammonia, am-

Table 3.14 Commonly Used Nitrogen Fertilizers [4]

| Compound | Formula | Primary nutrient (%) | | |
		N	P_2O_5	K_2O
Ammonia	NH_3	82.2		
Ammonium sulfate	$(NH_4)_2SO_4$	21.2		
Ammonium nitrate	NH_4NO_3	35.0		
Sodium nitrate	$NaNO_3$	16.5		
Calcium nitrate	$Ca(NO_3)_2$	17.0		
Urea	$CO(NH_2)_2$	46.6		
Calcium cyanamide	$CaCN_2$	34.9		
Monoammonium phosphate	$(NH_4)H_2PO_4$	12.1	61.7	
Diammonium phosphate	$(NH_4)_2HPO_4$	21.2	53.7	
Potassium nitrate	KNO_3	13.8		46.6

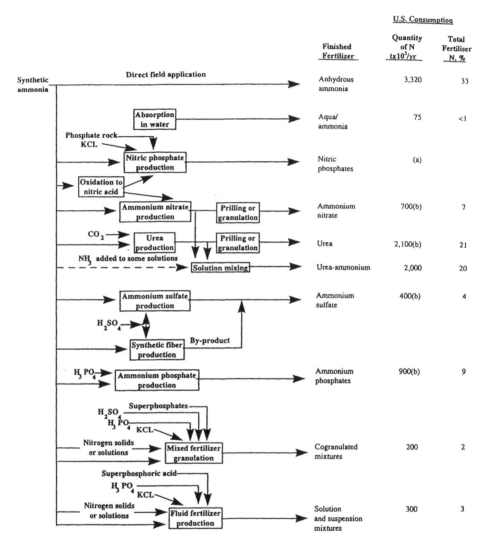

Figure 3.7   Routes for fertilizers from synthetic ammonia consumption (data based on year ending 30 June 1990). (a) significant quantities made in foreign countries; (b) includes quantities applied in dry blends. *Source*: [4].

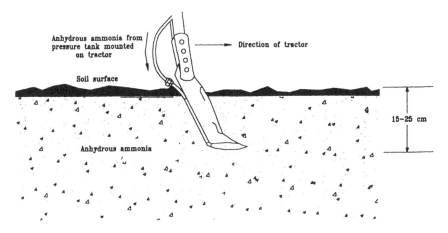

Figure 3.8 A schematic of soil injection of anyhdrous ammonia. *Source*: [5].

monium nitrate, or urea. Other compounds like ammonium sulfate and ammonium phosphate have long fertilizer histories but can no longer be mass produced economically.

Anhydrous Ammonia. Anhydrous ammonia is used in large amounts in North America particularly in the corn and wheat belts of the United States. This direct application accounts for about one third of the nitrogen fertilizer used in the United States. The advantage of the process is the extremely high nitrogen content of ammonia (82.2%) and the ability to compress or refrigerate into a liquid and ship via pipeline. The anhydrous ammonia is applied directly by the use of knives as shown in Figure 3.8 [4]. The ammonia is injected into the ground (15–25 cm) to limit loss.

Aqua Ammonia. Aqua ammonia is a solution of 23% ammonia (20% nitrogen) and water. This allows for a reduction of ammonia vapor pressure and thus easier storage and handling capabilities. However, this greatly increases the amount of storage necessary and complicates the handling requirements. The solution is applied similarly to anhydrous ammonia but at a more shallow depth of 8–12 cm. In the United States, less than 1% of the nitrogen fertilizer is delivered in this manner.

Ammonium Sulfate. Before 1950, ammonium sulfate [$(NH_4)_2SO_4$] was the most common nitrogen fertilizer. However, the low nitrogen content (21.2%) and lack of value in the secondary nutrient sulfur (24.3%) have reduced the demand. Today, this accounts for only 4% of the nitrogen fertilizer used in the United States.

Ammonium sulfate can be produced by the following processes: direct reaction between ammonia and sulfuric acid (crystalline form), reaction of ammonia with calcium sulfate in the presence of CO_2, and from the byproduct of other processes such as caprolactam and acrylonitrile.

The first process, direct production, is shown in Reaction 3.4. The process is carried out in monel reactors due to the highly corrosive environment.

$$2NH_3 + H_2SO_4 = (NH_4)_2SO_4 \tag{3.4}$$

The second process is a two-step reaction sequence (Reactions 3.5 and 3.6). The process is highly adaptable—one of three forms of calcium sulfate [gypsum (dihydrate), plaster of paris (hemihydrate), or anhydrate]—may be used. The advantage of this process is the avoidance of a sulfuric acid intermediary.

$$2NH_3 \text{ (g)} + H_2O \text{ (l)} + CO_2 \text{ (g)} = (NH_4)_2CO_3 \text{ (aq)} \tag{3.5}$$

$$(NH_4)_2CO_3 \text{ (aq)} + CaSO_4 \text{ (s)} = (NH_4)_2SO_4 \text{ (aq)} + CaCO_3 \text{ (s)} \tag{3.6}$$

The third and most common method is to use a byproduct from other processes and is presently the only process used in the United States. The low demand for ammonium sulfate fertilizer has made it uneconomical to produce by any other means.

Ammonium Nitrate.   Ammonium nitrate became an important fertilizer after World War II when chemical plants that had been producing munitions were converted to the manufacture of fertilizer. It has a high concentration of nitrogen (35%) which is immediately available to the plant in the form of nitrate. However, ammonium nitrate is hygroscopic and prone to have large density fluctuations with crystal structure changes at 32°C. Ammonium nitrate is explosive if stored improperly. Proper storage methods and safety considerations must be employed.

Ammonium nitrate is formed by the neutralization method which reacts nitric acid and ammonia. The process generally uses a 60% nitric acid solution to which ammonia is added (Reaction 3.7). This produces an ammonium nitrate solution of 8% water. Any higher ammonium nitrate concentration would be explosive since the heat of reaction could not be dissipated.

$$NH_3 \text{ (g)} + HNO_3 \text{ (aq)} = NH_4NO_3 \text{ (aq)} \tag{3.7}$$

The process may be operated at atmospheric pressure or at higher pressures. The advantage of low pressure is that atmospheric units can be used while at higher pressures the steam produced can be used elsewhere in the facility. The operating temperature for atmospheric operation is 145°C, while at higher pressures (4–5 atm) the operating temperature has a range of 175–180°C. A

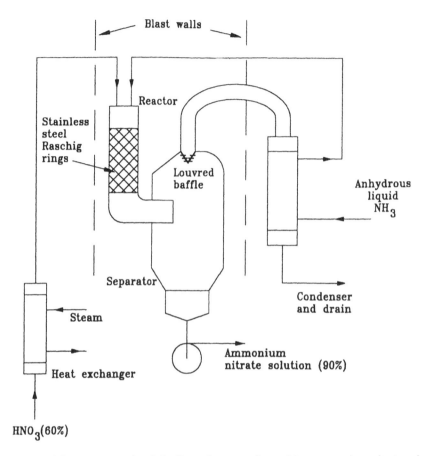

Figure 3.9 A schematic of the Stengel process for making ammonium nitrate solution. *Source*: [5].

commonly used process for manufacturing ammonium nitrate is the Stengel process shown in Figure 3.9.

Urea. Urea, $CO(NH_2)_2$, is the leading nitrogen fertilizer worldwide. It provides an estimated 41% of the world's total nitrogen fertilizer and 32% in the United States. Urea contains a large amount of nitrogen (46.6%), the largest of any solid fertilizer (ammonium nitrate has 35.0% and ammonium sulfate has 21.2% nitrogen). As a fertilizer, urea will not burn or explode and is less costly to produce per nitrogen unit. Urea first gained acceptance as a fertilizer for rice. Urea in rice paddies supplies ammonia nitrogen and can be easily retained in a flooded paddy. Consequently, fertilizers like ammonium nitrate are partially reduced to N_2O or N_2 which is volatized and lost. Rice is able to use ammonia directly as a fertilizer. For most crops, urea must undergo hydrolysis to ammonia and then nitrification to nitrate before it can be used by plants. In cooler climates these reactions are slow leading to a delayed effect of the fertilizer. In warmer climates the ammonia may be lost to the atmosphere, but this can be rectified by placing the urea in the ground. Urea also exhibits phytotoxicity, which is the poisoning of the seed during hydrolysis. Production of urea is discussed later.

Urea–Ammonium Nitrate Solutions.   An important fertilizer is the urea-ammonium nitrate solution which makes up 20% of the nitrogen fertilizer market. This fertilizer utilizes the unusually high solubility between urea and ammonium nitrate (Table 3.15), which allows for a high nitrogen content, low nitrogen cost, and ease of handling. The solution also can contain herbicides since they can be easily added to the solution. However, these solutions are corrosive, so corrosion inhibitors are added such as ammonium thiocyanate, sodium arsenite, sulfonate OA5, or trace ammonia.

Ammonium Phosphate.   Ammonium phosphates have a long history of use as a fertilizer. They are not as easily leached from the soil as ammonium nitrate, but cannot be made in high nitrogen concentrations. There are three forms of ammonium phosphate: monoammonium phosphate (MAP, $NH_4H_2PO_4$), diammonium phosphate (DAP, $[NH_4]_2HPO_4$), and triammonium phosphate (TAP, $[NH_4]_3PO_4$). Unfortunately, triammonium phosphate is unstable and loses ammonia easily. Ammonium phosphates are produced by reacting ammonia with phosphoric acid (Reaction 3.8).

$$H_3PO_4 \dashrightarrow NH_4H_2PO_4 \dashrightarrow (NH_4)_2HPO_4 \dashrightarrow (NH_3)_3PO_4 \qquad (3.8)$$

Table 3.15 Properties of Urea-Ammonium Nitrate Solutions [4]

Parameter	Solution		
Nitrogen (%)	28	30	32
Composition (wt%)			
Ammonium nitrate	40.1	42.2	43.3
Urea	30.0	32.7	35.4
Water	29.9	25.1	20.3
Specific gravity at 15.6°C	1.283	1.303	1.32
Salt-out temperature (°C)	−18	−10	−2

Table 3.16 Single Nutrient versus Multiple
Nutrient Application [4]

Nutrient	Single (%)	Multiple (%)
Nitrogen	80	20
Phosphate	8	92
Potash	65	35
Total	61	39

Both MAP and DAP are produced commercially as fertilizers and have a nitrogen-phosphorus-potassium (NPK) of 10–12:50–56:0 and 16–18:46–48:0, respectively.

Mixed Fertilizers

In many fertilization applications the soil requires more than one of the primary nutrients. This can be supplied by either multiple applications of single-nutrient fertilizers or by a single application of mixed fertilizer or multi-nutrient fertilizer. In many cases, the multinutrient fertilizer can be made of the levels necessary for the particular plant and soil type. This has led to over 21,358 registered grades of mixed fertilizers. The mixed fertilizer can be distributed in either nongranular, granular, or liquid forms. The single distribution of nutrient versus multiple nutrient applications are shown in Table 3.16 [4]. The majority of nitrogen and potassium fertilizers are distributed in the single nutrient form while phosphate is mainly distributed in the mixed form.

In many situations, economic and agronomic factors dictate that multiple applications of single nutrients are more desirable. An economic situation is the application of anhydrous ammonia with subsequent mixed fertilizer to fill the primary nutrient needs. Anhydrous ammonia is an inexpensive and effective method of nitrogen fertilization. In many areas it is widely available and affordable. Another example is shown in the growing of corn. Initially, the corn is fertilized with a mixed fertilizer to obtain the necessary primary nutrients; then by adding nitrogen at the proper stage of growth, the profitability can be greatly increased.

D. Urea and Resins

Urea is produced commercially by reacting NH_3 and CO_2 at high pressure and temperature. The process is usually near an ammonia synthesis plant which produces both reactants. The two-step process is shown in Reactions 3.9 and 3.10 [5].

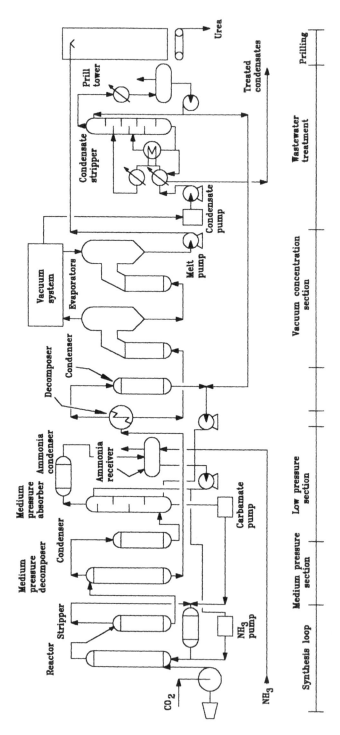

Figure 3.10 A schematic of the Snamprogetti ammonia stripping process. *Source:* [9].

$$2NH_3 + CO_2 = NH_2COONH_4 \qquad \Delta H = -117\,kJ/mol \tag{3.9}$$

$$NH_2COONH_4 = NH_2CONH_2 + H_2O \qquad \Delta H = 15.5\,kJ/mol \tag{3.10}$$

The first step is to react ammonia with carbon dioxide to form ammonium carbamate. The reaction is highly exothermic. The second step is the endothermic dehydration reaction of ammonium carbamate. The overall synthesis reaction is still highly exothermic.

Urea Synthesis Process

The most popular way of synthesizing urea commercially is producing it from ammonia and carbon dioxide using ammonia stripping process [6]. Figure 3.10 shows a schematic of Snamprogetti SPA process for urea synthesis. Ammonia and carbon dioxide react at 150 bar to produce urea and ammonium carbamate [9]. The operating temperature for the reaction is 185–190°C. The conversion in the reactor is very high due to the favorable NH_3/CO_2 ratio of 3.5:1. These conditions prevent corrosion problems.

In the first reactor, carbamate is formed; then the carbamate is decomposed in three stages at different pressure vessels [9]. The second vessel after the reactor, is under the same pressure, while vessel 3 and vessel 4 require different pressure conditions (i.e., vessel 3 is at 18 bar and vessel 4 is at 4.5 bar).

Reactants that are not transformed into urea are recycled to the reactor by means of an ejector. The plant is free from pollution problems. All vents are efficiently washed so that they are discharged to atmosphere practically free of ammonia and urea. Liquid discharge may have to be physically treated to meet local regulations or client requests. Also, discharge water can be reused as boiler feed water (BFW).

Table 3.17 shows the process economics for urea synthesis. Raw materials and utilities per 1,000 kg of urea are given. The finishing process can be coupled with synthesis, either prilling or granulation, both direct or via crystallization.

Table 3.17 Economics of Urea Synthesis Process (Prilling Case) [9]

Ammonia, kg	566
Carbon dioxide, kg	735
Steam, 110 bar, 510°C, kg	730
Electric power, kwh	21
Water, cooling, m^3	80
Condensate, kg	1,045

Urea is used as a feed supplement for cattle, up to 8 lb/yr per head of cattle. In liquid fertilizers, urea raises the upper limit for stable nitrogen solutions. In solids it prevents caking due to the formation of ammonium chloride in mixtures of ammonium and potassium fertilizers. "Ureaforms" which are condensates of urea and formaldehyde, can be used as time-release fertilizers when it is desired to release fertilizing urea at a controlled rate, such as on a golf course [8]. The main use of solid, prilled urea is as side-dressing.

Urea–Formaldehyde Resins

Straight urea-formaldehyde resins are used principally in the preparation of molding compositions and adhesives. As the name implies, the resin is made by reaction between urea and formaldehyde; however, the synthesis process somewhat depends upon the end-use envisaged. The following process is for a wood adhesive resin, and illustrates the general procedures involved [10].

Formalin is made slightly alkaline (pH of 8.0) by the addition of sodium hydroxide and then urea is added to obtain an urea to formaldehyde ratio of about 1:2 by mols. The resulting solution is boiled under reflux for 15 min and formic acid then is added to make the pH of the solution be 4.0. The resulting solution is boiled again for 5–20 min until the desired extent of reaction is attained. The product has to be neutralized with sodium hydroxide and then evaporated under mild vacuum until the desired content of solids is attained. For a liquid adhesive case, the typical solid content is about 70%.

Formation of Methylolureas. Under mildly alkaline conditions, both monomethylolurea and dimethylolurea are obtained in high yield and can be isolated as pure, crystalline solids: [10]

$$NH_2-CO-NH_2 \xleftarrow{\quad HCHO \quad} NHCH_2OH-CO-NH_2$$

$$NHCH_2OH-CO-NH_2 \xleftarrow{\quad HCHO \quad} NHCH_2OH-CO-NHCH_2OH_2$$

Trimethylolurea is also possible under these conditions, even though it does not appear to be formed under the resin preparation conditions.

Condensation of Methylolureas. In the second stage of the resin synthesis, the reaction is continued under acidic conditions by the following mechanism:

$$-CH_2OH + H^+ \longleftrightarrow -CH_2O^+H \longleftrightarrow -{}^+CH_2 + H_2O$$

$$-CH_2 + H_2N- \longleftrightarrow -CH_2-{}^+NH_2 \longleftrightarrow -CH_2-NH- + H^+ + H_2O$$

A typical resin solution includes several different species that contain methylol and amino groups; therefore, various types of condensation may result. For example [10],

$$NHCH_2OH-CO-NH_2 + NH_2-CO-NH_2 \xrightarrow{-H_2O}$$
$$NH_2-CO-NH-CH_2-NH-CO-NH_2$$

$$NHCH_2OH-CO-NHCH_2OH + NH_2-CO-NHCH_2OH \xrightarrow{-H_2O}$$
$$NHCH_2OH-CO-NHCH_2OH$$

The general structure of the methylene-containing polymers may be represented by:

$$HOCH_2-[-NH-CO-NH-CH_2]_n-NH-CO-NH-CH_2OH$$

Unmodified urea-formaldehyde resins are unsuitable for use in surface coatings formulations, due to their lack of solubility in common solvents and incompatibility with other resins. This limitation can be substantially overcome when the resins are modified by alcohols, such as *n*-butanol. Very commonly, butylated urea-formaldehyde resins are blended with alkyd resins to provide good flexibility and adhesion in coatings. Furfuryl alcohol-modified urea resins are used to bind foundry cores formed in preheated pattern boxes, due to their capability of rapid core production.

Urea formaldehyde molding powders are used primarily where particularly short molding cycles are desired for low unit costs (e.g., wiring devices and plastic closures) [8]. Important end uses of this resin can be found in industries dealing with plywood, particle-board, plastic closures, wiring devices, textile resins, paper resins, coatings, bondings, furfuryl alcohol modified foundry resins, etc.

Sulfamic Acid

Sulfamic acid is made by the reaction of oleum with urea:

$$NH_2CONH_2 + H_2SO_4 \dashrightarrow 2NH_2SO_3H + CO_2$$

The ammonium or amine salts of sulfamic acid are used as a fire-retardant for paper and synthetic fibers. This has economic advantages over chlorosulfonic acid, but gives off HCl. Sulfamic acid gives off ammonia and is less corrosive than chlorosulfonic acid. This compound is important as a S-N compound. The existence of N-S bonds in sulfamic acid ($H_3N^+SO_3^-$), the sulfamate anion ($H_2NSO_3^-$), and sulfamides, ($R_2N)_2SO_2$, is particularly important, since sulfur and nitrogen have similar electronegativities and strong tendencies to form single and multiple covalent bonds [7]. Other S-N compounds include disulfur dinitride (S_2N_2), polythiazyl, and S_xN_y compounds.

Melamine

Melamine, a trimer of cyanamide, was first prepared by Liebig in 1834. Melamine has become commercially popular due to its resins, also known as "Melmac."

In the traditional process route to melamine is produced by reacting calcium carbide with nitrogen to yield calcium cyanamide, and then to cyanamide, dicyandiamide, and melamine. The process requires a great deal of energy input due to its high reaction temperature.

$$CaC_2 + N_2 \xrightarrow{\sim 1100^\circ \text{ C}} CaCN_2 + C \qquad\qquad \Delta H = -297 \text{ kJ/mol}$$

calcium carbide　　　　　　　　calcium cyanamide

$$CaCN_2 \xrightarrow{H_2O} Ca(HCN_2)_2$$

$$Ca(HCN_2)_2 + H_2O + CO_2 \rightarrow 2NCNH_2 + CaCO_3$$

cyanamide

$$2NCNH_2 \rightarrow NH_2C(NHCN) = NH$$

dicyandiamide

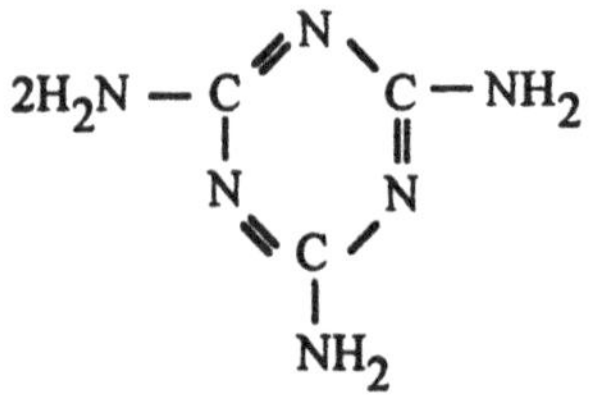

melamine

The last stage of the reaction is highly exothermic, and carried out in the presence of methanol for better heat removal.

The current standard process route to melamine uses urea as a starting material. Urea is heated in the presence of ammonia at 250–300°C and 4–20 MPa. The reaction probably involves the simultaneous dehydration and hydration of urea to form cyanamide and ammonium carbamate; trimerization of the cyanamide then leads to melamine [10].

$$2NH_2CONH_2 \rightarrow NH_2CN + NH_2COONH_4$$

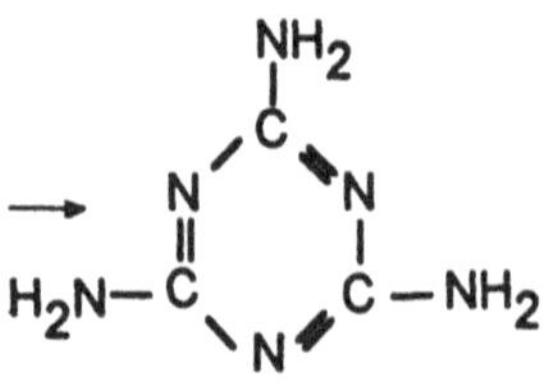

This process is based on the high pressure, high temperature reaction. Even though the once-through conversion of urea to melamine is merely 50%, this route is still the most economical of all the available processes.

Low pressure routes have also been developed by both BASF and Oesterreichische Stickstoffwerke (OSW). The low pressure process is carried out in two stages:

$$CO(NH_2)_2 \longrightarrow HNCO + NH_3$$

$$6HNCO \longrightarrow C_3H_6N_6 + 3CO_2$$

The second reaction is carried out catalytically above the sublimation temperature of melamine. The product is recovered by quenching and centrifuging, and the yield is reported to be 90% or higher.

The process used by the former Standard Oil of Ohio (SOHIO; currently BP America) involved electrolytic conversion of HCN to cyanogen bromide, and subsequent conversion to cyanamide by ammonolysis, then to melamine by trimerization:

$$HCN + Br_2 \longrightarrow BrCN + HBr$$

$$BrCN + NH_3 \longrightarrow NH_2CN + HBr$$

$$3NH_2CN \longrightarrow C_3H_6N_6$$

Since 1965, only the nonelectrochemical routes (i.e., starting from urea) have been commercially used.

Melamine is used primarily in the manufacture of amino resin. The resin formation reaction is analogous to the urea-formaldehyde system.

Formaline is made slightly alkaline, pH of 7.5–8.5, by adding sodium carbonate. Then, melamine is added in a 1:3 mole ratio of melamine to formaldehyde. Then the mixture is heated at 80°C for 1–2 hr until the desired extent of reaction is attained. The resulting syrup is stabilized by borax (pH buffer) and can be used without further processing. The reaction mechanisms and pathways are also believed to be analogous to the urea-formaldehyde resin.

Straight melamine-formaldehyde resins are used primarily in the preparation of molding compounds, laminates, and textile finishes. Nearly 90% of these molding powders are used to make dinnerware. Melamine dinnerware can be made very decorative for popularity. This resin has better physical properties than phenolics and it can be very easily colored, which phenolics cannot. Methylolmelamine also is used as a synthetic tanning agent for white leather.

Table 3.18 Nitric Acid Uses [11]

Percentage (%)	Use

76%	For ammonium nitrate of which
	   80% was for fertilizers
	   10% was for explosives
	   10% was for N_2O

9%	For adipic acid (fiber and plastics precurser)
3.5%	For dinitrotoluene (for toluene diisocyanate)
2%	For sodium, potassium and calcium nitrates
1%	For nitro-compounds for explosives
5%	For miscellaneous (pickling stainless steel, other metal-lurgical processes, and in the nuclear fuel cycle)

E. Nitric Acid and Nitrates

Nitric acid is a major industrial chemical and ranks 14th in chemical production with 17.24 billion lb produced in 1995 [1]. The large diversification of uses can be seen in Table 3.18 [11]. Although nitric acid initially was used industrially in the production of explosives and dyestuffs, the single largest use for nitric acid is for the formation of ammonium nitrate (NH_4NO_3), which is used by the fertilizer industry.

Nitric acid is a strong oxidizing agent that reacts with alkali and basic materials as well as oxides. Concentrated nitric acid reacts violently with many organic compounds. Nitric acid attacks most metals except for platinum, gold, aluminum, and chromium steel.

Most of the modern manufacture of nitric acid is done by the catalytic oxidation of ammonia (Ostwald process). Other now outdated processes include the reaction of sodium nitrate with sulfuric acid and direct synthesis from N_2 and O_2 by the arc process at temperatures in excess of 2,000°C. Once cheap ammonia became available these processes were no longer economical.

Although the process configuration varies widely there are three main steps common to each, they are: oxidation of ammonia to nitric oxide (NO), oxidation of NO to nitrogen dioxide (NO_2), and then absorption of NO_2 in water to produce nitric acid (Figure 3.11). The first step of the catalytic combustion of ammonia takes place over a catalyst consisting of platinum/rhodium (90:10) or platinum/rhodium/palladium (90:5:5). The reaction is very rapid and goes almost to completion as shown in Reaction 3.11. The reaction is one of the most efficient catalytic processes in industrial chemistry, having an extremely

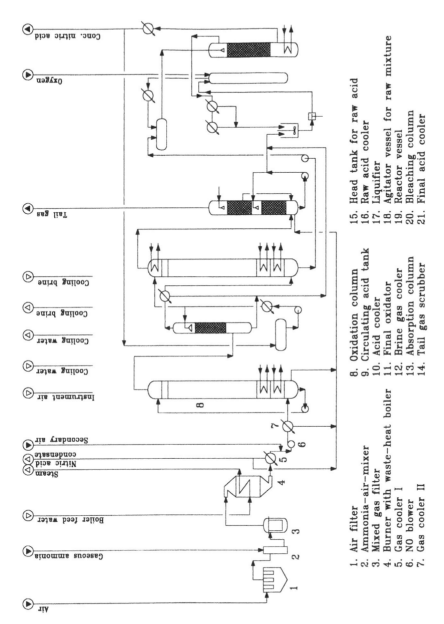

Figure 3.11 Nitric acid via oxidation of ammonia. *Source:* [3].

1. Air filter
2. Ammonia–air–mixer
3. Mixed gas filter
4. Burner with waste–heat boiler
5. Gas cooler I
6. NO blower
7. Gas cooler II

8. Oxidation column
9. Circulating acid tank
10. Acid cooler
11. Final oxidator
12. Brine gas cooler
13. Absorption column
14. Tail gas scrubber

15. Head tank for raw acid
16. Raw acid cooler
17. Liquifier
18. Agitator vessel for raw mixture
19. Reactor vessel
20. Bleaching column
21. Final acid cooler

Air

Gaseous ammonia

Boiler feed water

Steam
Nitric acid condensate
Secondary air

Instrument air

Cooling water

Cooling water

Cooling brine

Cooling brine

Tail gas

Oxygen

Conc. nitric acid

short residence time (10^{-11} sec) and a high selectivity. The reaction is carried out at temperatures of 820–950°C and pressures of 1–12 bar, although low pressures are preferred to limit side reactions that produce products such as NO and N_2O. The catalyst is a fine-woven wire mesh that typically contains rhodium and/or palladium for strength. The Pt oxidizes during the reaction to form PtO_2 which limits the life of the catalyst, although 80% of this can be recovered downstream by adsorption onto marble chips or palladium-gold gauzes.

$$NH_3 + 1.25O_2 \longrightarrow NO + 1.5H_2O \qquad \Delta H_{298K} = -226 \text{ kJ/mol} \qquad (3.11)$$

The next stage is the oxidation of nitric oxide as seen in Reaction 3.12. The reaction is slow and is favored at temperatures below 150°C and increasing pressures.

$$2NO + O_2 \longrightarrow 2NO_2 \qquad \Delta H_{298K} = -114 \text{ kJ} \qquad (3.12)$$

Nitrogen dioxide (NO_2) dimerizes almost instantaneously to an equilibrium mixture of NO_2 and N_2O_4 as shown in Reaction 3.13. This reaction is also favored at lower temperatures and increasing pressures.

$$2NO_2 = N_2O_4 \qquad \Delta H_{298K} = -57.4 \text{ kJ} \qquad (3.13)$$

The final stage is the absorption of nitric oxides in water are shown in Reactions 3.14 and 3.15. The important process parameters are temperature, pressure, and contact time. Low temperature and high pressure are preferred to increase the absorption. Increasing the contact time also increases the acid concentration. At atmospheric pressure nitric acid only can be made in 45–50% solutions, while at higher pressures 70% can be made. Although the exact mechanism of absorption is not understood, equipment design allows for a wide range of concentrations to be produced.

$$3NO_2 \text{ (g)} + H_2O \text{ (l)} \longrightarrow 2HNO_3 \text{ (aq)} + NO \text{ (g)} \qquad \Delta H_{298K} = -58.7 \text{ kJ/mol}$$
$$(3.14)$$

$$N_2O_4 \text{ (g)} + H_2O \text{ (l)} \longrightarrow HNO_3 \text{ (aq)} + HNO_2 \text{ (g)} \qquad \Delta H_{298K} = -65 \text{ kJ/mol}$$
$$(3.15)$$

The manufacture of concentrated nitric acid (98–99%) can be done by two methods, direct and indirect. The indirect method, mainly used in the United States, is performed industrially by two different systems, the sulfuric acid and magnesium nitrate processes. Both processes involve the dehydration of nitric acid with concentrated H_2SO_4 or $Mg(NO_3)_2$. The sulfuric acid process has acute corrosion problems. This process utilizes azeotropic rectification to produce concentrated nitric acid.

The direct manufacture of nitric acid, mainly used in Europe, uses a variation of the normal nitric acid process. The two industrial processes have evolved: the first uses the direct oxidation of N_2O_4 with pure O_2 in the presence of H_2O in pressures (Reaction 3.16).

$$N_2O_4 + H_2O + \frac{1}{2}O_2 \longrightarrow 2HNO_3$$

$$(3.16)$$

The second involves the absorption of NO_2/NO in concentrated HNO_3 (Reaction 3.17). This produces a superazeotropic acid which then is distilled to produce concentrated nitric acid. Examples of these processes are the Uhde CNA (concentrated nitric acid) process and Davy McKee's SABAR (Strong Acid By Azeotropic Rectification) process.

$$2NO + H_2O + \frac{3}{2}O_2 \longrightarrow 2HNO_3$$

$$(3.17)$$

Processes

Nitric acid plants all use one of two variations of the Ostwald process, which are monopressure and dual pressure. The monopressure systems can be further broken into monopressure low (1–2 bar), medium (3–7 bar), and high (8–12 bar). In the United States, 90% of the nitric acid plants are mono-pressure high, while in western Europe most plants are dual-pressure medium/high (medium pressure catalytic combustion and high pressure oxidation and absorption) systems. The high pressure plants are used because they have low plant costs, low space requirements, and satisfactory emissions; however, they have higher catalyst consumption, higher energy consumption, and lower NO yield. A schematic of the monopressure low, monopressure high, and dual-pressure systems is shown in Figure 3.12. In high pressure plants, the size of the equipment is greatly reduced, but the compressor cost can make up half of the total investment.

The manufacture of nitric acid by ammonia oxidation can be considered in six process sections: 1) filtering and mixing of air and ammonia; 2) converting ammonia to nitrix oxide; 3) oxidizing nitric oxide; 4) absorbing nitrogen dioxide dimer; 5) removing tail gases; and 6) storing and transporting nitric acid [5].

Filtering and Mixing. Since the reaction between ammonia and oxygen is catalytic in nature, it is very important that the effectiveness of the catalyst is maximized by keeping the catalyst surface area fully accessible by the gases. In particular, particles of metal or metal oxide may cause premature oxidation and affect the catalyst effectiveness. Therefore, air and ammonia passed to the mixer must be free from any suspended matter, dust, or mist (greater than 0.5 (μm) which could obscure or foul the catalyst surface [5].

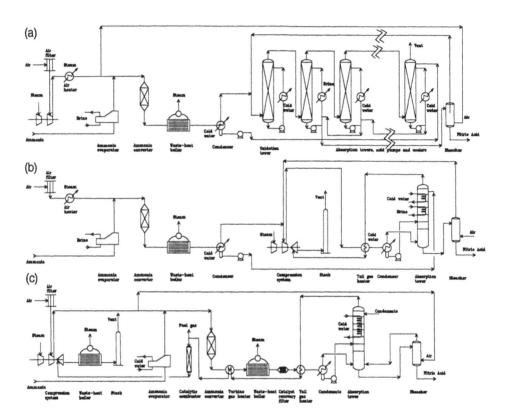

Figure 3.12   A schematic of a nitric acid plant. (a) Atmospheric, (b) Split pressure, (c) High pressure. *Source*: [5].

Typically, air is filtered through spun glass-fiber before and after mixing with ammonia, whereas ammonia liquid is passed through magnetic separators to remove iron and iron oxide particles and ammonia vapor through sintered metal fiber.

Oxidation of Ammonia.   The oxidation of ammonia is thermodynamically favored by low pressure:

$$4NH_3 + 5O_2 = 4NO + 6H_2O$$

However, pressure is applied in the manufacturing process due mainly to economizing the plant dimensions. Thus, the pressure used to some extent classifies the plant:

Low pressure:	1 atm
Medium pressure:	3–4 atm
High pressure:	7–10 atm

Because high pressure reactant flows are concentrated over the area of the catalyst gauze, there are more layers of gauze required to permit sufficient contact time for the reactant molecules on the catalytic surface. Typically, the air is compressed by a centrifugal compressor, either operated by a turbine driven by tail gases augmented by steam or by a compressor driven by an electric motor.

The temperature of the mixed gases entering the converter is at about 100°C and is achieved by a combination of the heat of compression of the air and the thermal energy added to the anhydrous ammonia. The liquid anhydrous ammonia must be completely vaporized in order to avoid explosions or damages to catalytic gauze. The boiling point of the ammonia must be considered based on the pressure at which the process is being operated. The boiling point of ammonia is: −33, −19, −1, 5, 21, and 27.5°C at 1, 2, 3, 5, 7, and 10 atmospheres, respectively [5]. For low pressure process, water used to vaporize the ammonia is thereby chilled and the cooled water can be used in the later parts of the process for cooling.

The stoichiometric composition of gas mixture yielding NO from NH_3 using dry air as oxygen source, is 14.35% by volume (8.969% by mass) of ammonia. The explosion limits for ammonia in air at atmospheric pressure are about 15–28% by volume of ammonia. Thus, the mixture could be explosive and is flammable.

Corrosion of material by the mixture gas also becomes an issue. Mild steel begins to decompose a 10% ammonia-air mixture at 200°C, 18–8 stainless steel at about 235°C, and aluminum-magnesium alloy at about 350°C. The compromise is to use titanium-stabilized 18–8 stainless steel.

The flow rates of gases are typically measured with orifice meters. Mixing of gases are achieved by various ways including venturi injectors, baffles or perforated plates, sparge pipes, and Raschig rings made of stainless steel and aluminum, etc.

Oxidation of Nitric Oxide.   The desired ammonia oxidation reaction is:

$$NH_3 \text{ (g)} + (5/4)O_2 \text{ (g)} = NO \text{ (g)} + (3/2)H_2O \text{ (g)} \qquad \Delta H^o = -226.5 \text{ kJ/mol}$$

The reaction is very fast and exothermic. Therefore, the contact between the reactants and the catalyst surface as well as the desorption of the product become crucially important. At the same time, reactions occurring homogeneously (not on the catalyst surface) must be controlled. In this regard, too high a temperature can lead to other homogeneous reactions [5].

The formation of nitric oxide is a stage in a progressive scheme of oxidation of ammonia as:

$$NH_3 + (5/4)O_2 = NO + (3/2)H_2O \qquad \Delta H^o = -226.5 \text{ kJ/mol}$$

$$2NO = N_2O + (1/2)O_2 \qquad \Delta H^o = 98.6 \text{ kJ}$$

$$N_2O = N_2 + (1/2)O_2 \qquad \Delta H^o = 82 \text{ kJ/mol}$$

Other associative and dissociative reactions also can occur and they include:

associations including NH_4OH, NH_2OH, HNO_2
decompositions including $NH_3 \longrightarrow N_2 + 3H_2$
                              $2NO \longrightarrow N_2 + O_2$

Busby et al. [12]. studied the effect of changes in mass flow rates of ammonia-air mixture per unit mass of catalyst at various temperatures on conversion. Their results showed the conversion ranged between 95.2% and 98.4% at temperatures of 850, 875, and 900°C. The optimal condition would still depend on the gauze structure and design (i.e., by the weave and its thickness, etc.).

The catalyst found most successful in attaining the desired conditions is platinum, with some addition of rhodium. The modern catalyst is platinum or platinum-rhodium wire made into gauze and then into wads of 4–30 gauzes, and used to form a bed. These gauzes are supported on coarser gauzes of chromium-nickel alloy. The catalyst can be poisoned by As, Bi, P, Si, S, Pb, and Sn. The life of catalyst is crucially important to the plant economics. A typical catalyst cycle is 60 days; then the catalyst is removed, pickled in hydrochloric acid, washed in distilled water, and dried by heating in a hydrogen flame. Further, catalyst evaporation is also a significant concern and the recovery of precious metals emitted from the catalyst is becoming more important [5].

Oxidation and Absorption. Reactions involved in the oxidation and absorption processes are:

$$NO\ (g) + (1/2)O_2\ (g) = NO_2\ (g) \qquad \Delta H^\circ = -56.57\ kJ/mol$$

$$NO_2\ (g) = (1/2)N_2O_4\ (g) \qquad \Delta H^\circ = -29.06\ kJ/mol$$

$$N_2O_4(g) + (2/3)H_2O(l) = (4/3)HNO_3(aq) + (2/3)NO(g)\ \Delta H^\circ = -34.31\ kJ/mol$$

All these reactions are exothermic and the energy is released at each stage. The stages involve nitric oxide to nitrogen peroxide, nitrogen peroxide to its dimer, dimer solution to nitric acid, and the release of nitric oxide, so that the cycle begins again.

The oxidation of nitric oxide to nitrogen peroxide involves heat generation and a decrease in the total number of moles. Therefore, the reaction is favored by a decrease in temperature and an increase in pressure. The dimerization reaction is very fast. The solid form (melting point −11.2°C; boiling point 21.1°C) consists entirely of dimer. The dimerization is favored by low temperature and high pressure, just as in the case of nitrogen peroxide formation.

The absorber consists usually of a single stainless-steel tower, although two or more may be necessary for low-pressure plants. Another kind of absorber used is a drum absorber. In general, the cool gases along with additional air are fed to the bottom of the tower, and demineralized water is fed to the top of the tower. Instead of demineralized water, clean condensed steam can be used. Nitric acid is taken out from the bottom, whereas tail gases leave at the top of the tower. The absorption towers can be packed with sieve plates, bubble cap plates, or turbo-grids.

The nitric acid from the absorption tower carries nitrous gases which color the liquid brown. Therefore, the acid must be stripped of these nitrous gases by air blowing through the liquor. The resultant gases are fed back to the absorber.

Removal of Tail Gases. The perfect removal of nitrous gases by the absorber would be ideal; however, it is realistic that a typical effluent from a plant consists of 0.15–0.25% nitrogen oxides (NO_x) and 2.5–5% oxygen, with the balance being nitrogen and water vapor. It must, however, be realized that emission standards are getting stricter and also vary from country to country. Nitrogen oxides are environmentally very hazardous, not only causing the well-publicized acid rain, but also contributing to smog formation; these gases are poisonous to human beings and other animals.

The nitrous gases can be eliminated by use of hydrogen peroxide or ozone; however, such treatments are very costly. A more practical method is burning the gases with hydrocarbons, using a catalyst at 300–400°C.

$$(3n + 1)NO + C_nH_{2n+2} = \frac{3n + 1}{2}N_2 + nCO_2 + (n + 1)H_2O$$

The catalyst used is vanadium oxide or platinum. Platinum is good for NO_2 to become N_2. Another catalyst, known as "honeycat," will convert NO_x to nitrogen when burnt with natural gas, hydrogen, naphtha, or LPG.

Another emission problem of concern is the formation of mist, which consists of droplets of acids. Mist may be removed by a fine mesh of stainless steel roll so that the gases have to pass through an intricate maze, or by electrostatic precipitation [5].

Uses of Nitric Acid

About 65% of all the nitric acid produced is reacted with ammonia to manufacture ammonium nitrate. About 80% of ammonium nitrate is used as fertilizer, the rest is as an explosive.

The other major use of nitric acid is in organic nitration. Nitration using mixtures of sulfuric and nitric acid is the first step in the synthesis of amino- and nitro-compounds such as aniline and trinitrotoluene (TNT). Many important dyes and pharmaceuticals are ultimately derived from these reactions. However, their quantities are quite small. Polyurethane polymers are also ultimately derived from nitrated toluene and benzene and this use accounts for 5–10% of nitric acid end uses.

F. Hydrazine

Hydrazine is a very important chemical and finds very versatile applications, including as an energy source in fuel cells and rocket propulsion, as a military missile fuel, in the synthesis of nitrogen compounds for agricultural and medical use, as blowing agents for polymers, and as an oxygen scavenger for boiler water.

Raschig Process

The most widely employed route until 1972 called the Raschig process involves indirect oxidation of ammonia:

$$NH_3 + NaOCl \longrightarrow NH_2Cl + NaOH$$

$$NH_2Cl + NH_3 \longrightarrow H_2NNH_2 + HCl$$

In the commercial production of hydrazine by the Raschig process, a large excess of ammonium hydroxide and fresh sodium hypochlorite are first reacted to make chloramine. Chloramine is then converted to crude hydrazine by rapid heating. The reactor effluent, containing about 2.5 wt% of hydrazine, is stripped of unreacted ammonia. The hydrazine solution is then concentrated to 3–4%. This solution is fed to an evaporator for removal of sodium chloride and boiling-off of hydrazine hydrate which can be dehydrated to anhydrous hydrazine by extractive distillation. The hydrazine yield based on ammonia is about 60%.

The drawbacks of the Raschig process include a large amount of byproduct chlorides (i.e., six times more than the product hydrazine). Further, both chloramine and hydrazine are unstable at higher concentrations. They decompose to nitrogen and ammonium chloride.

Schestakov Process

The ammonia of Raschig synthesis process is replaced by urea in Schestakov process.

$$NH_2CONH_2 + NaOCl \longrightarrow H_2NNH_2 + NaCl + CO_2$$

This process chemistry is analogous to the Hofmann degradation of amides to primary amines. However, due to several drawbacks including those of the Raschig process, the urea-based Schestakov process has not been employed commercially.

Bayer Process

The Bayer process uses ketone to form diaziridine in the presence of chlorine and excess ammonia [13].

$$2NH_3 + Cl_2 \longrightarrow NH_2Cl + NH_4Cl$$

The reaction between diaziridine and additional ketone produces azine, which is hydrolyzed to form hydrazine with the regeneration of ketone. There are two improvements over the Raschig process, namely, 1) use of ketone as a nitrogen carrier and 2) stabilizing the N-N bond of hydrazine.

ATOCHEM Process

The ATOCHEM process utilizes the reaction among ketone, ammonia, and hydrogen peroxide. The process is also called the Ugine Kuhlmann process. The process obtains azine in high yields via the hydrogen peroxide oxidation of ketone-ammonia system in the presence of nitrile (RCN). This synthesis route bypasses chlorination.

Properties and Uses

Hydrazine is a water-white, clear hygroscopic liquid with ammonia-like odor. The melting and boiling points of anhydrous hydrazine are $-2°C$ and $113°C$, respectively. Hydrazine hydrate ($N_2H_4 \cdot H_2O$), however, has melting and boiling points of $-52°C$ and $120°C$, respectively.

The most important nonmilitary application is synthesis of maleic hydrazide, which is used as a plant-growth regulator for tobacco suckering and tree pruning. Another important nonmilitary chemical is semicarbazide which is obtained by reacting hydrazine with urea:

$$H_2NNH_2 + NH_2CONH_2 \longrightarrow NH_2NHCONH_2 + NH_3$$
$$\text{semicarbazide}$$

Semicarbazide is used in pharmaceutical field in the manufacture of nitrofuran drugs. It is also used as an intermediate for azodicarbonamide,

$$2NH_2NHCONH_2 \xrightarrow{\text{(O)}} NH_2CON=NCONH_2 + 2NH_3$$
$$\text{azodicarbonamide}$$

Azodicarbonamide is mainly used as a blowing agent for making vinyl foams, in particular, calendered vinyl leather. This works as a blowing agent by releasing nitrogen and produces a foam.

II.   METHANOL AND DERIVATIVES

A.   Methanol Synthesis

The expanded production of methyl *tert*-butyl ether (MTBE), ethyl *tert*-butyl ether (ETBE), and *tert*-amyl methyl ether (TAME) as oxygenates has increased the global demand for methanol. The synthesis of methanol from syngas is a well-established technology. Syngas can be produced from natural gas or coal. The profitability of a methanol plant is made on a case-by-case basis to account for location-specific factors such as energy expenditure, environmental impact, and capital cost. Methanol plants exist where there are large reserves of competitively-priced natural gas or coal. The advent of methanol synthesis has given a boost to the value of natural gas. Conventional

Table 3.19 World Methanol Supply/Demand Balance

Demand	Forecast[a]				
	1991	1992	1993	1994	1995
Formaldehyde	7154	7242	7327	7512	7667
DMT	644	653	658	665	670
Acetic acid	1324	1407	1495	1517	1732
MTBE	3282	4075	4670	6616	8082
Methyl methacrylate	479	479	577	590	615
Gasoline/fuels	541	517	490	428	480
Solvents	730	749	776	796	813
Others	3931	4027	4149	4372	4228
Nontablulated countries	205	210	215	220	225
Total demand	18,290	19,359	20,357	22,716	24,500
Nameplate capacity	22,681	23,637	24,172	27,062	27,822
Capacity 90%	20,413	21,273	21,755	24,356	25,040
Percent utilization					
@Nameplate	80.6%	81.90%	84.20%	83.90%	88.10%
@90% nameplate	89.6%	91%	93.6%	93.3%	97.8%

[a] All quantities are in 1000 metric ton/yr.
Source: *Oil and Gas Journal*, pg 48, May 29, 1993.

steam reforming produces hydrogen-rich syngas at low pressure. However, this process is well suited to the addition of carbon dioxide which utilizes the excess hydrogen and hence increases the methanol productivity.

The global demand for methanol has increased about 8%/yr from 1991 to 1995. The global production capacity of methanol has expanded by about 5.1 million metric tons, or 23% in the same time period. Leading the growth is increased methanol demand for MTBE and formaldehyde production. The world methanol supply/demand balance is shown in Table 3.19.

All industrially-produced methanol is made by the catalytic conversion of synthesis gas containing carbon monoxide, carbon dioxide, and hydrogen as the main components. Methanol productivity can be enhanced by synthesis gas enrichment with additional carbon dioxide to a certain limit [14]. However, a CO_2–rich environment increases catalyst deactivation and shortens its lifetime, and produces water which adversely affects the catalyst matrix stability resulting in crystallite growth via hydrothermal synthesis phenomena [14]. Thus, a special catalyst has been designed to operate under high CO_2 conditions. This catalyst's crystallites are located on energetically stable sites that lower the tendency to migrate. This stability also minimizes the influence of water formed on the catalyst matrix so that it is only slightly affected. This catalyst preserves its higher activity due to a lower deactivation rate over long-term operations [14]. The basic reactions involved in methanol synthesis are:

$$CO_2 + 3H_2 = CH_3OH + H_2O \qquad \Delta H^o_{298} = -52.81 \text{ kJ/mol}$$

$$CO + 2H_2 = CH_3OH \qquad \Delta H^o_{298} = -94.08 \text{ kJ/mol}$$

$$CO_2 + H_2 = CO + H_2O \qquad \Delta H^o_{298} = -41.27 \text{ kJ/mol}$$

Of the three reactions, only two are stoichiometrically independent. The chemical mechanism of the methanol synthesis over the $Cu/ZnO/Al_2O_3$ catalyst is somewhat controversial [14,25]. However, evidence points toward the theory that methanol is produced predominantly via the CO_2 hydrogenation and the forward water–gas shift reaction [14,25].

$$CO_2 + 3H_2 = CH_3OH + H_2O$$

$$CO + H_2O = CO_2 + H_2$$

Different Reactor Systems for Methanol Synthesis

Conventionally, methanol is produced in two-phase systems: the reactants (CO, CO_2, and H_2) and products (CH_3OH and H_2O) forming the gas phase and the catalyst being the solid phase. Two reactor types are most popular: an adiabatic reactor containing a continuous bed of catalysts with quenching by cold gas injections (ICI system), and a multitubular reactor with an internal heat exchanger (Lurgi system). Both systems are operated at temperatures of 483–553 K and low pressures of 5–7 MPa using $Cu/ZnO/Al_2O_3$ catalysts. A more detailed account of the various reactor systems utilized for methanol synthesis is given below:

- Fixed-bed reactor
- Slurry reactor
- Trickle-bed reactor
- Gas-solid-solid trickle-flow reactor (GSSTFR)
- Reactor system with interstage product removal (RSIPR)

The reactor configuration most prevalent for methanol synthesis from syngas is the gas-phase fixed-bed reactor. It is a two-phase system in which the reacting gas flows through a bed of catalyst particles. ICI introduced its low pressure methanol process in low tonnage plants. This process typically operates at temperatures of 220–280°C and pressures of 5–10 MPa. The exothermic nature of the methanol synthesis reaction makes the temperature control difficult. The reactor operates adiabatically and the temperature rise in the bed is known to be high. Multibed quench reactors are necessary to counter this problem [14].

In 1975, Chem Systems developed the liquid phase methanol synthesis process that was later studied by Air Products and Chemicals, Inc. and The University of Akron [14]. The liquid phase methanol (LPMeOH) process uses a liquid-entrained reactor configuration which is a special type of slurry reactor. In slurry reactors, synthesis gas containing CO, CO_2, and H_2 passes upward into the reactor concurrently with the slurry which absorbs the heat liberated during the reaction. The slurry is separated from the vapor and recirculated to the bottom of the reactor via a heat exchanger, where cooling occurs by steam generation. The reactor effluent gases are cooled to condense the products and any inert hydrocarbon liquid which may be vaporized. Methanol and the inert hydrocarbon liquid are immiscible and are separated by a decanter. The methanol stream produced is suitable for use directly or can be sent to a distillation unit to produce chemical-grade product. Due to the excellent reaction temperature control, high conversions of syngas per pass can be achieved, which reduces the recycle gas flow and compression requirement. A small catalyst particle size can be used which prevents diffusional limitations. The near-isothermal temperature of the system allows favorable conditions to obtain the desired reaction kinetics. However, the slurry reactor has an upper limit for catalyst loading and hence operates at low space velocities. It has low conversion per pass due to the high extent of backmixing, and poses problems in separating the catalyst from the slurry due to catalyst attrition and agglomeration [15].

Methanol synthesis also has been investigated in a trickle-bed reactor and its performance has been compared to the conventional reactors. In a trickle-bed reactor, syngas and mineral oil flow concurrently over a fixed bed of catalyst. Trickle-bed reactors tolerate high catalyst loadings and hence can operate at high space velocities. They also operate at nearly plug flow conditions, leading to higher conversions per pass. The fixed bed causes no catalyst attrition which permits the use of costly catalysts. The trickle bed also consists of a liquid mineral oil phase which acts as a heat sink to absorb the heat generated by the exothermic reaction. Thus, the trickle-bed reactor incorporates the advantages of the gas-phase fixed-bed reactor and the slurry reactor. However, small catalyst particle size causes a pressure drop in trickle-bed reactors and large catalyst particle size causes mass transfer limitations due to intraparticle diffusion effects. Methanol synthesis reactions are carried out over a $Cu/ZnO/Al_2O_3$ catalyst at temperatures of 498–523 K and at a pressure of 5.2 MPa. The $H_2/(CO + CO_2)$ ratio varied at 0.5, 1, and 2, and the CO/CO_2 ratio was maintained at 9.0. Methanol productivities as high as 39.5 mol/(hr kg of catalyst) were obtained which were higher than the maximum methanol production rate of gas-phase methanol synthesis of 30 mol/(hr kg of catalyst).

Studies also showed higher methanol production rates in trickle-bed reactors as compared to slurry reactors at identical operating conditions. This is attributed to better gas/liquid/solid contact in trickle beds combined with its close proximity to plug flow conditions as opposed to high extent of back-mixing in the slurry reactors [16].

Two novel converter systems developed for the manufacture of methanol from synthesis gas are the gas-solid-solid trickle flow reactor (GSSTFR) and the reactor system with interstage product removal (RSIPR). In the GSSTFR system, a solid adsorbent trickles countercurrently over the catalyst bed and removes the product formed at the catalyst surface. Thus, the equilibrium limitation is alleviated thereby allowing high reactant conversions up to 100% in a single pass. The ratio of the adsorbent flow rate to the reactant feed flow rate determines the driving force for the forward reaction. As compared to the Lurgi low pressure process, the amount of catalyst used in this reactor system is reduced by 40%, the raw materials consumption is reduced by 10%, and the consumption of cooling water decreases by 50%.

In the RSIPR system, a series of adiabatic or isothermal fixed-bed reactors are used and the product is selectively removed in absorbers between the reactor stages. Here the methanol is removed selectively by absorbing in a liquid such as tetraethylene glycol dimethyl ether (TEGDME). This absorption takes place at the temperature level of the reactor inlet in order to achieve a high energy efficiency. The total solvent/methanol ratio is around 10.0. The specific raw material consumption of the RSIPR is similar to or lower than the conventional system. The cost of the converter and the catalyst are reduced because higher specific reaction rates are achieved. The recycle loop is eliminated completely. Thus, the investments are significantly lower than in the case of the Lurgi low pressure process. The catalyst amount decreases by 40% and accordingly a smaller reactor volume is needed [17].

Methanol synthesis using fluidized-bed technology exhibits better energy utilization, smaller reactor size, and higher per-pass conversion. In fixed-bed reactors, methanol productivity is limited by chemical equilibrium due to the reversible nature of the methanol synthesis reaction. In a fluidized-bed reactor, catalytic hydrogenation of carbon dioxide and carbon monoxide occurs in the dense phase, which contains the catalyst particles. The bubble phase is devoid of catalyst particles. The concentration gradients between the two phases induce the diffusion of methanol and water from the dense phase to the bubble phase while carbon dioxide diffuses in the opposite direction. The removal of methanol from the immediate vicinity of the catalyst particles enhances the rate of reaction and reactant conversion. The gas-solid trickle-flow reactor (GSTF) and the gas-solid-solid trickle-flow reactor (GSSTFR) have the advantage of shifting equilibrium in a favorable direction by removal of reaction products. However, these reactor types have the disadvantage of the presence of pore diffusion limitations. In the case of fluidized-bed reactors, the problem of pore diffusion resistances is largely eliminated because of the small catalyst particle sizes used. Industrial fixed-bed pellets are of the order of 6–12 mm diameter, whereas fluidizable catalyst particles can be smaller than 100 μm. Preliminary theoretical investigations show that the fluidized-bed reactor setup results in a CO conversion of about 72% which is higher than 63% attained in fixed-bed reactor setup. The methanol productivity is higher by 30% in the case of a fluidized-bed reactor setup [18].

Different Catalysts Used in Methanol Synthesis

In 1923, BASF found that methanol was the main product of carbon monoxide hydrogenation when mixed catalysts containing ZnO and Cr_2O_3 were used. In studying this catalytic synthesis reaction, it also was found that the presence of iron (or iron alloys) was detrimental to the process since it might react with CO to yield iron pentacarbonyl, $Fe(CO)_5$, which is a poison to the catalyst. This catalytic system was used until the early 1970s and is known also as "Zn-based" or "ZnO-based" catalytic systems. This catalyst is active at higher temperatures (350–400°C) and the process is operated under high pressure (250–350 atm), and is called "high pressure methanol synthesis." The methanol synthesis over ZnO/Cr_2O_3 proceeds via CO hydrogenation.

In 1963, ICI announced an innovative process system using $Cu/ZnO/Al_2O_3$ catalysts, later called "low pressure synthesis." The major constituents of this catalytic system are Cu (reduced form of CuO) and ZnO on Al_2O_3 support. The pressure and temperature conditions required are much lower than those for the high pressure process. Typical operating conditions are 220–270°C and 50–100 atm [14]. This catalyst has been found to be susceptible to sulfur and carbonyl poisoning, sintering, fouling, crystallite growth, and thermal aging. Nearly all of commercial methanol syntheses are carried out by low pressure processes.

In 1975, Chem Systems developed the liquid phase methanol (LPMeOH) process which is based on the low pressure synthesis technology except that the new process is carried out in an inert oil phase [79]. The catalytic system used is $Cu/ZnO/Al_2O_3$, that is modified for slurry operation (i.e., attrition resistant, finely powdered, and leaching resistant). The S3.85 and S3.86 catalysts of BASF and EPJ-19 and EPJ-25 catalysts of United Catalysts Inc. were developed for this process [14,19]. The process has been tested for commercial feasibility at a demonstration scale by Air Products and Chemicals, Inc [79].

Extensive research and process development work was conducted by Lee [14]. Cybulski [79] published an in-depth review of the liquid phase synthesis of methanol. Even though the chemistry of methanol synthesis over Cu-based catalyst has been controversial, more evidence point toward the CO_2 hydrogenation route as a predominant mechanism [14].

$$CO_2 + 3H_2 = CH_3OH + H_2O$$

$$CO + H_2O = CO_2 + H_2$$

A slurry phase concurrent synthesis of methanol using a potassium methoxide/copper chromite mixed catalyst has been developed. This process operates under relatively mild conditions such as temperatures of 100–180°C and pressures of 30–65 atm. The reaction pathway involves a homogeneous carbonylation of methanol to methyl formate followed by the heterogeneous hydrogenolysis of methyl formate to two molecules of methanol, the net result being the reaction of hydrogen with carbon monoxide to give methanol via methyl formate:

$$CH_3OH + CO = HCOOCH_3$$

$$HCOOCH_3 + 2H_2 = 2CH_3OH$$

$$2H_2 + CO = CH_3OH$$

The carbonylation of methanol to methyl formate is homogeneously catalyzed by alkali alkoxides like potassium methoxide, whereas the hydrogenolysis of methyl formate to methanol is carried out using a copper chromite catalyst. Thus, this concurrent synthesis involves two different catalysts in a single reactor, and methanol acts as a reactant and a solvent. The slurry-phase operation facilitates efficient heat dissipation resulting in isothermal operation. The products are methanol and methyl formate with traces of water and dimethyl ether from the dehydration of methanol. A conversion-per-pass of up to 90% and a selectivity for methanol of up to 94–99% have been achieved. This multistep single-stage methanol synthesis compares favorably with the more prevalent direct synthesis in terms of good heat transfer rates and high per-pass conversions [20]. This catalyst system has not been commercialized.

Studies on the kinetics of methanol synthesis confirm the fact that methanol can be formed from either carbon monoxide or carbon dioxide. Various syngas compositions were reacted over two catalysts Cu/ZnO and Cu/ZnO/Al$_2$O$_3$ at temperatures between 200–275°C, a pressure of 2.86 MPa, and in a differential packed-bed reactor constructed of copper tubing. When both CO and CO$_2$ are present in the syngas, the rates from either reactant were additive, with a further contribution to methanol arising from conversion of CO to CO$_2$ via the water-gas shift reaction, utilizing water formed in the methanol synthesis. At temperatures above 225°C, formation of methanol from CO$_2$ decreased over the Cu/ZnO catalyst, whereas a temperature greater than 250°C was necessary to affect the formation of methanol from CO$_2$ over the Cu/ZnO/Al$_2$O$_3$ catalyst [21]. This process has not been commercialized.

Studies on methanol synthesis using $CeCu_2$–derived catalysts show that methanol is synthesized by hydrogenation of CO and not CO_2. They also show that the active catalyst surface is extensively covered with a hydrogen-deficient methanol precursor under steady-state conditions and that the cerium oxide surface or its interface with the copper crystallites is intimately involved in the synthesis process. The hydrogenation of this precursor or methanol itself is rate limiting. Transient increases in the exit concentrations of methanol were achieved by displacing methanol from the catalyst surface using pulses of CO, CO_2, and O_2 into H_2–containing feed gas streams. Experiments were carried out in a single pass, fixed-bed microreactor which consisted of a 0.25 in. (o.d.) stainless steel tube contained within an aluminum-bronze heater block. The $CeCu_2$ alloy used as catalyst was prepared by high-vacuum electron-beam melting [22]. This process has not been commercialized.

A new kind of copper catalyst was prepared and optimized for methanol synthesis. The catalyst was prepared using a definite $La_2Zr_2O_7$ support. In the presence of carbon dioxide and water, Cu/La_2O_3 catalyst are quickly deactivated by the formation of lanthanum carbonates and hydroxycarbonates. However, the formation of these species can be prevented by incorporating La_2O_3 in a stable $La_2Zr_2O_7$ pyrochlore. After optimization, up to 600 g methanol/g_{cat}-hr can be produced over the $Cu/La_2Zr_2O_7$ catalysts. Experiments were carried out in a continuous-flow borosilicate glass reactor at standard conditions: 0.5 g catalyst, syngas flow rate of 4 l $g_{cat}^{-1}hr^{-1}$. Varying feed compositions were fed at 230, 250, 270, and 300°C, and a pressure of 6 MPa. Copper supported on a well-defined $La_2Zr_2O_7$ stable pyrochlore is an active catalyst for methanol synthesis even in the presence of carbon dioxide. The deactivation of the catalysts is caused by a high carbon dioxide content. This is mainly due to the formation of lanthanum carbonates and hydroxycarbonates and their spreading onto the metal particles, hence decreasing the catalytic activity. This carbon dioxide deactivating effect which is detrimental to methanol synthesis can be avoided in the presence of ZrO_2 or by promotion by ZnO [23]. This catalytic system is still in an investigational stage.

Another study investigates the hydrogenation of variable amounts of carbon dioxide and carbon monoxide in a synthesis gas mixture, using Raney copper, zinc oxide promoted Raney copper, and a commercial coprecipitated Cu-ZnO-Al_2O_3 methanol synthesis catalyst. Similar patterns of carbon monoxide and carbon dioxide conversion at various CO_2/CO ratios were observed for all the catalysts. The activity of the zinc-free Raney copper catalyst indicated that copper is capable of catalyzing the methanol synthesis reaction. However, zinc oxide-promoted catalysts show higher methanol activity than the Raney copper catalyst. This suggested that ZnO plays a role in enhancing the rate limiting step by interaction with carbon dioxide and carbon monoxide mole-

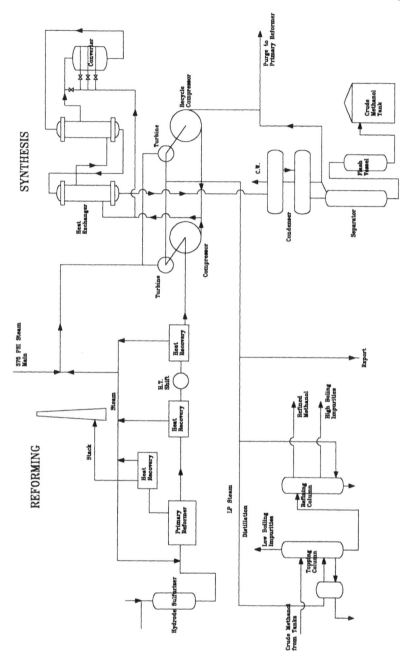

Figure 3.13 A schematic of the first ICI methanol synthesis process. *Source*: [26].

cules after initial adsorption on the copper surface [24]. These results are not necessarily in agreement with earlier findings by other investigators [14].

Methanol Synthesis Process Technology

The technology for low pressure methanol synthesis is quite well established. A number of companies now offer complete plants for the manufacture of methanol, starting from natural gas, naphtha, or coal as the feedstock. The catalytic systems are more or less standard and contain copper, zinc oxide, and alumina (or chromium oxide). The differences in these designs may be found mainly in the reactor configurations, use of different feedstock, energy integrations, arrangements of process units, recycle scheme, design of shift converter, etc. Some of these major plant designs are discussed.

The ICI Low-Pressure Methanol Synthesis Process. The first ICI unit was linked to a steam-naphtha reformer, the naphtha feed to which was first thoroughly desulfurized using zinc oxide and cobalt catalysts. A schematic flow diagram of the first ICI unit [26] rated for 300 tons of refined methanol per day is shown in Figure 3.13. The plant used a syngas feed that contained hydrogen, carbon monoxide, carbon dioxide, and methane. A shift converter was used to adjust the carbon monoxide/carbon dioxide ratio. The feed then was compressed to 50 atm in a centrifugal compressor and fed into a quench-type converter which was operated at 270°C. The product stream then was cooled and the methanol was condensed out. A purge gas stream was recycled to a reformer to convert the accumulated methanol into synthesis gas.

As the synthesis reaction proceeds with a decreasing number of moles, low pressure operation meant lower methanol concentration in the effluent stream and therefore higher recycle rates. However, high operating pressures involve bulkier equipment and sturdier compressors, and a balance must be achieved between energy costs and capital costs.

The design of the converter is very critical, since it should enable uniform gas distribution because the channeling of gases can cause local overheating and catalyst deactivation due to sintering. Reactor thermal stability and temperature runaway are of utmost importance in designing the converter. Provision must be made to warm the converter during start-up. Special attention must also be paid to ensure that the catalyst can be easily loaded and unloaded from the reactor to reduce downtime. It would also be a good safety measure, since the reduced catalyst is pyrophoric.

The crude methanol that is produced by the low-pressure process contains water, dimethyl ether, esters, ketones, iron carbonyl, and higher alcohols. However, these impurities, although similar to those formed in the high-pressure process, were much less in quantity.

ICI Katalco Low Pressure Methanol Synthesis Process. The ICI Katalco low pressure methanol process is divided into three main sections: synthesis gas preparation, methanol synthesis, and methanol purification. The principal feedstocks are natural gas, naphtha, heavier oil fractions, and coal. Synthesis gas is formed by the steam reforming of natural gas. The reformer effluent includes a mixture of hydrogen, carbon oxides, steam, and residual methane at 880°C and 20 bar. The methanol synthesis converter contains a copper-based catalyst and operates in the range of 240–270°C. Loop operating pressure is in the range of 80–100 bar. The methanol synthesis reaction is limited by chemical equilibrium and the unreacted gases are recycled to the converter. Crude methanol from the separator contains water and low levels of by-products, which are removed using a two-column distillation system. The first column removes light ends such as ethers, esters, acetone, and lower hydro-carbons. The second removes water, higher alcohols, and higher hydrocarbons.

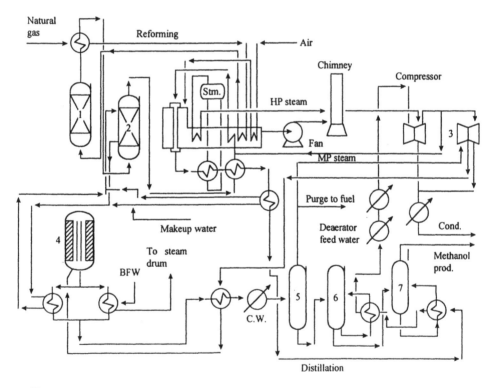

Figure 3.14 A schematic of the ICI Katalco low pressure methanol process. *Source:* [14,134].

Blends of gasoline with relatively low levels of methanol have been tested extensively worldwide. Blends of gasoline and methanol, plus high molecular weight alcohols, which are added to assist in dissolving the methanol, have been used extensively. Methanol-gasoline blends are subject to phase separation due to its high Reid vapor pressure, brought about by the inadvertent addition of small amounts of water. Concern over possible deleterious effects of methanol on engines and fuels systems in autos not designed for methanol use, also has been a negative factor for methanol acceptance. A schematic of the process flowchart is given in Figure 3.14.

The Lurgi Low Pressure Methanol Synthesis Process. In the process offered by the Lurgi Corporation [27] for the synthesis of methanol, the converter or synthesis reactor is operated at temperatures of 250–260°C. The operating pressure is 50–60 bar. The design of the converter is different from that of ICI. In the ICI design the catalyst forms a bed in which gas is introduced at various locations all along the length of the bed in order to get a uniform temperature distribution. The Lurgi design envisages a shell and tube reactor where the tubes are packed with catalyst and the heat of reaction is removed by circulating water on the shell side. Essentially, the reactor also plays a second role, that of a high pressure steam generator.

The integrated process also accepts gaseous hydrocarbons (e.g., methane) as well as liquid hydrocarbons (e.g., naphtha) as feedstock. The synthesis gas is generated either by the steam reforming route or by the partial oxidation route. The steam reformer is typically operated at 850–860°C where the previously desulfurized naphtha is contacted with steam to produce hydrogen and carbon oxides. The syngas is compressed in a centrifugal compressor to 50–80 bar and fed to the reactor.

In the second route, the heavy residues are fed into a furnace along with oxygen and steam. The feedstock gets partially oxidized at 1400–1450°C. Since the operating pressure of the partial oxidizer is 55–60 bar, there is no need for syngas compression. However, the gas needs to be cleaned to remove hydrogen sulfide and free carbon, and a shift conversion is necessary to adjust the hydrogen to carbon monoxide ratio. A schematic flow diagram is given in Figure 3.15.

The Mitsubishi Gas Chemical Low-Pressure Methanol Synthesis Process. A schematic flow diagram of the process developed by the Mitsubishi Gas Chemical Company [27] is shown in Figure 3.16. This process also uses a copper-based methanol-synthesis catalyst and is operated at temperatures of 200–280°C over a pressure range of 50–150 atm. The temperature in the catalyst bed is kept under control by using a quench-type converter design, as well as by recovering some of the reaction heat in an intermediate stage boiler. The process uses hydrocarbon feedstock. The feed is desulfurized and

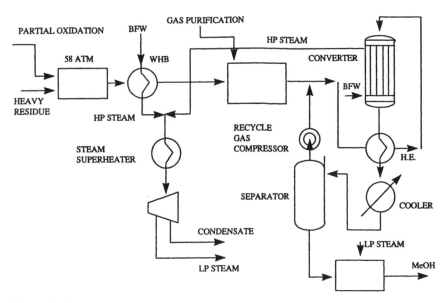

Figure 3.15   Flow sheet of the Lurgi low pressure methanol synthesis process.

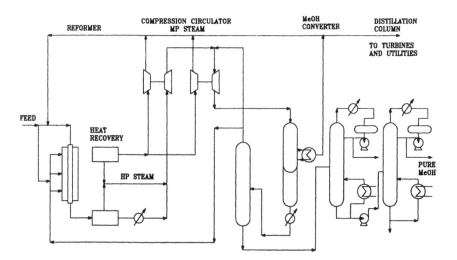

Figure 3.16   Flow diagram of the Mitsubishi Gas Chemical methanol synthesis process. *Source*: [27].

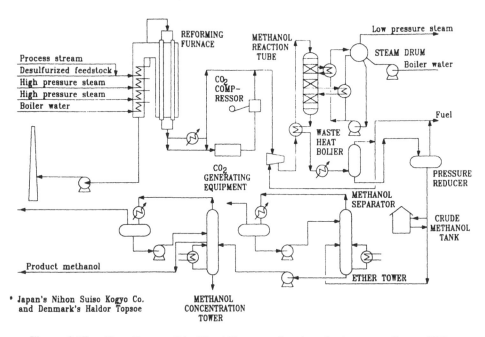

Figure 3.17   Flow diagram of the Nissui-Topsøe methanol synthesis process. *Source*: [29].

then fed into a steam reformer at 500°C. The exit stream from the reformer contains hydrogen, carbon monoxide, and carbon dioxide, and is at a temperature of 800–850°C. The gases then are compressed in a centrifugal compressor and mixed with the recycled stream before being fed into the converter.

Nissui-Topsøe Methanol-Synthesis Process.   A schematic flow diagram [29] of the process offered by Haldor Topsøe of Denmark and Nihon Suiso Kogyo Company of Japan is given in Figure 3.17. The process is based on a copper-zinc-chrome catalyst that is active at 230–280°C and at a pressures of 100–200 atm.

Japan Gas-Chemical Company Methanol Synthesis Process.   The Japan Gas-Chemical Company also came up with an intermediate-pressure process that operated at 150 atm. There is essentially not much difference among the various designs of methanol synthesis. A flow diagram of the process [29] is given in Figure 3.18.

Haldor Topsøe A/S Low Pressure Methanol Synthesis Process.   Figure 3.19 shows a schematic of Haldor Topsøe A/S process [9], starting from natural gas or associated gas feedstocks using two-step reforming. The syn-

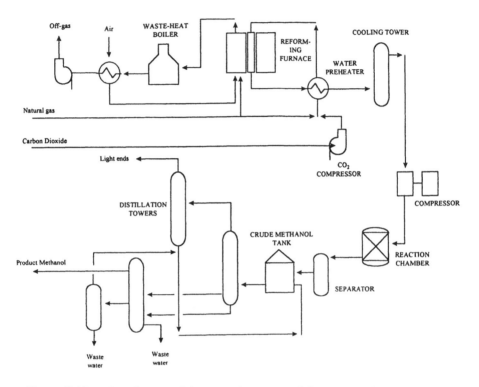

Figure 3.18   Flow diagram of the Japan Gas-Chemical Company methanol synthesis process. *Source*: [29].

thesis reactors are provided with heat exchangers between the reactors. Reaction heat is used to heat the saturator water. Effluent from the last reactor is cooled by preheating feed to the first reactor. The distillation section consists of three columns. The total energy consumption for this process is about 29.7 GJ/metric ton including energy requirements for oxygen production.

M. W. Kellogg Low Pressure Methanol Synthesis Process.   Figure 3.20 shows a schematic of the Kellogg methanol synthesis process [134], starting from hydrocarbon feedstocks using a high-pressure steam reforming process. The process uses BASF low-pressure synthesis catalyst. Desulfurized natural gas is converted to syngas via a single-stage steam reformer. Thermal efficiency is obtained by various schemes of energy integrations including heat exchange between reformer effluent and steam as well as operation of intercoolers and reactors. The reactor type is adiabatic.

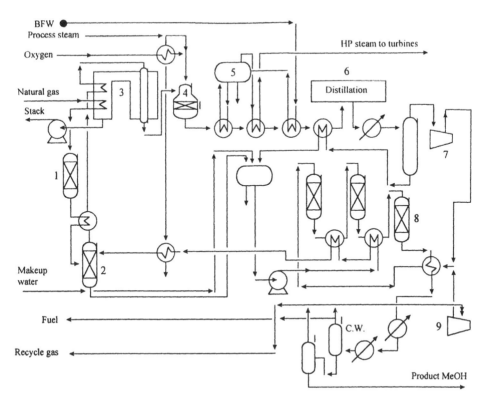

Figure 3.19   Flow diagram of the Haldor Topsøe A/S low pressure methanol synthesis process. (1) desulfurizer, (2) process steam generation unit, (3) primary reformer, (4) oxygen-blown secondary reformer, (5) superheated high-pressure steam generator, (6) distillation section, (7) single-stage syngas compressor, (8) synthesis loop, and (9) is recirculator compressor for recycle gas. *Source*: [9,14].

B.   Methanol Derivatives

The 1973 oil embargo and the ensuing petroleum and natural gas shortages forced the chemical industry to seek new resources for petrochemicals manufacture. About 86% of our domestic carbonaceous fossil fuel resources are coal and only about 2% each are petroleum and natural gas. Thus it is imminent to resort to a coal base for organic chemical products at this time. The manufacture of methanol from natural gas-based synthesis gas is a well-established technology and has been steadily improved over the years. Methanol also can be produced from coal-based synthesis gas in high yield at a

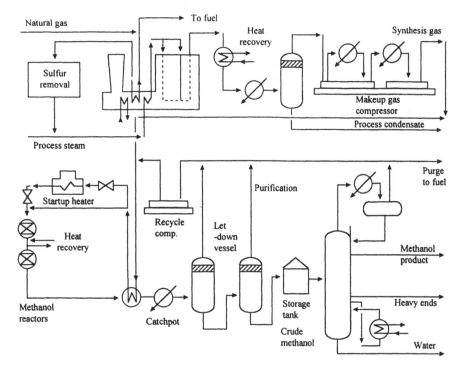

Figure 3.20 A schematic of the M. W. Kellogg low pressure methanol synthesis process. *Source*: [9,134].

relatively low cost. This makes methanol an attractive possibility as an inter-mediate to other chemicals. The greatest use of methanol in the chemical industry is as a fuel and a raw material for the synthesis of oxygenates like methyl *tert*-butyl ether (MTBE) and *tert*-amyl methyl ether (TAME). It is also used in the manufacture of other organic chemicals.

Formaldehyde

Before the advent of oxygenate ethers like MTBE, formaldehyde production was the largest single application of methanol, with at least 16 manufacturers and consuming over 30% of the methanol produced. Formaldehyde is used in the manufacture of urea-formaldehyde resins, phenol-formaldehyde resins, melamine-formaldehyde resins, acetal resins, acetylenic chemicals, etc. The reactions involved in formaldehyde synthesis are:

$$CH_3OH + \frac{1}{2}O_2 = HCHO + H_2O \qquad \Delta H = -154.8 \text{ kJ/mol, 900 K}$$

$$CH_3OH = HCHO + H_2 \qquad \Delta H = +121.3 \text{ kJ/mol}, 900 \text{ K}$$

Two processes are used mainly in this synthesis: an oxidation-dehydrogenation (silver catalyst) process and a purely oxidation (metal oxide catalyst) process. Both of these processes use air to oxidize methanol to formaldehyde or to supply in situ the endothermic heat of dehydrogenation (121.3 kJ) or both. The oxidation-dehydrogenation process developed by ICI runs at 600–700°C and uses a shallow bed of silver crystals. Other variants use layers of silver gauze or combinations of silver crystals and gauze. The typical reactor operates with 80% conversion. BASF uses a modification of this process where the reaction is carried out at higher temperatures and some dilute formaldehyde solution is recycled. The conversion achieved in this process is as high as 98.6%. In contrast, the metal oxide process operates at lower temperatures of 300–400°C and atmospheric pressure. Methanol conversion is 99%, with yields (based on methanol) in the range of 88–91%. The metal oxide catalyst commonly used is the iron oxide-molybdenum oxide type. The oxide process has the advantage of higher conversion, which makes distillation for methanol recovery unnecessary. Catalyst life is also longer because the oxide catalyst is less sensitive to impurities [30].

Methylamines

The production of mono-, di-, and trimethylamines has a number of applications in the chemical industry. Monomethylamine is used in insecticides and surfactants, dimethylamine is used in rubber chemicals, and trimethylamine is used for choline chloride and biocides. Dimethylamine is produced to the greatest extent, followed by mono- and trimethylamine. The methylamines are produced by reaction of methanol and ammonia in the vapor phase over a dehydration catalyst at a temperature of 450°C and from atmospheric to approximately 20.4 atm pressure [30]. The methylamine synthesis reactions are:

$$CH_3OH + NH_3 = CH_3NH_2 + H_2O$$

$$2CH_3OH + NH_3 = (CH_3)_2NH + 2H_2O$$

$$3CH_3OH + NH_3 = (CH_3)_3N + 3H_2O$$

Preferred catalysts generally are silica/alumina compositions which may be impregnated with promoters, such as silver phosphate, cobalt sulfide, etc.

Chlorinated Hydrocarbons

Chloromethanes have varied applications in the chemical industry. Methyl chloride is mainly used in the manufacture of silicones, tetramethyl lead, and as a solvent. Methylene chloride (often called methylene dichloride or dichloromethane, DCM) is a multipurpose solvent, a degreasing agent, and a blowing agent. Chloroform is an intermediate in the manufacture of refrigerants, aerosol propellants, and fluorocarbon resins, and carbon tetrachloride is an intermediate for aerosol propellants, and insecticides. The reactions involved in chloromethanes from methanol are:

$$CH_3OH + HCl = CH_3Cl + H_2O$$

$$CH_3Cl + Cl_2 = CH_2Cl_2 + HCl$$

$$CH_3Cl + 2Cl_2 = CHCl_3 + 2HCl$$

$$CH_3Cl + 3Cl_2 = CCl_4 + 3HCl$$

Most of the HCl is recycled and utilized. The catalysts used are alumina gel, cuprous chloride, or zinc chloride. The reaction temperature is in the range of 490–530°C [30].

Acetic Acid

The synthesis of acetic acid is one of the most rapidly growing chemical applications for methanol. The process for manufacture of acetic acid was developed by Monsanto. The reaction runs at a temperature of 150–200°C and a pressure of 30.62 atm. The catalyst used is rhodium salts with certain ligands and in the presence of an iodine compound. The reaction is:

$$CH_3OH + CO = CH_3CO_2H \qquad \Delta H = -123 \text{ kJ/mol}, 400 \text{ K}$$

The reaction usually takes place in a stirred-tank reactor and the product then flows into a flash tank and then to a refining train [30].

Acetic Anhydride

Acetic anhydride is used in the manufacture of cellulose acetate-based film, cigarette filters, and plastics. Eastman Chemical developed a process that is based on gasification of coal in a Texaco gasifier to make synthesis gas which then is converted to methanol. The methanol is converted to methyl acetate by esterification with acetic acid and then carbonylated. The carbonylation process uses rhodium salt catalysts with ligands and an iodine promoter [30].

Dimethyl Ether

Methanol can be converted to dimethyl ether (DME) by passing the vapors over a dehydration catalyst such as silica/alumina or γ-alumina at temperatures of 250°C or higher. Dimethyl ether can be used as an aerosol propellant and also as a precursor to a variety of hydrocarbons and organic chemicals; it also can be used as a refrigerant and a cold starting fluid for automobile engines. The reaction is [30]:

$$2CH_3OH = CH_3OCH_3 + H_2O \qquad \Delta H = -22.2 \, kJ, 600 \, K$$

Advances in the synthesis chemistry and technology allow the single-stage synthesis of DME from synthesis gas, which is also proven to be economically more favorable than the conventional two-stage routes.

Ethylene Glycol

The present commercial process for the synthesis of ethylene glycol (EG) involves the reaction of ethylene with oxygen over a silver catalyst. The ethylene oxide so formed is then hydrolyzed to ethylene glycol. Ethylene glycol also can be produced directly from synthesis gas but the operating pressures are too high for commercial use. Celanese and Redox Technologies developed an economically attractive process to ethylene glycol from methanol and formaldehyde. This process exhibits significant raw material advantages over the ethylene oxide process. Methanol and formaldehyde react in the liquid phase at 125–200°C and 21.4–41.8 atm using a free radical initiator. The selectivity to EG is high. The reaction is [31]:

$$CH_3OH + HCHO = HOCH_2CH_2OH$$

Styrene

Styrene (SM) can be synthesized in a single step via alkylation of toluene with methanol which offers significant advantages in raw material costs and energy consumption as compared to the benzene to styrene via ethylbenzene process. This single-step process is carried out at a temperature of 400°C and uses a cesium-boron type X zeolite catalyst. The reaction is [31]:

$$C_6H_5CH_3 + CH_3OH = C_6H_5CH=CH_2 + H_2O$$

Ethanol

Ethanol utilization in the chemical industry has grown significantly since the synthesis of oxygenate ethers like ETBE. Due to its relatively low blending vapor pressure, ethanol is very heavily used as an oxygenated gasoline blending material. In the United States, up to 10% of ethanol blending is very frequently seen. The Halcon SD Group has developed a new route to ethanol from methanol. The process flowsheet is shown in Figure 3.21 and the reaction chemistry is:

$$MeOH + CO = HOAc$$

$$HOAc + MeOH = MeOAc + H_2O$$

$$MeOAc + 2H_2 = EtOH + MeOH$$

Net reaction: $MeOH + CO + 2H_2 = EtOH + H_2O$

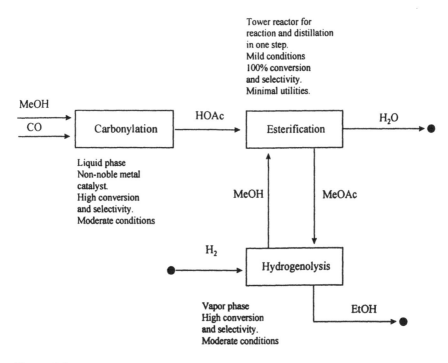

Figure 3.21   Flowsheet of the Halcon SD Process of methanol to ethanol. *Source*: [32].

In this process methanol first is carbonylated in the liquid phase to acetic acid at high yield. In the second step, acetic acid is esterified with methanol recycled from the earlier step. Mild conditions are used and the yield from methanol and acetic acid is essentially quantitative. The esterification is carried out in a novel manner that greatly reduces capital cost and steam consumption compared to conventional esterification technology [32].

C.   Formaldehyde

Formaldehyde (CH_2O) is an important compound in the synthesis of various chemicals on an industrial scale. One of the first industrial applications was in the production of artificial Indigo. The variety of end products produced from formaldehyde include resins or glues (produced by the condensation of formaldehyde with urea, phenol, or melamine) as well as rubber, paper, fertilizers, explosives, engineering plastics, and specialty chemicals like acrolein, methacrylic acid, methyl methacrylate, etc. Because it is nearly impossible to handle in its pure gaseous form, formaldehyde is almost exclusively produced and processed as an aqueous solution: formalin. Formalin is obtained commercially by absorbing the gases leaving the reactor in water. The industrial production of formaldehyde starts from methanol or methane. Thus, the production capacity of formaldehyde can be significantly increased by utilizing the natural gas and coal resources.

The selective oxidation of methanol to formaldehyde is very important for the chemical industry, for it allows the production of methanol-free formaldehyde and also prolongs the catalyst life. A lot of research was directed toward the synthesis of individual transition metal (Cr, Mn, Co, Ni, etc.) molybdates as well as the investigation of the properties of the classical iron-molybdenum oxide catalyst. In spite of having a high selectivity and thermal stability, these catalysts have a low specific activity. Studies were carried out to investigate the catalytic properties of multicomponent catalysts based on metal molybdates in methanol oxidation [33]. The results indicated that the oxidation to formaldehyde is the main reaction on these catalysts. Above 390°C, subsequent oxidation of formaldehyde to carbon monoxide takes place. However, the rate of this process is very low, which ensures the high selectivity to formaldehyde at high temperatures (350–390°C). The activity and selectivity both increase by adding a small amount of Tl and P to a Ni-Co-Fe-Bi molybdate catalyst. With catalysts having large specific surface areas and relatively small pore radii, the reaction rate is affected by transport phenomena.

Production of formaldehyde as an aqueous solution, for handling convenience, using the silver process gives high yields [34]. In this process, air and vaporized methanol are combined with steam and recycled gas, and then are passed over hot silver grains at ambient pressure. Methanol is converted to formaldehyde by partial oxidation and by reduction at a temperature of 600°C.

The reactions involved in formaldehyde synthesis from methanol are:

$$CH_3OH + 0.5O_2 = HCHO + H_2O$$

$$CH_3OH = HCHO + H_2$$

Product streams consist of nitrogen (50%), hydrogen (15%), water vapor (20%), formaldehyde (15%), and minor amounts of byproducts. Formaldehyde is condensed by passing the reactor effluent through a partial condenser. The resulting gas-liquid stream is subsequently fed to the absorber, such as to minimize the formaldehyde content in the tail gas and maximize the formaldehyde content in the product liquid leaving the absorber.

The dehydrogenation of methanol to formaldehyde has been studied in organic membrane reactors [35]. The combination of the membrane and catalyst holds significance in the industry, like the combination of reaction and separation. By using membrane reactors, the process operation is simplified and the conversion of the reactants is enhanced significantly. Deng and Wu

[35] utilized a palladium/ceramic membrane, a porous membrane, and a catalytic inorganic membrane. A conventional Cu-P/SiO$_2$ catalyst was also in the membranes. The dehydrogenation of methanol in membrane reactors using these membranes was compared to the same in a conventional fixed-bed reactor. The palladium/ceramic membrane reactor showed the highest conversion of methanol, followed by the catalytic membrane reactor. At a reaction temperature of 673 K and W/F (amount of catalyst per flow rate) of around 1000 g-cat min^{-1}mol^{-1}, the yield of formaldehyde increased by 15% in the palladium/ceramic membrane reactor, 8% in the catalytic porous membrane reactor, and 5% in the porous alumina membrane reactor above that obtained in a conventional single-pass fixed-bed reactor using the same catalyst. They concluded that inorganic porous membranes have a greater potential for industrial use than the palladium/ceramic membrane.

The partial oxidation of natural gas, consisting chiefly of methane, currently holds tremendous industrial potential. Possible routes for the direct synthesis of formaldehyde from methane, either via chlorine-based catalysts or with the use of chlorine-containing compounds in the gas feed (both using chlorine-modified supported palladium catalysts and at temperatures of 450–650°C) gave formaldehyde yields less than 7.7% under optimum conditions [36]. TiO$_2$–based catalysts, with the addition of small amounts of CH$_2$Cl$_2$ showed marked increase in the conversion of methane. Formaldehyde selectivities in excess of 80% were achieved over the anatase phase at a conversion level of greater than 1%. The function of anatase is to facilitate the decomposition of CH$_2$Cl$_2$ at low temperatures and to release chlorine-containing radicals into the gas phase. It also hinders the oxidation of formaldehyde, thereby maintaining the high selectivities of formaldehyde.

The partial oxidation of methane to formaldehyde by molecular oxygen on silica and silica-supported oxide catalysts at a pressure of 1.7 bar in the temperature range of 520–650°C was studied [37]. A batch reactor with external recycle was used to investigate the effects of reactor diameter, recycle flow rate, catalyst weight, and methane-to-oxygen ratio on the catalyst activity. The reactor is equipped with an external recycle pump and a liquid product condenser to prevent the further oxidation of the oxygenated products. A negligible reaction (gas-phase) at temperatures in the range of 500–650°C was observed. Increasing the methane to oxygen ratio from 2:1 to 15:1 showed only a slight increase in formaldehyde selectivities. Addition of V$_2$O$_5$ enhanced the activity of the silica catalysts by a factor of 2 and that of fumed silica catalysts by a factor of 70.

Figure 3.22 shows a schematic of the Haldor-Topsøe A/S Formaldehyde Process. The process starting material is methanol and the catalyst used is Haldor Topsøe A/S FK-series iron/molybdenum-oxide catalysts [9]. The process is carried out in a recirculation loop at near atmospheric pressure (1–

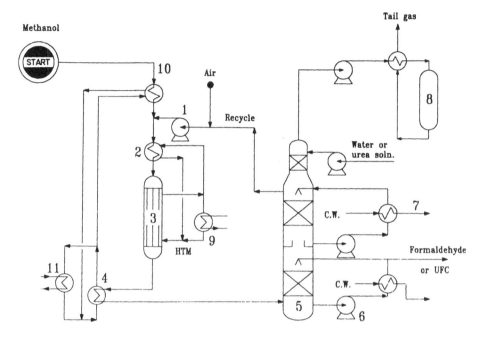

Figure 3.22   A schematic of the Haldor-Topsøe A/S Formaldehyde Process. (1) pump, (2) heat exchanger, (3) reactor, (4) low pressure steam boiler, (5) absorber, (6,7) cooling cycles, (8) tail gas scrubber, (9) reactor heater, (10) methanol evaporator, (11) boiler feed-water preheater.

1.4 atm). A mixture of vaporized methanol, air, and recycle gas may be preheated to about 250°C in the heat exchanger (2), or fed directly to the reactor (3). In the catalyst-filled tubular reactor, methanol and oxygen react to yield formaldehyde. Exothermic heat of reaction is removed by an oil heat transfer medium and the reactor exit gas is approximately at 290°C and is cooled in the low pressure steam boiler (4) to 130°C before entering the absorber (5). In the absorber, formaldehyde is absorbed in water or urea solution, and heat is removed by one or two cooling cycles (6,7). Tail gas quality is excellent, meeting strict environmental standards [9].

Energy integration is achieved by using steam produced by the reactor effluent heat exchanger (4) in the methanol evaporator (10) and in the boiler feed water preheater (11). Typical product specifications are given in Table 3.20. Twelve commercial units based on this process technology have been built, while three additional units are under construction [9].

Table 3.20 H-T Formaldehyde Product
Specifications

	AF	UFC
Formaldehyde	37–55	≤ 60
Urea	—	≤ 25
Methanol	< 0.7	< 0.3
Formic Acid	< 0.05	< 0.05

AF: aqueous formaldehyde.
UFC: urea formaldehyde precondensate.

The promoting action of K_2SO_4 on unsupported V_2O_5 catalysts used for the oxidation of methanol to formaldehyde has been studied in the temperature range of 550–750 K. The K_2SO_4 content of the catalysts was varied between 0 and 40 mol%. Experiments were carried out in a fixed-bed microreactor of 18-cm length and 2-cm inner diameter. The Pyrex reactor was installed in the oven of a gas chromatograph for accurate temperature control. The effect of K_2SO_4 addition on the catalytic properties of the vanadium catalysts was investigated using the vapor-phase oxidation of methanol as a test reaction. The only products formed were formaldehyde, water, CO, and CO_2. The activity of the catalysts for methanol oxidation falls rapidly when the K_2SO_4 content exceeds 20%. The higher activity of catalyst containing 10% K_2SO_4 as compared to pure V_2O_5 is attributed to the larger BET surface area of this sample. Selectivity to formaldehyde is improved for all samples containing K_2SO_4. With pure V_2O_5, the maximum selectivity was 0.85; the promoted catalysts exhibited a selectivity higher than 0.95. The oxygen-bond strength in the mixed catalysts is lowered if the K_2SO_4 content is increased from 0 to 40 mol%. The catalysts having lower oxygen-bond strengths are reduced to lower oxidation states during steady-state methanol oxidation. The activation energy for methanol oxidation decreases with increasing promoter content. The promoted catalysts show a higher selectivity to formaldehyde in the temperature range and composition range investigated [38]. This process is not in commercial operation.

D. Formaldehyde Derivatives

1,4–Butynediol

Ethynylation of formaldehyde is the major route for the commercial manufacture of 1,4–butynediol. This process is also known as the Reppe process. The stoichiometric reaction for this process is:

$$C_2H_2 + 2HCHO = HOCH_2CCCH_2OH$$

Investigations on the ethynylation of formaldehyde were carried out in a slurry reactor to study the reaction kinetics [39]. $Cu_2C_2/MgSiO_4$ was used as the catalyst for this study. Formaldehyde conversions of 98.6% were achieved in a 24-hr reaction run with a 98% selectivity toward the main product 1,4–butynediol. The gas/liquid mass transfer resistance can be neglected if the agitation rate exceeds 400 rpm. The particle sizes used, 0.21–0.25 mm and 0.15–0.18 mm, showed a negligible effect on reaction rate, and hence the intraparticle diffusion was not rate limiting. The rate was found to be dependent on the partial pressure of acetylene below pressures of 1.0 kg cm^{-2}. The effect of formaldehyde concentration on the rate showed a nonlinear relationship which indicated that the adsorption of formaldehyde plays an important role in the controlling step of the reaction [39].

A mixture of formalin and ethanol was passed at 240–320°C over various metal oxides supported on silica gel and metal phosphates. The main products were acrolein, acetaldehyde, methanol, and carbon dioxide. Acidic catalysts such as V-P oxides promoted the dehydration of ethanol to ethene. The best catalytic performances for acrolein formation are obtained with nickel phosphate and silica-supported tungsten, zinc, nickel, and magnesium oxides. With a catalyst with a P/Ni atomic ratio of 2/3, the yields of acrolein reach 52 and 65 mol% on ethanol basis with HCHO/ethanol molar ratios of 2 and 3, respectively. Acetaldehyde and methanol are formed by a hydrogen transfer reaction from ethanol to formaldehyde. Then acrolein is formed by an aldol condensation of formaldehyde with the produced acetaldehyde [40].

Methyl Methacrylate

Methyl methacrylate (MMA) is a basic chemical and is the raw material for a variety of synthetic processes, particularly for manufacturing plastics. It is used in sheeting for signs and building materials, surface coatings, and molding resins. The reaction between formaldehyde and propionaldehyde is:

$$CHOH + CH_3CH_2CHO = CH_2C(CH_3)CHO + H_2O$$

Ethylene and synthesis gas are used to produce propionaldehyde and formaldehyde. These aldehydes then are reacted in the presence of a catalyst to form methacrolein. The methacrolein reacts with atmospheric oxygen to form methacrylic acid in a catalytic multitubular reactor. The methacrylic acid is separated from water (a byproduct) and purified; it then reacts with methanol to form MMA. The byproducts are incinerated on-site to recover the energy in the process [41].

The vapor phase synthesis of methacrylic acid from propionic acid and formaldehyde was studied [42]. In particular, the choice of alkali metal cation and loading were evaluated for their effect on the activity and selectivity of silica supported catalysts. Experiments were carried out in 0.5 in. (o.d.) quartz reactors equipped with 0.125 in. thermowells. Alkali metal cations supported on silica are effective base catalysts for the production of methacrylic acid. Silica surfaces exchanged with alkali metal cations are capable of chemisorbing propionic acid yielding surface-bound silyl propionate esters and metal propionate salts. The alkali metal cation influences the temperature at which desorption of the ester occurs (Cs < Na < Li < support). For silica catalysts of equimolar cation loading, activity and selectivity to methacrylic acid show the opposite trend, Cs > K > Na > Li. Methacrylic acid selectivity reaches a maximum at intermediate cation loadings where interaction of adjacent silyl esters is minimized [42].

Hexamethylene Tetramine

Formaldehyde reacts with ammonia to yield hexamethylene tetramine (HMTA).

$$6HCHO + 4NH_3 \longrightarrow$$

This product is known as hexamine in the plastics industry and urotropin in the pharmaceutical industry. The reaction is rapid and the yield is above 96%. In order to carry out the reaction, formaldehyde in aqueous solution reacts with anhydrous ammonia in a cooled reactor. The reaction product is purified by evaporation, centrifuging, and drying. The centrifuge wash liquors are virtually all recycled.

Hexamethylene tetramine is an important chemical. Its most important application is as a source of formaldehyde for crosslinking phenolic molding powders, shell molding resins, and two-step curing resins for chip board. Quaternization of HMTA with an alkyl chloride (R-Cl) gives a family of bactericides (developed by Dow) for use in latex paints and as a dermatitis preventative in water-soluble cutting oil. Hexamine is used in pharmaceutical formulations to combat urinary tract infections and also as an intermediate in the production of chloramphenicol.

Pentaerythritol

Formaldehyde condenses with acetaldehyde in the presence of a basic catalyst to form pentaerythritol, $C(CH_2OH)_4$:

$$4HCHO + CH_3CHO + NaOH = C(CH_2OH)_4 + HCOONa$$

There are several side reactions forming byproducts. Etherification of two mols of pentaerythritol yields dipentaerythritol as:

$$2C(CH_2OH)_4 = (CH_2OH)_3C - CH_2-O - CH_2C(CH_2OH)_3 + H_2O$$

Byproducts include carbon dioxide and formic acid in addition to the condensation products. The commercial production of pentaerythritol, therefore, involves a long series of purification steps. The manufacturing process consists of crystallization–dissolution–filtration–redissolution–ion exchange–purification–vacuum crytstallization–redissolution–filtration–drying. A typical technical product contains about 10–11% dipentaerythritol. Depending upon the end uses, the two products may have to be obtained in pure forms, thus requiring further complicated purification schemes.

Pentaerythritol is used as a raw material for the explosive known as pentaerythritol tetranitrate (PETN). The most important end-use of pentaerythritol is in the manufacture of alkyd resins, in competition with glycerol. Due to its higher alcohol functionality, pentaerythritol gives alkyds somewhat better properties over glycerol. Other important end-uses include: pentaerythritol resin esters used as floor-polish, fire-retardant, high pressure lubricants. Pentaerythritol is also used to produce chlorinated polyethers by the following reactions:

$$C(CH_2OH)_4 + 4CH_3COOH \longrightarrow C(CH_2OOCCH_3)_4 + 4H_2O$$

$$C(CH_2OOCCH_3)_4 + 3HCl \longrightarrow (ClCH_2)_3CCH_2 - O - \overset{\displaystyle O}{\overset{\|}{C}} - CH_3$$

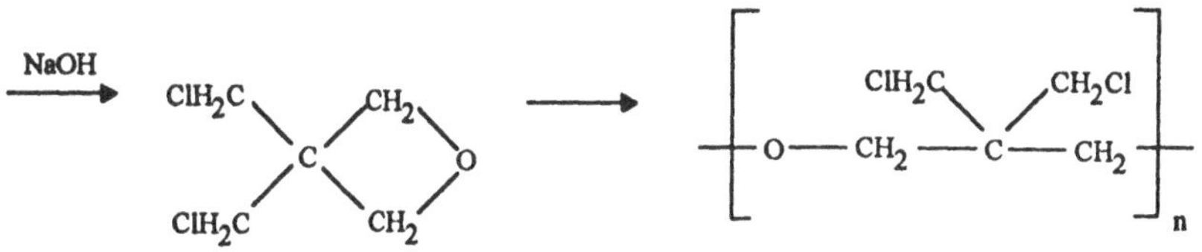

This resin is used primarily in corrosion-resistant coating and piping of process equipment, as a compromise between poly(vinyl chloride) (PVC) and the much more expensive fluorocarbons.

Polyacetals

Compounds containing carbonyl groups polymerize by forming polyacetals:

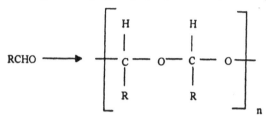

The most stable polyacetal polymer is polyformaldehyde (or polyoxymethylene, POM); this is the only polyacetal that has reached commercial production. This resin has unique properties (e.g., selflubrication) and is very widely used in automotive applications such as engineering plastics. Acetals are widely used engineering thermoplastics with high load-bearing characteristics and low coefficients of friction. Currently, over 200 million lb of acetals are molded and extruded in the United States.

There are two kinds of acetals: homopolymers and copolymers. The former were the first acetals made by DuPont in 1960. Strength-related properties of homopolymers are slightly higher than those of copolymers. The homopolymer has higher heat deflection temperatures and lower continuous use temperatures (85°C versus 100°C) than copolymers. The homopolymer has somewhat greater surface hardness and a slightly lower coefficient of friction. Other key properties of acetals are good mechanical stability and high fatigue strength. Acetals have high dielectric properties, but they are not flame-retardant. They have very low water absorption and moisture hardly affects their properties. Several specific grades are available, namely, 1) extrusion, 2) lubricated, 3) flame-retardant, 4) UV-stabilized, 5) impact-modified, 6) glass fiber-reinforced, 7) mineral filled, and 8) other grades with conductive carbon fibers and with aramid fibers (high strength and low friction coefficient).

As for processing of acetals, the type of polymer is important (i.e., homopolymer versus copolymer acetals). Injection molding is the most widely-used method for processing acetals, with conventional general-purpose screws of 16–20 L/D, and injection temperatures of 180–220°C. High molecular weight acetals are recommended for extrusion. Blow molding is rarely used and difficult to apply to acetals, because of the material's tendency to set up quickly.

End uses of acetals include industrial/mechanical and automotive products, appliances, plumbing, and electronics. Industrial and mechanical products are currently the largest end use, including molded rollers, bearings, gears, levers, rails, linkages, conveyor chains, etc. Automotive applications are the second largest area, typically including window lift mechanisms and cranks, ball sockets, interior fasteners, door handles, gear shift assemblies, etc.

Glycolic Acid

In the manufacture of ethylene glycol from formaldehyde, glycolic acid is formed as an intermediate:

$$HCHO + CO + H_2O = HOCH_2COOH$$

$$HOCH_2COOH + CH_3OH = CH_3OOCCH_2OH$$

$$HOCH_2COOCH_3 + 2H_2 = HOCH_2CH_2OH + CH_3OH$$

The significant end uses include raw material in chelating formulations, conversion to ethylene glycol, and use in leather industry.

Synthetic Gasoline

Methanol itself can be used as a transportation fuel just as liquefied petroleum gas (LPG) and ethanol; however, direct use of methanol as a motor fuel would require nontrivial engine modifications and substantial changes in lubrication system. This is why methanol to gasoline is quite appealing [69,73].

The Mobil Research and Development Corporation developed the methanol-to-gasoline (MTG) process. The process technology is based on the catalytic reactions using the zeolites of the ZSM-5 class [70,73]. Methanol-to-gasoline reactions can be written as:

$$nCH_3OH \dashrightarrow (-CH_2-)_n + nH_2O$$

The detailed reaction path is described by Chang [73]. The following simplified steps describe the overall reaction path:

$$2CH_3OH \dashrightarrow (CH_3)_2O + H_2O$$

$$CH_3OH(CH_3)_2O \dashrightarrow light\ olefins + H_2O$$

light olefins $\dashrightarrow$ heavy olefins

heavy olefins $\dashrightarrow$ paraffins

aromatics

naphthenes

Due to the shape-selective pore structure of the ZSM-5 class catalysts, the hydrocarbons fall predominantly in the gasoline boiling range. The product distributions are influenced by the temperature, pressure, space velocity, reactor type, and Si/Al ratio of the catalyst [73,74]. Paraffins are dominated by isoparaffins, while aromatics are dominated by highly methyl-substituted aromatics. The C_9+ aromatics are dominated by symmetrically methylated isomers reflecting the shape selective nature of the catalysis. The C_{10} aromatics are mostly durene (1,2,4,5–tetramethylbenzene), which has an excellent octane number but its freezing point is very high (79°C). Too high a durene content in the gasoline may impair automobile driving characteristics especially in the cold weather, due to its tendency of crystallization at low temperature [137]. Mobil's test found no drivability loss at −18°C using a gasoline containing 4 wt% of durene [137]. Mobil also developed a heavy gasoline treating (HGT) process to convert durene into other high-quality gasoline components by isomerization and alkylation. The MTG reactions are exothermic and go through the dimethyl ether intermediate route.

There are three basic types of chemical reactors developed for the MTG process: 1) adiabatic fixed-bed, 2) fluidized-bed, and 3) direct heat exchange. The first two were developed by Mobil and the last by Lurgi.

The adiabatic fixed-bed concept uses a two-stage concept, namely, 1) the first stage methanol reactor, and 2) the second stage methanol conversion to hydrocarbons. The first commercial plant of 14,500 bbl/day gasoline capacity was constructed in New Zealand [49,137]. The plant had been running successfully since 1985 start-up until a recent shut-down. The synthesis gas is generated via steam reforming of natural gas obtained from the off-shore Manifield. A HGT plant was also successfully run in New Zealand and reduced the durene content to 2 wt%. The successful operation of MTG in New Zealand was a very important milestone in the human history, since it made possible the chemical synthesis of gasoline from unlikely fossil fuel sources like natural gas and coal. Crude petroleum is no longer the sole source for gasoline.

A fluidized-bed MTG concept was concurrently developed by Mobil. The process research went through several stages involving bench-scale fixed fluidized-bed 4-bbl/day and 100-bbl/day cold-flow models, and a 100-bbl/day semiwork plant.

Typical MTG operating conditions in a fixed-bed reactor mode are: [137]

methane/water charge = 83/17 (by mass)
inlet/outlet temperatures of dehydration reactor = 306/404°C
inlet/outlet temperatures of gasoline conversion reactor = 360/415°C
pressure = 2,170 kPa
recycle ratio, mol/mol charge = 9.0
space velocity, WHSV = 2.0

Typical yields from fixed-bed operations (based on the original mass of methane input) include: 43.4% hydrocarbons, 56% water, 0.4% CO/CO_2, and 0.2% coke and others. Hydrocarbon products consist of 1.4% light gas, 5.5% propane, 0.2% propylene, 8.6% isobutane, 3.3% n-butane, 1.1% butenes, and 79.9% C_5+ gasoline [137]. Typical MTG operating conditions in a fluidized-bed mode differs from the fixed-bed mode and they are:

methanol/water charge = 83/17 (by mass)

reactor temperature of gasoline conversion reactor = 413°C
pressure = 275 kPa
space velocity, WHSV = 1.0

Typical fluidized-bed yields based on the percentages of methanol input include: 0.2% methanol + DME, 43.5% hydrocarbons, 56% water, 0.1% CO/CO$_2$, and 0.2% coke + others. Hydrocarbon products typically consist of 5.6% light gas, 5.9% propane, 5.0% propylene, 14.5% isobutane, 1.7% n-butane, 7.3% butenes, and 60% C$_5$+ gasoline [137].

During the MTG development, Mobil researchers also found that the hydrocarbon product distribution can be shifted to light olefins by increasing the space velocity, decreasing the MeOH partial pressure, and/or increasing the reaction temperature. Typical yields [137] from 4-bbl/day operation were: C$_1$–C$_3$ paraffins, 4 wt%; C$_4$ paraffins, 4 wt%; C$_2$–C$_4$ olefins, 56 wt%; C$_5$+ gasoline, 35 wt%. Using olefins from the MTO or FT (Fischer-Tropsch) processes, diesel and gasoline can be made via a process converting olefins to diesel and gasoline. Using acid catalysts, catalytic polymerization is a standard process and is being used at SASOL to convert C$_3$–C$_4$ olefins into (G+D). Recently, Mobil developed the Mobil Olefins to Gasoline and Diesel (MOGD) process using their commercial zeolite catalyst [49,135,136].

An innovative process enhancement was made under the sponsorship of the Electric Power Research Institute [138]. Their DTG (DTH, DTO) process is based on the conversion of dimethyl ether to hydrocarbon over ZSM-5-type catalyst [138]. Their process is based on the novel, economical, single-stage synthesis process of dimethyl ether (DME) from syngas. By producing DME in a single stage, the intermediate methanol formation is no longer limited by chemical equilibrium, thus increasing the reactor productivity substantially. This is especially true for the synthesis of methanol in the liquid phase. Furthermore, by feeding DME directly to the ZSM-5 reactor instead of methanol, the stoichiometric conversion and hydrocarbon selectivity increase substantially due to the less production of water. The difference between MTG and DTG, therefore, is in the placement of methanol dehydration reaction step (i.e., DME formation reaction). In the MTG process, methanol-to-DME conversion takes place in the gasoline reactor, whereas in the DTG process, this takes place in the syngas reactor. The process has not been tested yet on a large scale.

E.   Dimethyl Ether

Dimethyl ether (DME) is a useful chemical intermediate for the preparation of many important chemicals, including dimethyl sulfate. Dimethyl ether is synthesized via dehydration of methanol over a catalytic system:

$$2CH_3OH = CH_3OCH_3 + H_2O$$

Table 3.21 Reaction Equilibria for Methanol and DME Synthesis

---

1. Methanol synthesis

   (I) $CO + 2H_2$ $<\!\!-\!\!->$ CH_3OH

        $\Delta H_{600K} = -100.48$ kJ/mol $\Delta G_{600K} = +45.36$ kJ/mol

   (II) $CO_2 + 3H_2$ $<\!\!-\!\!->$ $CH_3OH + H_2O$

        $\Delta H_{600K} = -61.59$ kJ/mol $\Delta G_{600K} = +61.80$ kJ/mol

2. Water-gas shift reaction

        $CO + H_2O$ $<\!\!-\!\!->$ $CO_2 + H_2$

        $\Delta H_{600K} = -38.7$ kJ/mol $\Delta G_{600K} = -16.5$ kJ/mol

3. Dehydrogenation of methanol to dimethyl ether

        $2CH_3OH$ $<\!\!-\!\!->$ $CH_3OCH_3 + H_2O$

        $\Delta H_{600K} = -20.59$ kJ $\Delta G_{600K} = -10.71$ kJ

4. Direct synthesis of dimethyl ether

        $2CO + 4H_2$ $<\!\!-\!\!->$ $CH_3OCH_3 + H_2O$

        $\Delta H_{600K} = -35.31$ kJ $\Delta G_{600K} = +79.97$ kJ

Source: Sofianos, A., *Ind. Eng. Chem. Res.*, 30:2372-2378, 1991.

The catalyst proven effective is γ-alumina and the typical reaction temperature range is 300–350°C. The single-pass conversion exceeds 90%. The products are easily separated based on the boiling point difference. More recently DME has been increasingly used as an aerosol propellant to replace chlorofluoro-carbons (CFCs), which were believed to be culprits for destroying the ozone layer of the atmosphere. Also, DME can be used directly as a transportation fuel in an admixture with methanol or as a fuel additive. Dimethyl ether is the key intermediate in the synthesis of hydrocarbons from methanol. When-ever the methanol concentration is high in the reactor, methanol synthesis from syngas is limited by chemical equilibrium due to the reversible nature of the reaction. Coproduction of methanol and dimethyl ether offers an opportunity for increasing syngas conversion per pass [43–45]. The equilibrium constraints of methanol synthesis are alleviated by continuous conversion of methanol to dimethyl ether. These consecutive reactions essentially suppress the reverse reaction of the methanol synthesis, thus allowing syngas conversion to proceed further than limited by the methanol chemical equilibrium. The reaction equi-libria of methanol-DME coproduction are shown in Table 3.21.

Methanol from the first reaction is consumed in forming DME and water. The produced water reacts with carbon monoxide to produce carbon dioxide and hydrogen by the forward water–gas shift reaction.

$$CO_2 + 3H_2 = CH_3OH + H_2O$$

$$CO + H_2O = CO_2 + H_2$$

$$2CH_3OH = CH_3OCH_3 + H_2O$$

Carbon dioxide and hydrogen are reactants of methanol synthesis [11]. Thus, one of the products of each step is a reactant for another. This creates a driving force for the overall reaction and allows very high conversion of syngas in a single pass.

A study was carried out to investigate the complexity of the overall conversion, to show the importance of the choice of catalyst components, and to identify potential problems associated with the bifunctional character of the catalyst system such as deactivation and regeneration problems. The bifunctional catalyst system consisted of a Cu/Zn methanol synthesis catalyst and a methanol dehydration catalyst (e.g., γ-alumina, H-ZSM-5 zeolite with a high $SiO_2:Al_2O_3$ ratio of 90, amorphous silica-alumina, or Y zeolite in the H-form). The catalyst containing 50% active γ-alumina exhibited the highest catalytic activity, regarding both CO conversion and the production of oxygenates. For example, at 300°C the yield of DME/methanol was nearly 5 kg/kg_{cat}-hr which is significantly higher than the yield of the commercial methanol catalyst alone. The selectivity toward oxygenates over this system is high and stable over the temperature range of the test, whereas the water–gas shift activity is comparatively low. The system that has amorphous silica-alumina as its acid component initially displays a behavior similar to the one based on γ-alumina. An increase of the reaction temperature to above 250°C, however, effects a drastic increase in the yield of oxygenates obtained over the γ-alumina-containing system which retains its high selectivity. Carbon dioxide also is formed in relatively large quantities since the water–gas shift reaction is kinetically enhanced by an increase in the reaction temperature. The CO_2 formation is highest over the catalyst containing the H-Y zeolite, whereas it remains very low over the catalyst based on γ-alumina, even at high reaction temperatures. Methanation plays almost no role in all cases, since the experimental conditions adopted are rather moderate. Carbon monoxide conversion and DME yield decrease slightly with an increase in the space velocity. The selectivity remains stable, while the overall yield of the oxygenates increases proportionally to the space velocity. Thus, the DME synthesis reaction is favored much more than the methanol synthesis reaction in terms of thermodynamics. Hence, high CO conversions and high product yields could be attained at relatively low pressures and high space velocities [42].

Methanol and dimethyl ether have been coproduced by CO hydrogenation over Pd/NaY catalysts prepared by ion exchange. By controlling the calcination program, Pd/NaY catalysts can be tuned to selectively produce either branched lower hydrocarbons or methanol or dimethyl ether. CO hydrogenation at various temperatures and 10–11 atm total pressure was carried out in a continuous-flow fixed-bed reactor system. In the case of Pd/NaY(500/350) the rates of methanol and dimethyl ether formation increase sharply and reach a maximum by the end of the first hour of reaction. For the remainder of the reaction there is little deactivation in higher hydrocarbons production while rates of methane, methanol, and dimethyl ether formation decrease steadily. In the case of Pd/NaY(250/250), there are increases of at least one order of magnitude in the rates of methanol and dimethyl ether formation during the first two hours of reaction. After this period there is little deactivation in the formation of these two products while methane production decreases steadily. For Pd/NaY (500/350) increasing the temperature by 23°C increases the overall CO conversion by 18%. In the case of Pd/NaY(250/250) increasing the reaction temperature from 269 to 285°C increases the overall CO conversion by 40%. However, while methanol and dimethyl ether yields increase by 49 and 13%, respectively, the yield of methane increases by 127%. For Pd/NaY(500/350) the total decrease in the space velocity from 1800 to 400 hr^{-1} leads to an increase in the CO conversion by three times. For Pd/NaY(250/250) the same decrease in the space velocity results in a doubling to tripling of the CO conversion and in increases of at least 65% in the methane selectivity at the expense of methanol and dimethyl ether [46].

A single-step liquid-phase dimethyl ether synthesis from syngas was successfully developed [43–45]. Similar efforts have been made by researchers at Haldor Topsøe [28]. The conversion of syngas to DME involves three key reactions (i.e., adjusting syngas composition via the water-gas shift reaction, methanol synthesis, and methanol dehydration). All three reactions are exothermic. The catalysts for this process are combinations of the catalysts for methanol synthesis (e.g., Cu/ZnO type) and the methanol dehydrating catalysts (e.g., alumina, silica-alumina, zeolites, solid acids, and solid-acid ion-exchange resins). For the liquid system, the process uses powdered slurry catalysts. There may be a single reactor or multiple staged reactors. Three phases are present and the DME synthesis occurs in a single stage. Process conditions include a pressure range of 30–120 atm, a temperature range of 250–325°C, and space velocities of 1,000–10,000 l of syngas per kg of catalyst per hour. This process in the liquid phase is particularly useful for syngas with high concentrations of carbon monoxide. Carbon monoxide concentrations of more than 50% by volume have been used. Dimethyl ether can be used as a clean-burning automotive fuel and as a feedstock for the synthesis of hydrocarbon and oxygenates synthesis. The potential of DME as a feedstock over methanol is shown in Table 3.22 [47].

Dimethyl ether reacts with oleum to give dimethyl sulfate:

$$CH_3OCH_3 + SO_3 = (CH_3)_2SO_4$$

Table 3.22 DME Has Potential for Ether Synthesis

$$CH_3OCH_2CH_2OC(CH_3)_3 \xleftarrow{MTBE} \quad H_3COCH_2CH_2OCH_3$$

$$CO \longrightarrow H_3COCH_3 \xrightarrow{DME}$$

$$H_3CCO_2CH_3 \longleftarrow \quad H_3COCH_3 \xrightarrow{CH_3OH}$$
$$DME$$

$$H_3COCH \!=\! CH_2 \xrightarrow{H_2} CH_3OCH_2CH_3$$

$$(CH_3)_3COCH \!=\! CH_2 \xleftarrow{CH_3OH} H_3COC(CH_3)_3 \xrightarrow{MTBE} (CH_3)_3COCH_2CH_2OC(CH_3)_3$$
$$MTBE$$

$$\Big\downarrow H_2$$
$$(CH_3)_3COCH_2CH_3$$

The main application of dimethyl sulfate is as a methylating agent, for example, for making aryl methyl ethers.

F. Methyl *tert*-Butyl Ether

Oxygenates

Oxygenates are liquid organic compounds that are blended into gasoline to increase its oxygen content. During combustion, the additional oxygen in the gasoline reduces the output of carbon monoxide and limits emissions of ozone-forming materials. Two new programs called the oxygenated fuel program and the reformulated gasoline program were initiated by the Clean Air Act Amendments of 1990. The oxygenated fuels program requires gasoline sold in 39 metropolitan areas that do not meet federal air quality standards for carbon monoxide to contain 2.7% oxygen by weight. The program runs for at least four winter months, although areas can choose to extend it. The reformulated gasoline program, starting 1 January 1995 requires all gasoline sold during the summer season in nine large metropolitan areas with most serious ozone problems to contain 2.0% oxygen by weight. Aliphatic alcohols and ethers, in specific volumes and combinations, have similar octane values as conventional gasoline and therefore they are able to provide the oxygen that gasoline will need under the oxygenated fuels program. It is in this context that compounds such as methyl *tert*-butyl ether (MTBE) are considered oxygenates, even though they may have other properties and uses [48].

Blends of gasoline with relatively lower levels (5–10 vol%) of lower alcohols such as methanol have been tested extensively worldwide. Although they extend gasoline supplies, enhance octane rating and so improve engine performance while decreasing pollution, methanol-gasoline blends are subject to phase separation that is brought about by the inadvertent addition of small

Table 3.23 Oxygenate Properties

Property	MTBE	ETBE	TAME
Blending octane (R + M)/2	110	111	105
Blending octane (RON)	112–130	120	105–115
Blending octane (MON)	97–115	102	95–115
Reid vapor pressure (psi)	7.8	4	2.5
Boiling point			
(°C)	55	72	88
(°F)	131	161	187
Density			
(kg/l)	0.742	0.743	0.788
(lb/gal)	6.19	6.2	8.41
Energy density			
(kcal/l)	89.3	92.5	98
(kBtu/gal)	93.5	96.9	100.8
Heat of vaporization (kcal/l)	0.82	0.79	0.86
Oxygenate requirement (vol% @2.7 wt% oxygen)	15	17.2	16.7
Solubility in water (wt%)	4.3	1.2	1.2
Heat of reaction			
(kcal/mol)	9.4	6.6	11
(kBtu/lb-mol)	17	12	20

RON, research octane number.
MON, motor octane number

amounts of water. Concern over possible deleterious effects of methanol on engines and fuel systems in autos not designed for methanol use, also has been a negative factor for methanol acceptance. Mixed alcohols are less suitable for reasons of evaporation and phase separation [49].

Ethers like MTBE and ETBE have advantages over lighter alcohols (e.g., methanol, ethanol, etc.), especially with lower Reid vapor pressures and lower vaporization latent heats, etc. Methyl *tert*-butyl ether has a number of advantages over alcohols, including its compatibility with gasoline, its high blending octane number, its established efficient manufacture from methanol and isobutylene, the need for octane enhancement occasioned by the removal of lead from gasoline, and the perceived problems with gasoline blends containing methanol. Commercial growth of MTBE manufacture has been phenomenal [49]. The properties of the different oxygenates are shown in Table 3.23.

The Need for Oxygenates

The production and distribution of clean-burning gasoline has been mandated in cities not being able to comply with carbon monoxide (CO) levels. A large proportion of hydrocarbons contributing to the ozone problem comes from emissions from automobile exhausts. The options available to alleviate this problem are alternative fuels and oxygenate blending in gasoline.

The alternative fuels consist of:

- compressed natural gas (CNG)
- neat alcohols
- methanol fuel (M85; 85% methanol + 15% gasoline)

However, the utilization of alternative fuels requires a separate fuel distribution infrastructure and vehicle fuel systems need to be modified for these fuel properties. In contrast, blending oxygenates in existing gasoline fuel provides higher energy equivalent than conventional gasoline, its distribution cost is little or none as compared with alternative fuels, and the automotive fuel system modification is not needed. Thus, the net difference between the usage of oxygenate-blended gasoline and alternative fuels is $4 to $8 per million Btu in favor of the former. This is not only a summary of the present situation, but is a very reliable trend for the future [50].

Thus, blending aliphatic alcohols and ethers in gasoline not only expands existing gasoline volume and octane supplies, it utilizes gasoline's existing distribution infrastructure, and reduces emissions from existing vehicle fleets. In essence, the high octane and environmental benefits make oxygenates more attractive [50].

The raw materials for tertiary ethers are the alcohols, methanol or ethanol, and the C_4 or C_5 olefin fractions. Methanol and isobutylene react to form methyl *tert*-butyl ether (MTBE), methanol and isoamylene form tertiary amyl methyl ether (TAME), ethanol and isobutylene form ethyl *tert*-butyl ether (ETBE), and ethanol and isoamylene form tertiary amyl ethyl ether (TAEE).

MTBE Demand and Supply

According to a study conducted by engineers at Bechtel Corp., the U.S. MTBE capacity at the end of 1991 was 128,970 bbl/day. This number reached 209,170 bbl/day by the end of 1992. This rapid growth continued in 1993, with 84,900 bbl/day coming on stream, bringing the U.S. total to 294,070 bbl/day. Looking a little further out, projections for 1995, according to the study, are 389,270 bbl/day; this total includes projects in the engineering phase and two projects, totaling 37,500 bbl/day, under study [51]. The estimated worldwide production capacity for MTBE is shown in Figure 3.23 [49].

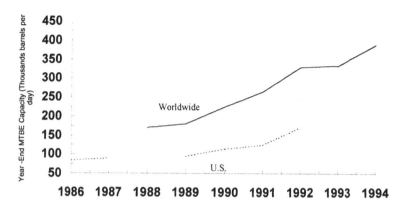

Figure 3.23   Worldwide production capacity of MTBE. *Source*: [49].

Synthesis of MTBE

The conventional MTBE synthesis consists of a reaction of isobutene and methanol over an acidic sulfonated cation-exchange catalyst. This reaction is highly selective, equilibrium-limited, and exothermic in nature. Several types of industrial reactors such as tubular reactors, adiabatic reactors with recycle, and catalytic distillation configurations have been utilized to carry out the MTBE synthesis reaction. The factors considered in the optimal design of a MTBE unit include the following items [52].

Reaction temperature: Lower temperatures enhance isobutene conversion, prolong catalyst life, decrease formation of byproducts like diisobutene and *tert*-butyl alcohol. Higher temperatures cause catalyst deterioration, favor decomposition of product, and reduce yield.

Methanol/Isobutene ratio: Lower ratios facilitate recycling of unreacted methanol and also allow its removal as an azeotrope with butenes.

Isobutene conversion: Conversion of isobutene depends on the number of stages in the process. A single-stage reaction achieves conversions of 90–96%, whereas two-stage reactions (where product is selectively removed before the second stage) achieve 99%. Higher conversions facilitate the fractionation of pure 1–butene from the process raffinate.

Catalyst life: Lower operating temperatures favor prolonged catalyst activity, due to the exclusion of catalyst poisons. The catalyst life is a more important factor than the cost of the catalyst.

MTBE product purity: Diisobutene (DIB) and *tert*-butyl alcohol are the primary byproducts in this process. In spite of these impurities, the

MTBE product purity is 97–99%. These byproducts, in small quantities, have no adverse effect on the MTBE product because they possess similar blending properties.

Different MTBE Synthesis Processes

EC Erdolchemie GmbH developed a MTBE synthesis process deviating from the conventional process in the areas of reactor design, conversion levels, and energy consumption. A schematic of this process is shown in Figure 3.24. This process utilizes two simple-shaft furnace-type reactors in succession, to enhance the isobutene conversion. The reactants are C_4 raffinate feed from a naphtha steam cracker, containing 45 wt% isobutylene and chemical grade methanol that are reacted over a special Bayer catalyst. The process setup includes a distillation column to separate the MTBE product from the butenes and methanol. High purity MTBE (> 99 wt%) is produced and the plant capacity is 2,500 metric ton/yr [53].

A typical single-stage MTBE synthesis process is shown in Figure 3.25. Isobutene conversion levels achieved by this single-stage process range 90–96%. This process utilizes a packed-bed reactor containing an ion-exchange catalyst, typical feeds being methanol and a C_4 fraction containing isobutene. A debutanizer tower separates the reactor effluent, giving the MTBE product at the bottom and a methanol-butene azeotrope overhead. The overhead stream is separated further into methanol (which is recycled to the synthesis reactor) and a C_4 raffinate saturated with water [52].

Higher levels of conversions (> 99%) can be achieved by a two-stage MTBE synthesis process (Figure 3.26). The first reactor is a typical MTBE synthesis, using isobutene and methanol as feeds over a packed-bed ion-exchange reactor. The product is separated in a debutanizer tower and the overhead of this reactor is charged to another synthesis reactor to achieve higher conversion of isobutene. A secondary debutanizer is used to separate the additional MTBE produced in the secondary packed bed reactor. Methanol removed from the overhead stream is recycled back to the primary synthesis reactor [52].

The synthesis of MTBE also can be carried out using methanol and n-butenes or mixed butanes, or n-butane as the C_4 feed. These feeds are typical of Middle East situations, where there is an abundantly higher supply of LPG as compared to isobutene. Although these substitute C_4 feeds are not commercially used for MTBE synthesis, their usage is feasible (Figure 3.27) [52].

Conventional MTBE synthesis is carried out over acidic ion-exchange catalysts. The feasibility of the use of shape-selective zeolite catalysts (e.g., ZSM-5 and ZSM-11) for MTBE synthesis has been studied [54]. These zeolite catalysts are thermally stable at high temperatures, give no acid effluent, and are less sensitive to the methanol-to-isobutene ratio. The high selectivity of the zeolite catalysts is attributed to its well-defined geometry of channel structures

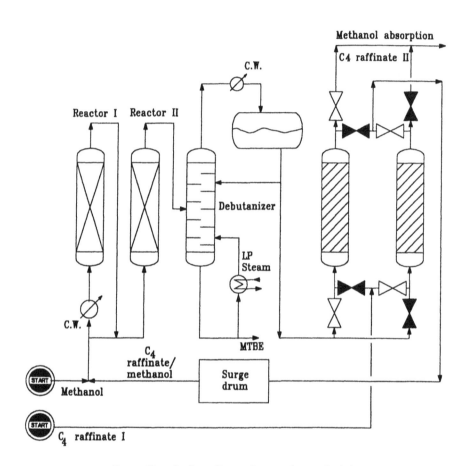

the methanol absorption columns shown shaded

Figure 3.24   A schematic of the EC Erdolchemie GmbH process of MTBE synthesis (the methanol absorption columns are shaded). *Source*: [53].

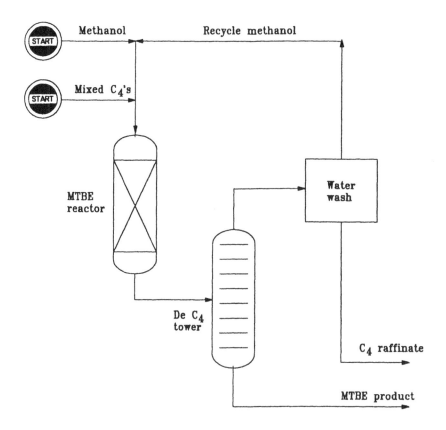

Figure 3.25   Typical single-stage MTBE synthesis process. *Source*: [52].

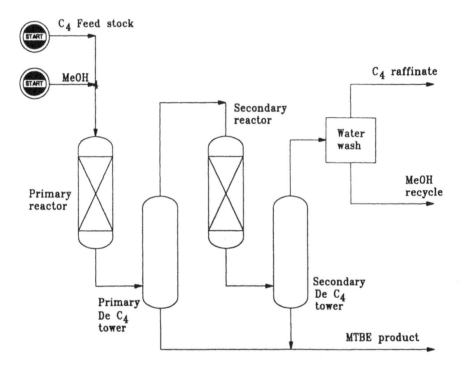

Figure 3.26   Typical two-stage MTBE synthesis process. *Source*: [52].

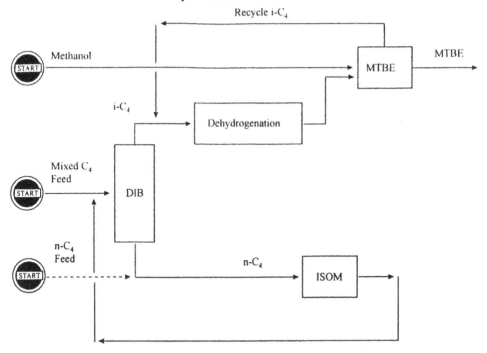

Figure 3.27   MTBE synthesis using substitute C_4 feeds. *Source*: [52].

of molecular dimensions. This property greatly diminishes the formation of byproducts like diisobutene and tert-butyl alcohol. Methanol and isobutene, after passing through the preheater, were charged to the reactor at atmospheric pressure. Both ZSM-5 and ZSM-11 give high selectivities (> 99%) towards MTBE, and conversions of isobutene in the range of 25–30%. In the liquid phase, the zeolites and the conventional ion-exchange catalyst (Amberlyst-15) were used separately and also in conjunction with each other. Experiments performed over HZSM-5 (SiO_2/Al_2O_3 ratio of 40) showed MTBE yields in the range of 90–94% (expressed as % of theoretical). Diisobutene byproduct was minimized to as low as 0.025% of isobutene converted. Selectivities of 100% toward MTBE were obtained by using HZSM-11 catalyst (SiO_2/Al_2O_3 ratio of 25). This increase of selectivity was due to the lower silica-alumina ratio or, in other words, due to the higher acidity of the catalyst. Experiments conducted with Amberlyst-15 as the catalyst showed comparable selectivities only at lower temperatures, and at high methanol to isobutene ratios. Thus, an excess of methanol is required to achieve comparable selectivity when Amberlyst-15 is used as the catalyst. Thus, zeolite catalysts display superior selectivity to MTBE over the entire range of methanol-to-isobutene ratios as compared with

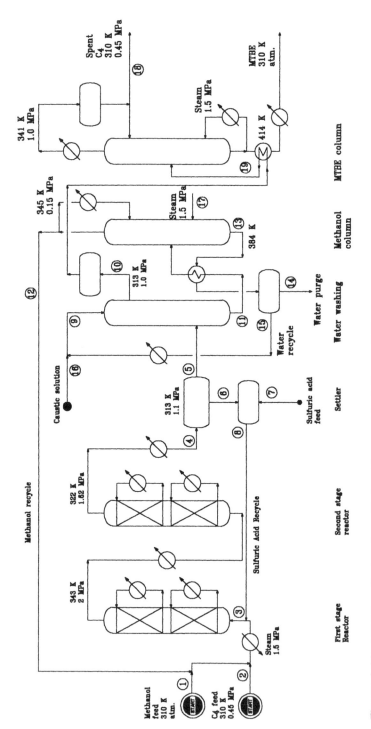

Figure 3.28 MTBE synthesis using sulfuric acid as a catalyst. *Source:* [55].

Amberlyst-15 catalyst. Zeolite catalysts allow operation at higher temperature, and higher space velocities, without causing deactivation of the catalyst.

Sulfuric acid is a new catalyst [55]. The process is carried out in the liquid phase and is exothermic, yielding 42.69 kJ/mol of MTBE. A flowsheet of the process is shown in Figure 3.28. Feeds include methanol and mixed butenes containing 45 wt% isobutene. After the reaction mixture is preheated to a temperature of 70°C, it enters the primary reactor where it is heated to 120°C in the presence of sulfuric acid (5 wt% of the reaction mixture). Conversion of isobutene is 90% in this reactor. The reactor effluents are cooled to 40°C and fed to the secondary conversion reactor where the isobutene conversion of 98% is achieved. Effluents from this reactor are sent to the settler where 89% of the sulfuric acid is recovered and recycled. The organic phase from the settler enters the water washing column where the remaining acid is neutralized by washing with caustic soda. The extract phase containing 15 wt% methanol is directed to the methanol recovery column and the overhead raffinate phase is sent to the MTBE column. In the methanol column, the water-methanol mixture is distilled, and methanol from the distillate is recycled; the bottom fraction, consisting mainly of water, is recycled and purged. The raffinate from the water washing column, consisting mainly of C_4 hydrocarbons and MTBE product, is separated in the MTBE column at a preheated temperature of 85°C. The C_4 mole fraction is collected as the distillate and stored, whereas the MTBE product is obtained as the bottom fraction. Typical column efficiency is about 50%. MTBE product composition consists of 99% MTBE, and minor amounts of diisobutene and tertiary butanol.

In 1992, refiners began to choose a variety of routes to the synthesis of MTBE [51]. Valero Refining & Marketing, in its MTBE synthesis plant, uses a butane/butylene mixture from the heavy oil cracker vapor recovery unit which on hydrogenation converts butadiene to butylene. This is then mixed with methanol in the MTBE synthesis unit, the MTBE product is separated and the butane/butene stream is charged to the alkylation unit. The butadiene is removed from the alkylation unit. This improves alkylate quality and reduces acid consumption. A block diagram of this unit is shown in Figure 3.29.

Valero also has a process to upgrade butane to butylene on the way to the synthesis of MTBE, a block diagram [51] of which is shown in Figure 3.30. The C_4 stream is treated in a caustic treatment unit to remove sulfur, and then butane is then fed to the Butaner unit to produce isobutane. The isobutane is fed to the Oleflex unit, where it is contacted with platinum catalyst to give isobutene. The heavy bottoms in this unit consist of isopentane which is directed toward gasoline blending. The liquid isobutene product is then contacted with methanol in the MTBE synthesis reactor. A butene column then separates the MTBE product from the C_4 raffinate. The MTBE product is directed toward storage, whereas methanol is removed from the C_4 raffinate.

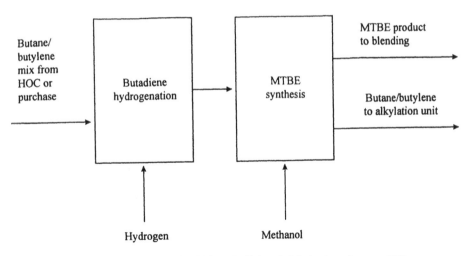

Figure 3.29   MTBE process by Valero Refining & Marketing. *Source*: [51].

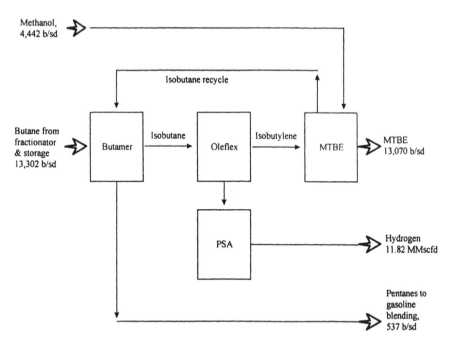

Figure 3.30   Valero processing of butane to butylene. *Source*: [51].

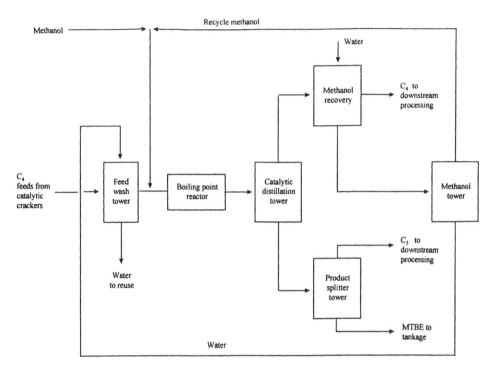

Figure 3.31 Flowsheet of the CDTech MTBE process. *Source*: [51].

CDTech uses catalytic distillation to convert isobutene and methanol to MTBE, where the simultaneous reaction and fractionation of MTBE reactants and products takes place [51]. A block diagram of this process is shown in Figure 3.31. The C_4 feed from catalytic crackers undergoes fractionation to extract deleterious nitrogen compounds. It is then mixed with methanol in a BP reactor where 90% of the equilibrium reaction takes place. The reactor effluent is fed to the catalytic distillation (CD) tower where an overall iso-butene conversion of 97% is achieved. The catalyst used is a conventional ion-exchange resin. This process selectively removes MTBE from the product to overcome the chemical equilibrium limitation of the reversible reaction. The MTBE product stream is further fractionated to remove pentanes, which are sent to gasoline blending, whereas the raffinate from the catalytic distilla-tion tower is washed with water and then fractionated to recover the methanol.

Conventional raw materials in the synthesis of MTBE are methanol and isobutene. However, it is believed that there may not be enough supply of isobutene to meet the increasing MTBE demand. One of these alternatives is the use of isobutanol. Research has been carried out to study the synthesis of MTBE from methanol and isobutanol [56]. This study employed two stainless steel tubular fixed-bed flow microreactors. The first reactor was used as an isobutanol dehydration reactor. Isobutanol and methanol were reacted at a temperature of 225°C, over silica-alumina catalyst. The product stream from the dehydration reactor then was passed to the second reactor . This reactor was maintained at a temperature of 50°C and ion-exchange resin catalyst (Amberlyst 15) was used in this reactor to convert the butenes and methanol to MTBE. An attempt was made to directly synthesize MTBE from methanol and isobutanol over Amberlyst 15, but extremely low conversions were achieved. Thus, it was accepted that the presence of isobutene is mandatory for MTBE formation. Therefore, isobutanol was first dehydrated to isobutene. Silica alumina was used as the dehydration catalyst because it offers higher degrees of isobutanol dehydration than methanol. This catalyst was preferred over zeolite H-ZSM-5 and γ-alumina. At a temperature of 50°C, MTBE + MIBE yields of 27.8% by mass were obtained from the second reactor. And, most significantly, the ratio of MTBE:MIBE was found to be as high as 11.7:1.

In MTBE synthesis, conversion of isobutene and selectivity towards MTBE has exceeded 90% using conventional ion-exchange resin catalysts. Productivity per given reactor, however, can be increased by utilizing a higher acid group density and higher acid capacity in the conventional acidic cation-exchange resin catalyst. This has been the focus of a study to induce an enhanced conversion of methanol and isobutylene to MTBE in a fixed reactor [57]. The catalyst used was Amberlyst 35 which has a higher acid content (5.5 mequiv/g) than the conventional Amberlyst 15 (4.7 mequiv/g). Two pilot scale reactors, fabricated from stainless steel, were placed in series. Effluents from the first reactor were introduced in the second reactor. Adiabatic operation was simulated in the lead reactor and a 90% conversion occurred. Amberlyst 35 gives higher conversion and liquid hourly space yields than Amberlyst 15. Higher conversions were obtained for Amberlyst 35 even when the reactor system was operated at higher space velocities. Space yields were 16–24% higher than those for Amberlyst 15. Thus, Amberlyst 35 is more productive than Amberlyst 15 [57].

G. ETBE, TAME, and DIPE

Can ETBE Compete with MTBE?

The use of oxygenated fuels for environmental reasons and increasing octane levels of gasoline has encouraged a great deal of research in the field of oxygenates. Tertiary ethers are preferred to lighter alcohols as oxygenates because of their lower Reid vapor pressures and lower latent heat of vaporization. The most common ether to date is methyl *tert*-butyl ether (MTBE) because of the availability of methanol as a raw material. However, a lot of current research has been directed toward ethyl *tert*-butyl ether (ETBE), which is made from ethanol and isobutene. The attractions about ETBE are its lower Reid vapor pressure and higher octane values. The lower vapor pressure will gain importance in 1995 when the Federal reformulated gasoline (RFG) specifications get more stringent. The RFG legislation mandates gasoline supplied in some United States locations during the summertime to have a blending Reid vapor pressure (bRvp) of 7.8 psi. The lower bRvp of ETBE (4 psi) as compared to MTBE (8–10 psi) makes it a suitable candidate as an oxygenate.

Another advantage of ETBE is that it can be made from renewable ethanol. This would in turn reduce the dependence on oil imports and create additional markets for agricultural wastes and biomass. The possibility of utilization of azeotropic ethanol (96.5 wt%) or subazeotropic ethanol (80–85 wt%) could increase further the availability, and reduce the cost of ethanol for ETBE production. To sum the advantages of ETBE production:

- ETBE's high octane values replace octane lost from lower aromatic levels required for reformulated gasoline. ETBE can provide more octane than other oxygenates at equal oxygen levels.
- ETBE's lower bRvp (4 psi) may reduce or totally eliminate capital investment for vapor pressure reduction. ETBE blending can reduce the volatility of a reformulated gasoline stock 0.5–0.6 bRvp.
- ETBE's clean volume allows significant dilution of undesirable benzene, olefin, and sulfur compounds in reformulated gasoline.
- ETBE is made from corn ethanol, a domestic renewable oxygenate. This could drastically cut down the dependence on oil imports. As an ether, ETBE displays superior water tolerance properties and can be shipped by pipeline.

Thus, the reasons for research for ETBE synthesis seems quite positive. MTBE producers seem to be shifting the focus of new plants towards flexibility in MTBE/ETBE production.

Synthesis of ETBE

The absorption of isobutene in subazeotropic aqueous ethanol, to give ETBE and *tert*-butyl alcohol (TBA) was investigated [58]. The experiments were conducted in a stainless steel autoclave of one liter capacity. The catalyst used was Amberlyst-15, which is an acidic, macroporous cation exchange resin, in the form of spherical beads. In the ETBE synthesis reaction using ethanol and isobutene, the side reactions are the dimerization of isobutene to form diisobutene and the formation of diethyl ether. These byproducts show a tendency to increase with an increase in reaction temperature. Hence, the reactions were carried out in the temperature range of 303–333 K to shift the selectivities towards ETBE and TBA. The experiments were carried out with agitation above 900 rpm to make the reaction free from external mass transfer resistance. The usage of cumene and *n*-heptane as solvents increased the conversion of ethanol to ETBE. The hydration and etherification of isobutene are reversible and exothermic reactions. Operating at lower temperatures (about 303–333 K) ensured that the equilibrium was in favor of the desired products. Thus, it was observed that isobutylene can be hydrated and etherified simultaneously by absorption in aqueous subazeotropic ethanol, in the presence of an acidic ion-exchange resin as catalyst to form ETBE.

The liquid-phase addition of ethanol to isobutene to give ethyl *tert*-butyl ether on the ion-exchange resin Lewatit K2631 was investigated [59]. This catalyst is a macroporous, sulfonic cation-exchange resin. The experiments were carried out in continuous flow stainless steel packed-bed microreactors, with a temperature range of 40–90°C, at a pressure of 1.6 MPa, and at liquid hourly space velocities exceeding 60 hr^{-1}. Preliminary experiments carried out at the highest temperature (89°C) assured that the mass transfer limitations were not involved in obtaining rate data. Ethanol inhibits the reaction, whereas isobutene enhances it. The reaction rate decreased on increasing ethanol or ETBE concentration, but it increased with that of isobutene. The presence of ETBE increased the rate of decomposition of the ether, at the same time as ethanol and isobutene concentrations in the feed decreased. Reaction temperature highly influences the rate. It decreased by 2.5 times when temperature decreases by 10°C.

Tertiary ethers have the advantage of high octane levels and low vapor pressures as compared to lower alcohols. However, they suffer from a major drawback in that they may increase aldehyde emissions of substantially, which have a high atmospheric reactivity [60]. According to the reformulated gasoline legislation, emissions of aldehydes and other toxic compounds (e.g., benzene) have to decrease by 15% by 1995. This drawback may be circumvented to a large extent by mixing the tertiary ethers with tertiary alcohols such as *tert*-butyl alcohol (TBA) and *tert*-amyl alcohol (TAA), which can be obtained by adding water to isobutene or isoamylene, respectively. *tert*-Butyl alcohol has a very low atmospheric reactivity and low aldehyde emission. Thus, the aldehyde emissions from ethers are lower in the presence of these alcohols. As compared to lighter alcohols, the water susceptibility of both TBA and TAA is much lower. Also, TBA- and/or TAA-blended gasoline is shippable through pipelines. The vapor pressure of these alcohols is also lower than that of lighter alcohols. Thus, adding tertiary alcohols to tertiary ethers lowers the emissions and has the advantage of using water as a raw material. The potential drawback of excess water also can be circumvented by using azeotropic ethanol which reduces the amount of residual water significantly.

The effect of water content of 0–5 wt% in the alcohol on ETBE or MTBE synthesis equilibrium composition and on the reaction rates has been studied [60]. The catalyst used is a macroporous sulfonated copolymer of styrene-divinyl benzene in H^+ form. The experiments were conducted in a stainless steel-jacketed batch reactor. The reaction medium was agitated at 500 rpm by a magnetic-drive turbine at a temperature of 313–353 K and the pressure was set at 1.6 MPa. The ethanol-isobutene ratio ranged from 0.81 to 1.44, and the water percentage in ethanol was in the range of 1.39–4.52 wt%. Diisobutene (DIB), which is the first byproduct formed in isobutene polymerization, is detected only at higher temperatures, at lower initial water amounts, and for an initial ethanol-isobutene ratio less than unity. However, its influence on the activity coefficients of ethanol, isobutene, ETBE, water, and TBA is negligible. The presence of water in amounts equal to or lower than that in azeotropic ethanol does not affect the equilibrium constants of ETBE or MTBE synthesis reactions, and is able to increase the isobutene conversion. Water shows an inhibitor effect in production rates at the beginning of the reactor. This effect disappears accordingly as water is consumed to yield TBA, which is produced rapidly. ETBE (or MTBE) and TBA can be simultaneously produced from azeotropic ethanol or methanol impurified with water, respectively, in existing MTBE units if the reactor length is long enough to get a high isobutene conversion.

A flowsheet of the ARCO MTBE/ETBE/TAME process [9] is given in Figure 3.32. The feed to the process includes methanol or ethanol and hydrocarbon streams containing *tert*-olefins such as isobutylene and isoamylene. The process can be designed to handle feedstocks from steam crackers, catalytic crackers, isobutane dehydrogenation, *n*-butene or amylene isomerization. The primary oxygenates manufactured are MTBE, ETBE, TAME and TAEE, or mixtures thereof. The hydrocarbon stream is preheated (1) to remove trace contaminants. It is reacted with methanol or ethanol over the ion exchange resin in two fixed-bed adiabatic reactors (2,3). The reactive olefin that can be handled range from 8% to 95%. MTBE, ETBE, TAME, or other ethers are separated from the nonreactive hydrocarbons and unreacted alcohol in a distillation column (4). The product ether is recovered at high purity from the bottom of the column. The excess alcohol is removed in the overhead stream. It is recovered in an optimized water wash system consisting of a small reactor (5) and alcohol stripper (6), and is recycled [9].

Figure 3.33 shows a schematic of the CDTech MTBE/ETBE/TAME process. This is essentially the same as the CDTech MTBE process presented earlier. The process is unique in the sense of using a "boiling point" reactor and catalytic distillation (CD) [61]. The C_4 feed and methanol is fed to the boiling point reactor (1). This is a fixed-bed downflow adiabatic reactor, in which the liquid is heated to its boiling point by the heat of reaction and

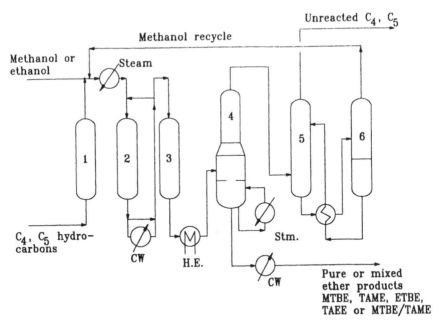

Figure 3.32 Flowsheet of the ARCO MTBE/ETBE/TAME process. (1) preheater, (2,3) fixed-bed adiabatic reactors, (4) distillation column, (5) reactor, (6) alcohol stripper. *Source*: [9].

limited vaporization occurs. The equilibrium-converted reactor effluent flows to the CD column (2) where the reaction is continued, and MTBE is concurrently separated from unreacted C_4 hydrocarbons as the bottom product. As the boiling reactor conversion decreases, the CD-reaction column recovers lost conversion, so that a high overall conversion is sustained. The CD column overhead is washed in an extraction column (3) with a countercurrent water stream to extract methanol. The water extract stream is sent to a methanol recovery column (4) for recycle. For MTBE, the isobutylene conversion is 99%; it is slightly lower for ETBE. For TAME and TAEE, isoamylene conversion is 95+%.

Figure 3.34 shows a schematic of the Ethermax process by Hüls AG and UOP [61]. Feed for the process includes methanol or ethanol and hydrocarbon streams containing reactive tertiary olefins such as isoamylene and isobutylene. Typical hydrocarbon streams are FCC light gasoline, steam cracker C_4 hydrocarbons, or product from a butane dehydrogenation unit. In the production of MTBE, the feed first passes through an optional water wash system (1) to remove resin contaminants. The majority of the reaction is carried out in a

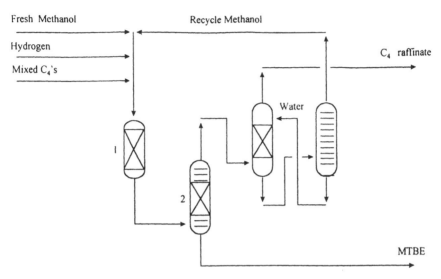

Figure 3.33 Flowsheet of the CDTech MTBE/ETBE/TAME process. (1) boiling point reactor, (2) CD column, (3) extraction column, (4) methanol recovery column. *Source*: [61].

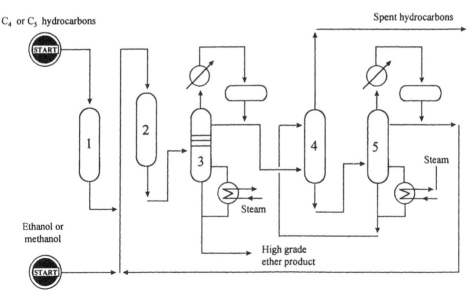

Figure 3.34 A schematic of the Ethermax process by Hüls AG and UOP. (1) water wash system, (2) simple fixed-bed reactor, (3) distillation (RWD) column, (4,5) methanol-water distillation column. *Source*: [61].

Olefins

Methanol recycle

Methanol/
ethanol

Unreacted
C_4's

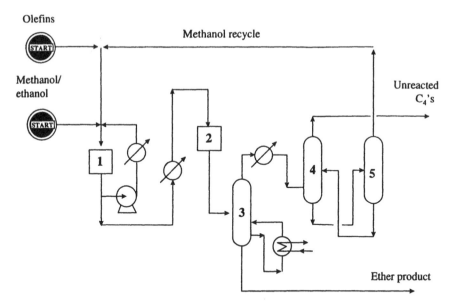

Figure 3.35   A flowsheet of the Phillips MTBE/ETBE/TAME process. (1,2) reaction
section, (3) MTBE fractioner, (4) methanol extractor, (5) methanol fractionator. *Source*: [61].

simple fixed-bed reactor (2). The reactor effluent feeds the reaction with
distillation (RWD) column (3) containing a proprietary packing where simultaneous reaction of the remaining isobutene and distillation occur [61].
Overhead from the RWD column is sent to methanol recovery, a simple
counter-current extraction column using water and a methanol-water distillation column (4,5). Methanol is recycled to the reactor section. Hydrocarbon
raffinate is typically sent to a downstream alkylation or oligomerization unit.
Optional treatment of raffinate includes removal of residual oxygenates in a
UOP oxygenate removal unit (ORU) and selective hydrogenation of diolefins
and monoolefins in a Hüls SHP Unit. The Ethermax process is environmentally compatible and does not produce any hazardous wastes [61].

Figure 3.35 shows a process flow diagram of Phillips MTBE/ETBE/TAME
process. This process is often called the Phillips Etherification Process. The
reaction section (1,2) which receives methanol and isobutene concentrate,
contains an ion exchange resin. The isobutene concentrate may be mixed olefins from a Fluid Catalytic Cracking Unit (FCCU) or steam cracker or from
the on-purpose dehydration of isobutene (Phillips STAR process). High purity
MTBE (99 wt%) is removed as a bottoms product from the MTBE fractionator
(3). All of the unreacted methanol is taken overhead, sent to a methanol

extractor (4), and recycled back to the reactor. Extract from the methanol extractor is sent to the methanol fractionator (5). The process achieves 99+% conversion of isobutylene in the cases of MTBE and ETBE, and 90+% conversion of isoamylene from TAME. The process uses down-flow fixed-bed reactors, but not catalytic distillation.

Tertiary Amyl Methyl Ether

MTBE is the ether most sought after to fulfill the demands set for lower fuel emissions by increasing the oxygen content of the gasoline. However, if MTBE is the only oxygenate used for this purpose, refineries would be required to buy more MTBE from outside sources since they do not have enough isobutylene for feed purposes. Synthesis of ETBE also competes with MTBE for isobutylene as a raw material. Fluid catalytic cracker (FCC) gasoline has a high olefin content which could be used for producing heavier ethers. Thus, the idea of production of tertiary amyl methyl ether (TAME) from isoamylenes has gained significant importance. Olefins with tertiary bonds selectively form carbonium ions and react with alcohols to produce ethers. Of the three isoamylenes, 2–methyl butene-1 and 2–methyl butene-2 are reactive while 3–methyl butene-1 is nonreactive. Although TAME has a slightly lower octane rating, it has favorable comparisons for vapor pressure, boiling point, energy density, and water miscibility. A few units have been built to etherify light FCC gasoline to TAME.

The world's first commercial TAME plant to produce high octane blending components was commissioned at the 190,000 bbl/day Lindsey refinery, Immingham, U.K. The TAME process combines an isoamylene charge from the refinery's fluid catalytic cracker (FCC) with methanol. In the TAME unit the isoamylene fraction of the light catalytically-cracked spirit (LCCS) from the FCC is reacted with methanol and hydrogen in the presence of a resin catalyst. The process achieves a conversion of 63%. The reactor effluents are distilled to yield a blend of TAME gasoline and methanol. This blend is mixed with water in an extractor to remove the methanol, leaving the TAME gasoline to be recombined with LCCS. The high octane blending component is then stored. The TAME plant capacity is 50,000 metric tons/yr.

EC Erdolchemie, Germany, developed a catalyst for the synthesis of TAME from isoamylene and methanol [62]. Instead of reacting the isoamylene with the aid of a conventional strong-acid cation-exchange resin that is capable only of proton catalysis, strongly acidic resin that is doped with a noble metal is used in the presence of hydrogen. This new bifunctional catalyst is a macroporous strong-acid cation exchanger with sulfonic groups and with less than 5 g/l palladium. It can catalyze simultaneously the isoamylene reaction and the hydrogenation of the troublesome substances. The palladium is fixed in a special process to the resin beads in a finely dispersed form. By selective hydrogenation of unsaturated compounds, the gum formation that is usually occurring with conventional ion-exchange resins can be prevented. TAME-gasoline produced in this way can be added directly to the fuel as a component with high octane number. A further advantage of this process is better heat dissipation which results from the metal coating of the catalyst resin. Consequently, fixed-bed reactors do not suffer from any peak temperatures that could lead to resin damage.

The commonly tapped sources of isoamylenes are FCC light gasoline and steam-cracker pyrolysis gasoline. A TAME unit was developed where all reactive olefins of FCC gasoline are etherified. Methanol was introduced with the hydrocarbon feed to the etherification prereactors, where methanol reacted with tertiary olefins and ethers were formed. Typical conversions were 70% for isoamylenes, 40% for C_6 tertiary olefins, and 20% for C_7 tertiary olefins. This mixture of ethers, hydrocarbons, and excess methanol was distilled in a column. Ethers and heavy hydrocarbons were collected at the bottom of the column, and methanol, unreacted C_4–C_8 paraffins, and olefins went to the top of the column. To achieve enhanced conversions another reactor system was attached to the column. Feed to this reactor was taken as a side stream from the column and passed through the side reactor system back to the column. The octane number of etherified light gasoline increased by 2–3 over the feed light gasoline. Typically, 90% of isoamylenes were converted to TAME, 60% of C_6 tertiary olefins were converted to tertiary hexyl methyl ethers (THxME), and 40% of C_7 tertiary olefins were converted to tertiary heptyl methyl ethers (THpME). The degree of olefin reduction with this process is significant, which gives important environmental benefits [63].

The BP Etherol process is a conventional two-reactor fixed-bed etherification process employing strong-acid macroporous ion exchange resins as catalysts. This two-reactor arrangement in series is capable of using a palladium-loaded ion exchange resin catalyst as a partial charge in the first etherification reactor, thus enhancing the catalyst life and producing an aqueous clear product. In another system provided by the same licensor, catalytic distillation is effected in a fractionation tower containing a patented catalytic packing that allows simultaneous catalytic reaction and distillation. Integration of fixed-bed reactor technology capable of using palladium-loaded ion exchange resin catalysts followed by catalytic distillation achieving isoamylene conversion in excess of 95% was developed as the CDETHEROL process for TAME synthesis. The overhead stream from fractionation is a C_5 stream containing a small amount of unreacted methanol which is recovered by extraction with water. The water/methanol mixture is fractionated to recover methanol which is recycled to the reactor. Water is recycled to the methanol extractor. The process combines hydroisomerization and selective hydrogenation with the previously mentioned high isoamylene conversion, which allows the refinery to produce a significant portion of its oxygenate demand within the refinery. The raffinate is upgraded by removal of diolefins and gum formers, thus providing a longer catalyst life [64].

ARCO Chemical is the world's largest producer of MTBE and ETBE with four plants in the United States, two plants in Europe, and 27 additional units licensed worldwide, resulting in a total ethers capacity of 140,000 bpsd, as of 1993 [9]. Their technology produces various oxygenates for octane enhancement of gasoline and making reformulated gasoline to meet U.S. Clean Air Act requirements. The primary oxygenates manufactured are MTBE, ETBE, TAME, and TAEE (or mixtures thereof). Typical feedstocks are methanol or ethanol and hydrocarbon streams containing tertiary olefins such as isobutylene and isoamylene. The process can be designed to handle feedstocks from steam crackers, isobutane dehydrogenation, and n-butane or amylene isomerization. The hydrocarbon stream after pretreatment reacts with methanol or ethanol in the presence of an ion-exchange resin catalyst in two fixed-bed adiabatic reactors in series, as described in an earlier section. The ethers formed are separated from the nonreactive hydrocarbons and unreacted alcohol in a distillation column. The ethers are recovered from the bottom of the column, whereas excess alcohol is removed from the overhead stream [65].

Diisopropyl Ether

A new process that converts propylene and water to diisopropyl ether (DIPE) was developed by Mobil Research & Development Corp. DIPE is a high-octane gasoline blending agent which, unlike other ethers, utilizes propylene in its synthesis. The DIPE reaction takes place in a fixed-bed catalytic reactor via a series of reaction steps. Isopropyl alcohol (IPA) is an intermediate which is recycled within the process. A propane/propylene splitter is included in the feed purification section to increase the concentration of propylene in the feed and maximize the DIPE production. DIPE utilizes propylene from the refinery and does not depend on an outside supply of alcohol. DIPE has similar octane blending values of RON and MON as other ethers like MTBE and TAME. DIPE also has a lower Reid vapor pressure than that of MTBE. DIPE is virtually nontoxic and has not caused adverse systemic effects or tissue toxicity [66].

H. Synthetic Gasoline

Fischer-Tropsch Process

The most highly developed commercial technology for converting coal-derived syngas into a light syncrude is a hybrid gasification-liquefaction process known as Fischer-Tropsch technology. The Fischer-Tropsch process catalytically converts synthesis gas into a variety of hydrocarbons, alcohol, ketone and organic acid products [67]. The Fischer-Tropsch technology is used by Sasol Corporation, S. Africa, to produce liquid fuels from coal on a commercial basis. In this type of reaction, the coal is first gasified in a high pressure Lurgi gasifier. Gaseous streams of synthesis gas are passed through purification stages where hydrogen sulfide and ammonia are removed. Then, the resulting gases are passed over iron, nickel, or cobalt catalysts yielding products ranging from methanol to hydrocarbons of high molecular weights. Of the three metal catalysts mentioned here, cobalt has the greatest tendency to produce hydrocarbons with more than one carbon atom per molecule. Iron is less active. Nickel, especially at higher temperatures, shows high activity toward the formation of methane.

There are currently two reactor types in use, the Lurgi/Ruhrchemie-developed Arge fixed-bed reactor, and the Kellogg-developed Synthol fluidized-bed reactor. Each of these two reactor types produces a different product distribution and operates under different conditions of temperature, pressure, and feed composition. The fixed-bed reactor operates at temperatures of 220–255°C, at a pressure of 25 bar, and with a hydrogen/carbon monoxide ratio of 1.7 in the feed. It attains a conversion of 65% of the syngas, yielding 33.4% gasoline (C_5–C_{10}), 5.6% LPG (C_3–C_4), 16.5% diesel oils, and the rest is heavy hydrocarbons. On the other hand, the fluid-bed reactor operates at temperatures of 320–335°C, at a pressure of 23 bar, and with a hydrogen/carbon monoxide ratio of 2.8 in the feed. It achieves conversions of 85% of the syngas with a product distribution of 72.4% gasoline, 7.6% LPG, 3.4% diesel oils, and the rest is heavy hydrocarbons. The range of reactions may be summarized as follows:

At a hydrogen/carbon monoxide ratio of 2.0,

$$nCO + 2nH_2 = (-CH_2-)_n + nH_2O$$

At a hydrogen/carbon monoxide ratio of 0.5,

$$2nCO + nH_2 = (-CH_2-)_n + nCO_2$$

These are linked by the shift reaction,

$$CO + H_2O = H_2 + CO_2$$

However, the limitation of this technology is that its product distribution spectrum is too diverse. It ranges from light gases like methane to very high molecular weight compounds like waxes. The product is composed mostly of straight-chain hydrocarbons with light aromatic hydrocarbons. A significant amount of oxygenates are also produced. It is possible to increase the olefin content of the hydrocarbons by increasing the carbon monoxide concentration of the syngas. At a hydrogen to carbon monoxide ratio of 2, the olefin content of the gasoline is 20%, while at a ratio of 0.5 the olefin content is almost 70%. Because of the wide product distribution obtained by this process, a large number of subsequent unit operations are necessary to separate and refine the product. The need for additional separation and refining adds both to the capital investment for a plant and its operating costs. Removal of wax is also technologically significant. Although it is ideally suited for a diesel or burner fuel, it cannot be used as a high octane gasoline fuel. Extensive processing is required to render it suitable for marketing as gasoline. A major thrust of catalyst research has been to find catalysts which will break the wide distribution and provide better selectivity to a narrow range of products.

Indirect Hydrocarbon Synthesis (via Methanol)

An advanced two-stage reactor system was proposed, where syngas is converted first to a mixture of methanol and dimethyl ether which then is converted to hydrocarbons in the presence of syngas unchanged in the first stage. The hydrocarbon synthesis was carried out in a fixed-bed reactor. A hybrid catalyst consisting of equal weights of Pd/SiO_2 for methanol synthesis and γ-alumina for methanol dehydration was used in the first reactor. The hydrocarbon synthesis was carried out using a zeolite ZSM-5 catalyst. The reactions were carried out at 300°C for methanol synthesis and dehydration and at 350°C for hydrocarbon synthesis. The operating pressures ranged from 1–5 MPa, with a hydrogen to carbon monoxide ratio of 2:1 in the syngas feed. This process achieves syngas conversions and hydrocarbon product yields of 90.8% and 57.3% respectively, at an operating pressure of 5 MPa. The hydrocarbon product profile obtained was essentially the same as that obtained from methanol conversion to hydrocarbons [68]. This process is not in commercial operation.

The conversion of methanol to hydrocarbons in the boiling range of gasoline is a highly exothermic reaction, releasing 398 cal/g of methanol converted at 371°C. This amount of heat release translates into an adiabatic temperature rise of about 650°C in the reactor, which causes higher catalyst deactivation rates. Further, such high temperatures make it difficult to dictate the product distribution spectrum. A process to accomplish the sequential restructuring of vaporized methanol to hydrocarbons in the gasoline boiling-point range is utilizing a fixed-bed reactor. Dimethyl ether was an intermediate formed during the process. Aromatics and isoparaffins in the C_5–C_9 range make up the product distribution. The purpose of this process was to control the large amount of heat released. Methanol is first dehydrated to dimethyl ether over γ-alumina catalyst, during which 20% of the total heat is released and utilized to bring the methanol-dimethyl ether charge to a sufficiently high temperature to entail the subsequent conversion to gasoline-range hydrocarbons. The catalyst used for hydrocarbon synthesis is a ZSM-5 crystalline zeolite, ZSM-11, ZSM-12, ZSM-21, or a TEA mordernite. The process conditions for methanol conversion include temperatures ranges of 205–310°C and LHSV of 10 or lower for methanol. Low pressures (in the range of 2–12 atm) are preferred for hydrocarbon synthesis in order to restrict undesirable products like durene from being formed [69].

A process was devised for hydrocarbon synthesis from methanol, wherein the large amount of heat released during the reaction was efficiently controlled. The process also forms dimethyl ether as an intermediate product. This process utilizes two separate reaction-temperature control mechanisms. The heat released preheats the methanol-dimethyl ether charge to the subsequent temperature required for the hydrocarbons synthesis reaction. The second mechanism relies on the use of light hydrocarbon gases (C_5 or lower) in conjunction with water formed in the methanol synthesis reaction to dilute the dimethyl ether passed to the ZSM-5 catalyst system. These gases can be separated easily from the higher boiling point components in the product stream. Aromatics and isoparaffins are again the primary products of the overall reaction [70].

Mobil Research and Development Corporation has developed a new process which can be used to upgrade C_2–C_4 olefins in fuel gas or LPG, to gasoline-range hydrocarbons. It utilizes a fluid-bed reactor and the shape-selective characteristic of the ZSM-5 catalyst to carry out a number of light olefin reactions, yielding a high octane gasoline product [69,70]. This Mobil Olefins to Gasoline (MOG) process converts all light olefins including ethene which is the most difficult of the light olefins to upgrade. The light olefins are efficiently oligomerized, redistributed, hydrogen-transferred, cracked, and aromatized. The fluidized-bed design controls the temperature rise which results from the high exothermicity of the MOG reaction. The MOG reaction section consists of a fluidized-bed reactor and regenerator vessels which are operatively connected for continuous regeneration of the spent ZSM-5 catalyst. The MOG operating conditions are relatively mild, thereby eliminating erosion and the need for expensive alloys in construction of the equipment. The product recovery section is highly simplified, owing to the fact that the reactor effluent consists only of gasoline and lighter components. A conversion of 90% of olefins per single pass can be accomplished, yielding 80–90% higher hydrocarbons. The type of olefin feed has a small effect on the product selectivity. The product gasoline has octane numbers of 92–97. Thus, this process is highly efficient in upgrading olefin-containing fuel gas streams such as the FCC and coker fuel gas to high octane gasoline product. This process also can be integrated with the MTBE process, allowing efficient upgrading of the unconverted butanes and methanol in addition to other light oxygenates leaving the MTBE reactor.

An M2 Forming process for converting light olefins and paraffins to BTX aromatic hydrocarbons, namely, benzene, toluene, and xylenes. The catalyst used in this study was a HZSM-5 type catalyst with a silica to alumina ratio of 70. Feedstocks used in this study included propene, n-pentane, n-hexane, and a number of commercial streams like light FCC and paraffinic naphthas. The reaction conditions were temperatures of 425–575°C, flow rates of 1–75 LHSV, and atmospheric pressure. The reactions involved in the M2 Forming process are complex, consecutive, and acid catalyzed; they include conversion of olefinic and paraffinic molecules to small olefins via acidic cracking and hydrogen-transfer reactions, formation of C_2–C_{10} olefins via transmutation, oligomerization cracking, and aromatic formation via cyclization and hydrogen transfer. A light FCC gasoline which contains 41% olefins and 10% aromatics yields 54% aromatics and a virgin naphtha which contains 47% paraffins, 41% naphthenes, and 12% aromatics produces 44% aromatics. Under the reaction conditions used, all the nonaromatic cracked products are aromatizable. Only methane and ethane are not aromatized and their concentrations increase as the reaction severity is increased. The feed composition affects the reaction in different ways. The more unsaturated and the higher the carbon number of the feed, the more it is reactive for conversion to the aromatics. The aromatics yield increases with decreasing hydrogen content of the feed and parallels with the reactivity of the feed. The nature of the feedstock affects the rate of catalyst deactivation. The hydrogen-deficient feedstocks are more prone to coking, leading to a higher catalyst-aging rate. The aromatization of paraffinic feeds is a highly endothermic reaction, while the aromatization of olefinic feeds can be an exothermic reaction. The endothermicity of the reaction increases as more hydrogen is produced in the reaction. Increasing the reaction temperature at a constant space velocity increases the aromatic yield. Increasing the temperature also changes the distribution of the aromatics toward more benzene and toluene. The product distribution varies with contact time and aromatic yields are maximized at space velocities of one or lower. Aromatization is not adversely affected by operating at higher pressures [71].

The conversion of methanol to aromatic hydrocarbons over ZSM-5 zeolites is the basis of the Mobil MTG (methanol-to-gasoline) process. Although the Mobil team considered a variety of configurations, including tubular heat-exchanger reactors, staged fixed-bed reactors, and fluid-bed reactors, the fixed-bed process and the fluid-bed route were selected for the initial development studies [69,70].

The fixed-bed reactor system was comprised of two reactors. The first reactor was a methanol dehydration reactor where a methanol/dimethyl ether/water mixture was produced, and the second reactor was a hydrocarbon-forming reactor which converted this mixture over ZSM-5 catalyst to a gasoline-range hydrocarbon product. As noted earlier, the conversion of methanol to hydrocarbons is a highly exothermic process. At 375°C, the heat of reaction is 398 cal/g of methanol converted which translates to an adiabatic temperature rise of 650°C. In this system, 20% of the heat is released in the methanol dehydration reactor. The methanol dehydration is carried out at temperatures of 320°C, pressures of 13–23 atm, at a space velocity of 20 hr^{-1} WHSV. The hydrocarbon synthesis reaction was carried out in a fixed-bed reactor at a temperature of 360°C, pressures of 13–23 atm, at space velocities of 1.5–5 hr^{-1} WHSV. The light gases produced in this reactor are recycled to the second-stage reactor to control the temperature rise. The product yield consisted of 43% hydrocarbons, with the rest was water, carbon monoxide, and carbon dioxide. The product distribution consisted of 22% lower paraffins (C_1–C_4) and isoparaffins, 3% lower olefins, 37% C_5+ nonaromatics, and 38% aromatics. Thus, the yield of gasoline-range hydrocarbons is as much as 75%. The fixed-bed process was subsequently scaled up to a 4 bbl/day pilot plant. For the final gasoline product, alkylate could be produced from the isobutane and lower olefins, so that the total gasoline selectivity was about 85%. The rest is used as LPG and fuel gas.

Development of the fluid-bed process followed a course similar to the fixed-bed process. A fraction of the ZSM-5 catalyst from the fluid bed is removed continuously, regenerated, and returned so that the selectivity is not affected. The average fluid-bed reactor temperature is 413°C, the operating pressure is 275 kPa at a space velocity of 1 hr^{-1} WHSV. The product yield is comprised of 44% hydrocarbons and the rest is water, carbon monoxide, carbon dioxide, and unconverted methanol-dimethyl ether. The hydrocarbon product distribution consists of 60% C_5+ gasoline-range compounds, and the rest is lower olefins and paraffins. The final gasoline yield, including alkylate, is about 88%. The rest is again used as LPG and fuel gas. The fluid bed produces more olefins and less gasoline-range hydrocarbons than the fixed-bed process. However, the light olefins, when alkylated with isobutane and blended back, give higher gasoline yields than the fixed bed. The one hydrocarbon in MTG gasoline not found in significant amounts in conventional gasoline is durene (1,2,4,5–tetramethylbenzene). Although durene has a very high octane number, owing to its high melting point (79°C), its content in gasoline must be controlled to about 3%, depending on the ambient temperature, so as not to crystallize out of solution. The formation of solid crystals by deposition from solution during cold weather is called carburetor icing and is undesirable because it blocks the flow of fuel to the engine.

The selectivity of a novel kind of zeolite was investigated for methanol conversion to gasoline-range hydrocarbons by Le Van MaO and McLaughlin (1989) [72]. Two new categories of ZSM-5 zeolites were prepared: nontoxic zeolites derived from chrysotile asbestos fibers, and Zn- and Zn-Mn-modified ZSM-5 zeolites. Feeds in this study included methanol, 1–butanol, 2–methylpropanol, and a mixture of C_1-C_4 alcohols. The experiments were carried out in a vertically mounted, stainless-steel, fixed-bed reactor 2.5 cm in diameter and 30 cm in length. The reaction conditions included three different temperatures: 390, 430, and 470°C; and pressures of 50, 100, and 150 psi. With the ZSM-5 zeolite, the maximum C_5-C_{11} liquid hydrocarbon yield was observed at about 100 psi for all three reaction temperatures. As the temperature was increased from 390 to 470°C, decreasing yields of liquid hydrocarbons and increasing aromatic content were observed. At 390°C, the aromatic contents in the product gasoline increased with an increase in the pressure. The BTX content in the product aromatics always decreased with increasing pressure. At 390°C, in the case of Zn-modified ZSM-5, the total aromatic contents in the liquid hydrocarbons as well as the percentage of BTX in the aromatic fraction were much higher, coupled with a significant reduction in the production of durene. At higher temperatures, the liquid hydrocarbons yield decreased slightly. However, these decreases were less than those observed in the parent zeolite. The BTX fraction of the aromatic product was noticeably higher. At lower reaction temperatures, the H-asb-ZSM-5/Zn catalyst showed a slight decrease in the activity and selectivity when compared to those of H-ZSM-5/Zn. However, at a higher temperature of 470°C, the H-asb-ZSM-5/Zn had higher amounts of liquid hydrocarbons as well as total and BTX aromatics. The same trend was observed for methanol, 1–butanol, 2–methylpropanol, and C_1-C_4 alcohol feeds. The best catalyst in terms of yield of liquid hydrocarbons and light olefins and also contents of total aromatics and BTX was Zn-modified Asb-zeolite [72]. This process is not in commercial operation.

I. Synthetic Olefins

Selective conversion of methanol or dimethyl ether to lower olefins is a process of commercial significance. It is generally accepted that lower olefins are intermediates in the conversion of methanol or dimethyl ether to higher hydrocarbons over ZSM-5-type zeolite catalysts. The reaction pathway can be simplified as follows:

DME --> C_2-C_4 olefins --> Paraffins + Aromatics

In contrast to lower olefins from diminishing resources like naphtha or associated gas, the lower olefins from methanol or dimethyl ether produced from abundant coal or natural gas are attracting attention. Of particular interest is the synthesis of ethylene and propylene from dimethyl ether because of their growing demand as raw materials for polyethylene and polypropylene. The usage of these polymers in everyday life is diverse (e.g., molded plastic items, plastic packaging films, etc.). Increasing demand for isobutene is inevitable since isobutene is used as the raw material for MTBE, MMA (methyl methacrylate), and isoprene. 1–Butene has growing significance as the raw material for maleic anhydride and as an ingredient for linear low-density polyethylene (LLDPE). 1– and 2–Butenes are important as ingredients in the synthesis of methyl ethyl ketone (MEK). Thus, lower olefins have varied usage in everyday life, so the research devoted toward their production from nonpetroleum sources is of economic interest.

Olefins are the essential intermediates in the conversion of methanol to gasoline range hydrocarbons. The selectivity of methanol conversion can be markedly directed toward olefin products by suppressing the aromatization reaction. This is accomplished by any of the following approaches: utilizing shape-selective catalysis over zeolites, reacting methanol at subatmospheric partial pressures, conducting partial conversion of methanol with recycling, using high temperatures for methanol conversion, or by reacting methanol over structurally modified zeolite catalysts.

Operation of the methanol to gasoline process at partial conversion of methanol gives high yields of olefins which can be separated from the product stream, and the unconverted feed can be recycled. This is the basis of a fluid-bed study carried out to entail the conversion of methanol to an olefins-rich hydrocarbon product [73]. Methanol feeds with varying amounts of dilution with water is used in this process. Increasing the water dilution was found to improve ethene selectivity. Using methanol containing 17% water, 48% C_2–C_5 olefins selectivity was obtained at 52% conversion. The fluid-bed was operated at a temperature of 300°C and pressure of 6 psig, at a methanol weight hourly space velocity of 0.4 hr^{-1}. Ethylene and propylene were the principal olefins identified in the product and they were formed in substantial amounts of 21.3 wt% and 17.2 wt%, respectively. Butenes and pentenes were the other olefins formed as products. The product spectrum included lower paraffins, aromatic, and higher (C_5+) nonaromatics. Operating the system at a higher temperature of 343°C gives a per-pass conversion of 84%. The product yield consists of 37% hydrocarbons with an olefins selectivity of 40%. Ethylene and propylene are produced in lesser amounts, while the aromatic and higher nonaromatic fractions are produced in larger amounts [73].

A catalytic process was developed which selectively produces a hydrocarbon product, comprised principally of ethylene and propylene. The catalyst used for this process was a crystalline aluminosilicate zeolite of the erionite-offretite family, designated as ZSM-34. The conversion conditions included temperatures of 260–540°C, pressures of 0.1–30 atm, and WHSV of 0.1–30. The feed consisted essentially of methanol and dimethyl ether, with at least 0.25 moles of water per mole of organic charge, owing to the fact that although ZSM-34 is a very effective catalyst for selectively converting methanol to lower olefins, its selectivity is markedly improved by the presence of added water to the feed. This process was directed toward enhancing the growth of production of plastics and synthesis fibers by providing inexpensive raw materials other than petroleum (e.g., ethylene and propylene). A major use of ethylene and propylene is to produce polyethylene and polypropylene. They are plastic materials which are virtually ubiquitous in everyday life, and are as diverse as molded plastic items, plastic packaging films, and electrical insulation [74].

The production of olefin can be enhanced by modifying the structural parameters of the zeolite catalyst and controlling the process operating conditions. A crystalline aluminosilicate zeolite having a Si/Al ratio greater than 30, and containing phosphorus incorporated within the crystal structure in an amount greater than 0.78% by weight was developed for olefin synthesis. Contrary to the strong-acid sites present in zeolites, the phosphorus-containing zeolite contains more acid sites than the parent zeolite but these sites are of lesser acid strength. As the strong-acid sites are believed to be responsible for the aromatizing activity of zeolites, replacement of these sites with a larger number of relatively weak-acid sites could explain the selectivity of phosphorus-containing zeolites toward olefins. Butter and Kaeding [75] observed that the use of the phosphorus-containing zeolite resulted in a reduction of the formation of aromatic hydrocarbons to 20.4%, and the production of olefin-paraffin ratios as high as 10. This compares to the use of zeolite without phosphorus which had aromatic yields and olefin-paraffin ratios of 40.46% and 0.2, respectively. Associated with the low aromatic yield was the absence of ethane and a low yield of propane and butane. The high olefin selectivity was attributed to the change in surface acidity and channel accessibility of the zeolite catalyst modified with phosphorus. The experiments using phosphorus-containing zeolite were conducted at a methanol weight hourly space velocity of 3.8, whereas the ones using an unmodified zeolite were conducted at a methanol weight hourly space velocity of 1.33. Thus, operating the process at high methanol space velocities shifts the selectivity toward synthesis of olefins [75].

The synthesis of olefins from methanol using aluminophosphate molecular sieve catalysts was studied [76]. Process studies were conducted in a fluidized-bed bench-scale pilot plant unit utilizing small-pore silicaluminophosphate catalyst synthesized at Union Carbide. These catalysts are particularly effective in the catalytic conversion of methanol to olefins, when compared to the performance of conventional aluminosilicate zeolites. The process exhibited excellent selectivities toward ethylene and propylene, which could be varied considerably. Over 50 wt% of ethylene and 50 wt% propylene were synthesized on the same catalyst, using different combinations of temperatures and pressures. These selectivities were obtained at 100% conversion of methanol. Targeting light olefins in general, a selectivity of over 95% C_2–C_4 olefins was obtained. The catalyst exhibited steady performance and unaltered molecular sieve crystallinity after a 1200-hr life test. These particularly high selectivities were attributed to the crystal structure and metallic elements incorporated in the framework of the silicaluminophosphate catalyst. The openings formed by the eight-member oxygen rings hinder the egress of any aromatic that may have formed inside the sieve cages. Thus the product spectrum from the methanol to olefins reaction is terminated at the synthesis of C_2–C_4 olefins. The product spectrum was devoid of any aromatics.

Another technique was used to increase the shape-selectivity of zeolites, thereby maximizing the yield of olefins from methanol. The enhanced yields were achieved by controlled deposition of materials like silica in the zeolite channels and cavities, thus increasing the diffusion path of guest molecules and introducing new steric constraints. By depositing 7.4% silica in the pores of zeolite ZSM-5, the aromatization reaction in methanol conversion was suppressed. At a temperature of 370°C and liquid hourly space velocity of 1 hr^{-1}, the selectivity toward olefins was significantly enhanced. The product distribution was comprised of 26.2% ethene, 18.7% propene, 10.8% butenes, 16% higher (C_5+) aliphatics, and only 19% aromatics. These yields are significantly higher when compared with those attained with untreated ZSM-5 which consist of 17% olefins, 41% paraffins, and 42% aromatics.

III. OXOCHEMICALS

A. Octanols and Heavy Oxochemicals

There are primarily two types of C_8 alcohols (octanols) that bear commercial significance: 2–ethylhexanol, and isooctanol. The total U.S. demand for these two octanols in 1965 was 285 million pounds. The industrial applications of octanols are as follows:

- Plasticizers: phthalates, phosphates, adipates, azelates, sebacates, epoxidized plasticizers. Plasticizers are used mainly in PVC compounding, the most important being the phthalates. Esters of aliphatic acids are used in footwear and household refrigerator parts where low-temperature flexibility is required.
- Surface-active agents
- Pesticides
- Octoic acid

Although isooctanol is slightly cheaper than 2–ethylhexanol, it is still considered more important due to the slightly higher volatility of diisooctyl phthalate, the unavailability of heptene, and the better odor of di-2–ethylhexyl phthalate. 2–Ethylhexanol is produced by the aldolization of *n*-butyraldehyde and subsequent hydrogenation, whereas isooctanol is made by the oxo reaction

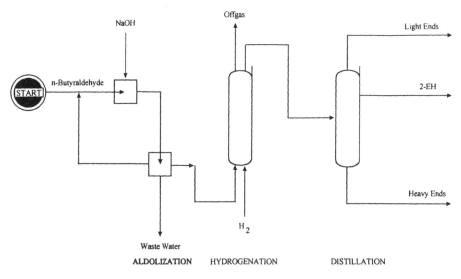

Figure 3.36   2-Ethylhexanol synthesis. *Source*: [8].

from a heptene feedstock [9]. The *n*-butyraldehyde can be obtained from two sources (i.e., partial hydrogenation of crotonaldehyde from aldolization and from propylene by the oxo process). A schematic diagram of 2–ethylhexanol process is shown in Figure 3.36. Octanols are obtained as a byproduct in the manufacture of sebacic acid from castor oil. Also, some normal octanol is made by manufacturers of detergent alcohols via fatty acid reduction of the Ziegler-type processes.

Isooctanol is a common name and is actually a mixture of several C_8 isomers. It consists of the following isomers: 3,4–dimethyl-1–hexanol (20%), 3,5–dimethyl-1–hexanol (30%), 4,5–dimethyl-1–hexanol (30%), 3–methyl-1–heptanol and 5–methyl-1–heptanol (15%), and other unidentified alcohols (5%). Isooctanol is soluble in all the common solvents but essentially insoluble in water. These characteristics of the oxo chemicals point to the use of isooctanol as solvents for fats, oils, waxes, gums, and resins. The alcohols may be used as lacquer solvents and plasticizers, in perfumes, and as industrial odorants.

2–Ethylhexanol can be converted to octoic acid by liquid-phase oxidation. Octoic acid is used in the form of its metal salts, mainly for odorless paint driers. The development of octoate paint driers represents an attempt on the part of oil-paint makers to retain some of the indoor market since one of the chief selling points of emulsion paints is their less unpleasant odor.

Heavier oxochemicals such as isodecyl alcohol and diisodecyl phthalate also find commercial significance in the industry. Tridecyl alcohol is made from propylene tetramer. Its main application is di-tridecyl phthalate (e.g., in PVC wire coating where low volatility is desirable). Diisodecyl phthalate is used mainly in PVC wire coatings because of its low volatility. It is also used to make synthetic lubricants and plasticizer esters from acids other than phthalic anhydride. The higher oxoalcohols are important raw materials for the production of both nonionic and anionic detergents and wetting agents. Ethylene oxide and propylene oxide react with isooctyl, decyl, and tridecyl alcohols to form polyester of any desired molecular weight. The ethylene oxide derivatives are especially well suited for use as household and industrial detergents. The propylene oxide products are oil soluble and may be used as lubricant oil additives or plasticizers for synthetic resins and rubber. Sodium tridecyl sulfate, which is formed from decyl and tridecyl alcohols, is an anionic detergent. These products have the advantage that it is easy to adjust the solubility balance [77]. Oxo tridecyl alcohol is particularly suitable for the production of nonionic detergents. It is uniform in color and free of objectionable odor. These detergents can be used in both textile and dishwashing applications. However, in recent years, the relative importance of protease enzyme in detergents has been steadily increasing.

B. n- and Isobutyraldehydes and Propionaldehyde

Butyraldehyde is used mainly as an intermediate in the production of synthetic resins, rubber vulcanization accelerators, solvents, and plasticizers. It reacts with polyvinyl alcohol to form polyvinyl butyral which is used as an interlayer for safety glass as well as for coating fabrics. Isobutyraldehyde is used as a solvent or plasticizer in the plastics industry, or as a raw material in the synthesis of isobutyl alcohol, methacrylic acid, neopentyl glycol, etc.

The Ruhrchemie AG oxo process is a commercial one for the manufacture of n-butyraldehyde from propylene and synthesis gas. The reaction is:

$$2CH_3CHCH_2 + 2CO + 2H_2 = CH_3CH_2CH_2CHO + (CH_3)_2CHCHO$$

Propylene reacts with syngas in the presence of a water-soluble rhodium complex forming n-butyraldehyde and a small amount of isobutyraldehyde. The catalyst is separated from the oxo product and recycled back to the reactor. The unconverted propylene is removed from the product by stripping with fresh syngas and is recycled to the oxo reactor. n- and isobutyraldehyde are obtained in a single-stage distillation. The flowsheet of this process by Davy Process Technology is shown in Figure 3.37 [134].

Another type of reaction giving butyraldehyde is the dehydrogenation of 1–butanol is:

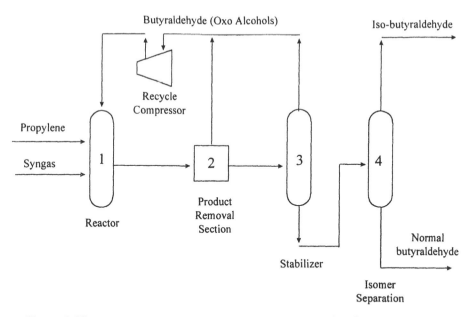

Figure 3.37   Davy process technology oxo process. *Source*: [134].

$$CH_3CH_2CH_2CH_2OH = CH_3CH_2CH_2CHO + H_2$$

The catalyst used for this reaction contains copper and chromium. More recent catalysts include zinc and aluminum in addition to copper and chromium. Conversion of 50% with a selectivity to butyraldehyde of 90% in a temperature range of 573–603 K was obtained [78].

A new catalyst, bismuth phosphate pure or with a small amount of Mo, has been developed for 1–butanol dehydrogenation. Experiments were carried out in a conventional flow apparatus with a glass U-tube reactor. The main products of this reaction are butyraldehyde, butenes, and carbon dioxide. Butyraldehyde is formed during dehydrogenation of 1–butanol, and butenes are the products of 1–butanol dehydration. CO_2 results from an oxidation process. The product distribution is dependent on the catalyst composition and the presence of oxygen in the reactant stream. The most selective catalyst for butyraldehyde is pure bismuth phosphate. An increase in the dehydrogenation activity is observed after introduction of oxygen into the reactant stream for a catalyst mixture of bismuth phosphate and MoO_3. The presence of MoO_3 along with bismuth phosphate improves the activity compared to pure bismuth phosphate. However, MoO_3 has a negative influence on selectivity to butyraldehyde, even in the presence of oxygen, when too high amounts of molybdenum are introduced. When selectivity is considered, the best catalyst for butyraldehyde formation is bismuth phosphate. It is better, with respect to conversion and selectivity, than the other catalysts but working with much higher liquid hourly space velocity. This catalyst is highly selective to butyraldehyde provided that oxygen is present, in a small amount, in the reactant stream [78].

The hydroformylation of propylene gives a mixture of *n*- and isobutyraldehyde in the presence of cobalt carbonyl catalyst. *n*- and isobutyraldehyde are produced in varying ratios between 1.6 and 4.4 depending on the temperature, catalyst concentration, and partial pressures of carbon monoxide and hydrogen. An investigation was carried out to study the effect of these parameters on the ratio of *n*- and isobutyraldehyde produced by this reaction over unmodified cobalt carbonyl catalyst. Experiments were carried out in an oil-thermostated 2–L capacity autoclave, with three propeller-type stirrers. The propellers in the reactor were fixed at positions in which an improved gas distribution was observed leading to intensive gas-liquid contact with gas bubbles reaching all parts of the liquid. Conversion levels up to 95% were obtained under different operating conditions in the temperature range of 383–423 K. The *n*-/isobutyraldehyde ratio decreased with increased propylene concentration, which suggests that increasing propylene concentration increases the isobutyraldehyde formation faster than that of *n*-butyraldehyde. However, the propylene dependencies of these individual reactions are different, isobutyraldehyde formation having a higher order of propylene dependence than that for *n*-butyraldehyde formation. At a partial pressure of syngas of 100 bar and a temperature range of 383–423 K, the *n*-/isobutyraldehyde ratio increased with an increase in catalyst concentration. The CO concentration in the liquid phase was not influenced by the rate of CO absorption. The *n*-/isobutyraldehyde ratio increased with an increase in the partial pressure of CO. This was attributed to the fact that a five–fold species, $HCo(CO)_4$ is more stable at higher CO partial pressures and is responsible for the formation of *n*-butyraldehdye. The partial pressure of hydrogen has a reproducible effect on isomer distribution. The higher partial pressure of hydrogen gave higher percentages of *n*-butyraldehyde. Low temperatures favor higher percentages of the *n*-butyraldehyde, indicating a higher activation energy for isobutyraldehyde formation. The rates of isomer production showed a fractional order of dependence with respect to catalyst concentration and partial pressure of hydrogen, but a complex dependence on the partial pressure of carbon monoxide showing substrate-inhibited kinetics at higher CO pressures.

Isobutyraldehyde can be synthesized selectively from methanol and ethanol in one step by using titanium oxide-supported vanadium oxide (V_2O_5/TiO_2) as a catalyst. This catalyst is also capable of synthesizing isobutyraldehyde from methanol and *n*-propyl alcohol. The reactions of methanol and *n*-propyl alcohol were carried out in a downflow continuous fixed-bed reactor. The reactor is a vertical quartz tube with an inside diameter of 1.8 cm. At 350°C, the conversion of n-propyl alcohol decreased and the selectivity to isobutyraldehyde decreased slightly to 60%, while as a compensation to that the selectivity to propionaldehyde increased with reaction time. Among the metal oxides tested, TiO_2 was found to be the only one which is effective in the reaction of n-propyl alcohol with methanol to give isobutyraldehyde. Improved results in both conversion and selectivity were observed by loading vanadium on these metal oxides. The reactions over V/TiO_2 catalysts yield isobutyraldehyde selectively with minor amounts of propionaldehyde, propane, and isobutane at high conversion. Almost all the excess methanol was recovered, together with a small amount of methane, after the reaction. The activities of V/TiO_2 strongly depend on the concentration of vanadium. The catalyst properties can be enhanced only by the addition of a fairly small amount of vanadium. Excess addition, however, results in a decrease in activity and selectivity to isobutyraldehyde. Catalysts with higher concentrations of vanadium show lower activities for the formation of isobutyraldehyde but produced hydrocarbons as main products. The formation of isobutyraldehyde and isobutane increased with n-propyl alcohol concentration, passed through a maximum, and then decreased. The optimum concentration is about 30 mol% n-propyl alcohol. However, the formation rate of propionaldehyde and propane, which are formed by the unimolecular reaction of n-propyl alcohol, increased directly with the concentration of n-propyl alcohol. The formation of isobutyraldehyde is more appreciable both at higher reaction temperature and at longer contact time. The selectivity of isobutyraldehyde and isobutane increased, with increasing contact time, while that of propionaldehyde and propane decreased with the contact time. Based on the reaction scheme, it is concluded that isobutyraldehyde is formed via propionaldehyde as an intermediate. It is a reasonable assumption that isobutane is formed from a propane-like species which is formed from n-propyl alcohol on the catalyst surface and then converted to propane [142].

The reaction of adsorbed CO on Rh/SiO_2 and $S-Rh/SiO_2$ with propylene and hydrogen was studied using an in situ infrared technique, and the effect of sulfur on the n-/isobutyraldehyde ratio was investigated [80]. Sulfided Rh/SiO_2 exhibits a higher CO insertion selectivity, a lower hydrogenation activity, and a lower n-/isobutyraldehyde ratio than Rh/SiO_2 during steady-state propylene hydroformylation at 513 K and 0.1–1 MPa. Propylene hydroformylation is a reaction of propylene with syngas that is catalyzed by rhodium and cobalt carbonyls and their phosphine-modified complexes in organic solvent. However, the main disadvantage of this homogeneous process is that an energy-intensive separation stage is required to recover the catalyst from the product and solvent stream. The effect of sulfur on steady-state propylene hydroformylation has been identified as a decrease in the CO conversion and a suppression of hydrogenation activity. Sulfur decreased the *n*-/isobutyraldehyde ratio at 513 K and 0.1–1 MPa. The decrease in the ratio of *n*-/isobutyraldehyde is attributed to the effect of phosphine ligands in homogeneous hydroformylation [80].

Studies have been carried out to synthesize butyraldehyde in a single step by dehydrogenation of butanol, using zinc oxide as catalyst in a fixed-bed glass flow reactor. The kinetics of the reaction using zinc oxide catalyst showing maximum yield of butyraldehyde have been studied. The reactor was housed in a 9-in. long heated ceramic tube. The design of the reactor facilitated uniform heat and mass transfer between vapor and the catalyst particles. The main products of the reaction were butene and hydrogen in the gas phase; the liquid product contained butanol and butyraldehyde. All types of zinc oxide prepared by different precursors showed very good activity as well as selectivity giving yields of 67–90%. In the temperature range of 350–450°C considered, conversion in all grades of zinc oxide increased with increase in temperature. In the case of zinc oxide prepared by decomposition of carbonate and oxalate, though the conversion kept on increasing with an increase in temperature, at higher temperatures beyond 450°C, a lot of carbonization made the catalyst lose its activity. Conversion increased rapidly with weight-time increase in the range of 3.4–19.6 g cat. min dm^{-3} and became steady at higher W/F_{AO}. Thus, zinc oxide has shown very good activity and selectivity in the synthesis of butyraldehyde. The zinc oxide calcined from zinc hydroxide gave maximum activity and selectivity of 90% towards butyraldehyde at a temperature below 400°C [81].

The Reppe hydroformylation of ethylene to produce propionaldehyde and 1–propanol in basic solutions containing $Fe(CO)_5$ as a catalyst was studied under carefully controlled conditions at a temperature range of 110–140°C. Propionaldehyde is the main product formed when NaOH is used as the base. The reaction is shown below:

$$CH_2 = CH_2 + 2CO + H_2O = CH_3CH_2CHO + CO_2 \qquad \Delta G = -80.8 \text{ kJ/mol}$$

The experiments were carried out in small stainless steel autoclaves having an internal volume of 700 mL. The autoclaves, having been charged with a particular catalyst solution and gas mixture of interest, were mounted vertically in electrically heated ovens. The factors affecting the rate of the reaction are: partial pressure of carbon monoxide, partial pressure of ethylene, catalyst concentration, temperature, base concentration/pH, and the nature of the base. Carbon monoxide has an inhibitory effect upon the reaction. The rate of reaction increases linearly with ethylene pressure in the low-pressure regime but exhibits saturation at ethylene pressures exceeding 17 atm. The reaction is second order with respect to catalyst concentration. The nature of the base used determines whether an aldehyde is produced, as in the experiments where KOH is the base, or whether the reaction system is sufficiently reducing in nature to cause the aldehyde product to be reduced to the corresponding alcohol [82].

The synthesis of acetaldehyde and propionaldehyde from CO hydrogenation over Na-Mn-Ni catalysts was studied. Coprecipitated Na-Mn-Ni catalysts exhibited high activities for the synthesis of acetaldehyde from CO hydrogenation and the synthesis of propionaldehyde from addition of ethylene to CO hydrogenation. The Ni/SiO_2 catalyst produced methane as a major product with C_{2+} hydrocarbons as minor products. Addition of Na and Mn to the Ni/SiO_2 resulted in increases in the selectivity to C_2+ hydrocarbons. Both Na and Mn promoted the formation of a small amount of acetaldehyde and propionaldehyde. The coprecipitated Na-Ni catalyst produced methane as a major product, while the coprecipitated Na-Mn-Ni catalyst exhibited high activity and selectivity for the formation of acetaldehyde and propionaldehyde. Addition of ethylene to syngas over the Ni/SiO_2 led to increases in the rate of ethane formation as well as the production of propionaldehyde. The significant increases in the rates of formation for these products indicate that the major competitive reactions for the added ethylene are ethylene hydrogenation and CO insertion. The Ni/SiO_2 causes a marked decrease in the hydrocarbon formation, but remains active for catalyzing the formation of propionaldehyde [83].

2–Ethyl-1,3–hexanediol is made by aldolization and subsequent hydrogenation of *n*-butyraldehyde. It is used in the manufacture of polymeric plasticizers, and is well known as an insect-repellent.

$$2C_3H_7CHO \longrightarrow C_3H_7C(OH)HCH(C_2H_5)CHO \xrightarrow{\text{H}_2}$$
$$C_3H_7C(OH)HCH(C_2H_5)CH_2OH$$

Trimethylolpropane is produced by reacting butyraldehyde with three mols of formaldehyde:

$$3HCHO + C_3H_7CHO \xrightarrow{\text{NaOH}} C_2H_5C(CH_2OH)_3 + HCOONa$$

The product is called 1,1,1–Trimethylolpropane (TMP) and is used mainly in the manufacture of triols by adduction with propylene oxide (PO), for flexible polyurethane foams and of synthetic lubricants by esterification with fatty acids. The allyl ethers of trimethylolpropane are used as crosslinking agents, for example, in acrylic resin systems. The world-wide capacity for TMP production is in an increasing trend.

n-Butyric acid is produced by liquid-phase oxidation of *n*-butyraldehyde:

$$C_3H_7(CHO) + 0.5O_2 \longrightarrow C_3H_7COOH$$

It is used mainly in the manufacture of cellulose acetobutyrate, the preferred cellulose ester for molding powders. Cellulose acetobutyrate (CAB) and cellulose acetate (CA) are commonly used for eyeglass frames. Cellulose acetobutyrate is used for automotive applications such as steering wheels.

n-Butyronitrile is made from butyric acid and ammonia. This is one of the raw materials for the widely-used coccidiostat "Amprolium."

IV. PHOSGENE AND DERIVATIVES

A. Phosgene

Uses and Production

Phosgene ($COCl_2$) is a colorless gas, of unpleasant odor that is a severe respiratory irritant [84–89]. This chemical has a very interesting past but an endangered future. Phosgene was used by Germany during World War I as a poison gas and caused numerous casualties. The toxicity of phosgene is a major reason to replace its synthetic applications by less noxious chemicals. The Bhopal, India, disaster in December 1984 resulted from a massive release of methyl isocyanates (MIC); these are prepared from methylamine and phosgene. This tragic event provided an additional stimulus to avoid the use of phosgene and the production and transportation of MIC and to arrive at the desired methyl carbamates [MeN=CH-C(=O)-R], by alternate means. The issues of guilt, the specific cause of the release of MIC, indemnities for the families of casualties and surviving victims, the degree of the manufacturer's responsibility, and related issues remain undecided before the courts at the time of writing [84].

The annual capacity of the United States to manufacture phosgene is about 2 billion lb, and the unit price is $0.55 (per pound 1993). The gas is packaged in 1-ton returnable containers that cost about $3,000 each. Most of the demand of about 1.6 billion lb is consumed in the production of polyurethane (PUR) followed by the production of polycarbonates (PC) [85].

Phosgene is produced by means of a typical free-radical chain reaction between chlorine and carbon monoxide [85]:

Initial step:	$Cl_2 \longrightarrow 2Cl\cdot$
Propagation steps:	$Cl\cdot + CO \longrightarrow ClCO\cdot$
	$ClCO\cdot + Cl_2 \longrightarrow Cl-CO-Cl + Cl\cdot$
Termination steps:	$2Cl\cdot \longrightarrow Cl_2$
	$Cl\cdot + Cl-CO\cdot \longrightarrow Cl-CO-Cl$
Minor byproduct:	$2Cl-CO\cdot \longrightarrow Cl-CO-CO-Cl$

While phosgene was discovered and was given its name because the reaction is catalyzed by sunlight, it is currently manufactured at a minimum temperature of 50°C, at a pressure of 5–10 atm, and in the presence of activated carbon. In the subsequent applications of phosgene, the presence of free chlorine is undesirable and thus excess carbon monoxide has to be used in order to make sure the chloride is completely converted. Purification is accomplished by condensing out the product, which is a liquid below 7°C. The off-gases are washed with water to destroy any remaining phosgene, and elaborate precautions are necessary to prevent any loss of product.

Recent phosgene plants constructed in the Gulf Coast of the United States obtain their CO from synthesis gas by low-temperature fractionation [8]. Early units derived their raw material from coal, directly or indirectly. Most phosgene plants have their own caustic chlorine facilities. Since most phosgene is consumed captively, the hydrogen chloride is recovered in an oxidation unit.

Consumption of Phosgene in the United States

Most of the current demand of 1.6 billion lb of phosgene is used by the polyurethane (PUR) industry, and most of it is produced for captive use because of its hazardous nature and the consequently high shipping and insurance costs [85].

The PUR polymers represent an extremely versatile family of materials that can be manufactured in the form of rigid but impact-resistant solids, semiflexible or flexible foams, thermoset or thermoplastic elastomers, coatings, adhesives, elastomeric fibers known as Spandex, etc. The products range from bedding and furniture cushions, flooring and roofing materials, automotive components such as bumpers and window gaskets, components of appliances, ingredients of paints, binders for structural or decorative boards and foundry molds, thermal and sound-insulating materials, etc, and even include bioengineering parts. Flexible foam products constitute about half of PUR consumption. In 1986 production of PURs in the United States was about 2.7 billion lb and was valued at about $2.5 billion; it is expected to grow in the foreseeable future at a rate that exceeds the growth of the GNP, especially because of the rapidly expanding reaction-injection molding (RIM) and reinforced-reaction-injection molding (RRIM) technologies employed in the highly productive manufacture of large objects such as components of automobiles. The RIM and RRIM technology is not limited to PURs and its worldwide application is growing at an average annual rate of 17%.

The discrepancy between the isocyanate and polyurethane production levels signifies that components other than the isocyanates contribute to the final PUR production materials. These other materials, often referred to simply as "polyols," are double- and higher-functionality alcohols.

The formation of a single urethane bond by the addition reaction of hydroxyl and isocyanate functional groups is represented simply by:

R–N=C=O + R'–OH --> R–NH–CO · O–R'

It may be useful to recall that the somewhat confusing traditional nomenclature calls structures such as H_2N–CO·O–R *carbamates* and Cl–CO–O–R *chloroformates*, while substitution at both N and O terminals gives *urethanes*.

The great variety of PUR-based products, and the nearly endless variations of the fundamental PUR chemistry, call for a separate, concise summary of the parameters that can be manipulated to control the nature of the end products and their processing technology.

The other important family of polymers dependent on the phosgene building block is that of the polycarbonates (PCs). Their production (about 350 million lb) consumes approximately 6–7% of the phosgene demand. The high-performance characteristics of the PCs also promise continued growth albeit at a lower level than that of the PURs, because the PCs are not used to the same extent as the PUR in the manufacture of large objects. On the other hand, special grades of PCs are utilized in the manufacture of compact disks (CDs) and other popular electronic devices.

The remaining few percentage points of the current phosgene demand are for the production of valuable pesticides of the carbamate family and other fine chemicals. The use of carbamate pesticides is likely to continue in order to protect agricultural and silvicultural productivity, but the impact of the Bhopal disaster has become a strong incentive for utilizing, wherever possible, nonphosgene technology, especially when this takes place in industrially less-developed countries [85].

Polyurethane (PUR) and Phosgene Derivatives

Phosgene is used as a building block of isocyanates and is being challenged by dimethyl carbonates (DMC), an intermediate obtained without intervention of chlorine and by direct use of carbon dioxide. A possible breakthrough in the use of phosgene is the announcement of an allegedly safe facility developed by Rhone-Poulenc [85]. It is used for the synthesis of hexamethylene diisocyanate, one of the polyurethane building blocks.

The synthetic uses of chloroformates (Cl–CO·OR) for the preparation of fine chemicals is one area of phosgene chemistry that is likely to survive, although admittedly, it represents a small fraction of the current phosgene demand.

The principal variables and parameters that can be chosen for the production of the above-mentioned great variety of PUR products are enumerated in the following section.

Choice of PUR Processing Technology.   This depends on the nature of the end product, convenience, and safety considerations [104]. It can consist of a one-step process in which the major building blocks of the desired

PUR (isocyanate and alcohol or some other ingredient that reacts with iso-cyanate) are mixed together with catalysts, fillers, reinforcing and coloring agents, blowing agent (if a foam is to be produced), and other minor constitu-ents of the reaction mixture to give the final solid product (most commonly as blocks, pads, and otherwise-shaped objects of thermoset elastomers or rather rigid foams). An alternative is a two-step process in which a relatively high molecular weight PUR prepolymer is assembled in such a fashion that it con-tains reactive terminal groups (most commonly isocyanate-terminated or hy-droxyl-terminated structures obtained by the use of an excess of one or the other component). The prepolymer can then be subjected to a reaction with a chain extender, a rather reactive and relatively low molecular weight diol or diamine that links molecules of the prepolymer to give larger two- or three-dimensional PUR end products. The two-step process is the basis of reaction-injection molding technology, but it is also used to avoid the handling of hazardous isocyanates—particularly tolylene or toluene diisocyanate (TDI)—during shipment or during final processing operations. The control during the formation of PUR prepolymer is not as simple as it may sound: only recently Air Products announced a new technology that produces a "perfect prepolymer."

A modification of the two-step reaction concept is the use of blocked isocyanates. This entails the conversion of isocyanates to a derivative that liberates the isocyanate under relatively mild thermal conditions and allows it to react with the partner component, while the blocking agent is either volatilized or also incorporated into the end product. Blocked isocyanates are obtained by the addition of a reagent that offers a good leaving group on thermal activation of the adduct. For example:

$$R-N=C=O + Q-O-H \longrightarrow R-NH-CO-O-Q \longrightarrow R-N=C=O(+Q-O-H)$$

where Q-O-H represents a phenol, $R'_2C=N-OH$, an oxime, and so on [85].

Choice of Isocyanate Component.

Aromatic Diisocyanates. Tolylene or toluene diisocyanate (TDI) is the traditional "work horse" of the PUR industry [94]. Because of its origin, which starts with nitration of toluene, ordinary TDI is a 80:20 mixture of 2,4– and 2,6–diisocyanatotoluene ($1.01). The minority component of this mixture reacts slower than the 2,4–isomer because both isocyanato groups are sterically inhibited, and hence 100% pure 2,4–isomer is also available at a premium price of $1.60 (Mobay's Mondur TDS). The pure 2,4–isomer of TDI is likely to give an overall more linear PUR end product. The consumption of TDI is affected strongly by the consumer demand for furniture (cushions, mattresses, etc.) and carpet pads, and thus it is sensitive to the general state of the economy and the construction sector, in particular [85].

Methylene di-*p*-phenylene diisocyanate, diphenylmethane 4,4'-diisocyanate, or simply MDI, and its polymeric analog, PMDI ($0.91) have recently caught up and exceeded the 760 million lb demand for TDI and are consumed in the United States at a level of about 900 million lb. Both are derived from aniline and formaldehyde and are listed together in most statistical reports. The growing demand for MDI and PMDI by the PUR industry is responsible for the strong aniline market since they consume about 60% of the latter. Both MDI and PMDI give rise to more rigid PUR end products than does TDI and are used for the manufacture of insulating panels for refrigerators and other appliances, automotive components such as bumpers and fascia, and in the constantly expanding RIM and RRIM operations. The polyfunctionality of PMDI is responsible for crosslinking of the resulting PURs and the formation of rigid end products. On the other hand, because of the nearly linear structure and a relatively restricted freedom of rotation about the central methylene group, MDI alone is used to assemble the elastomeric textile fiber, Spandex, known to many consumers of support hosiery and many types of elastic garments. This elastomeric textile fiber was commercialized by Du Pont in 1962 under the trade name "Lycra."

The structures of these large-volume aromatic isocyanates are shown below.

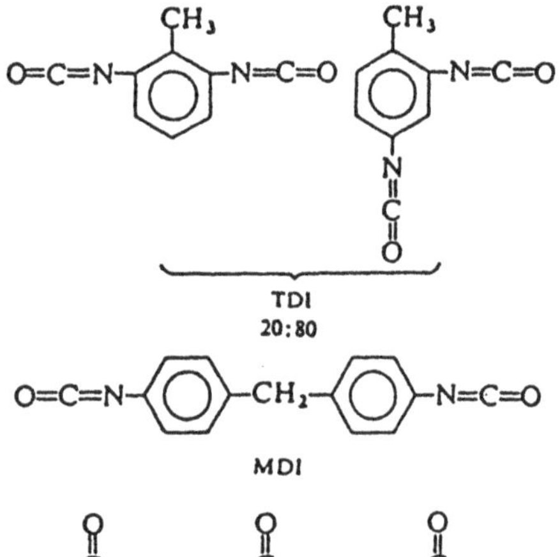

TDI
20:80

MDI

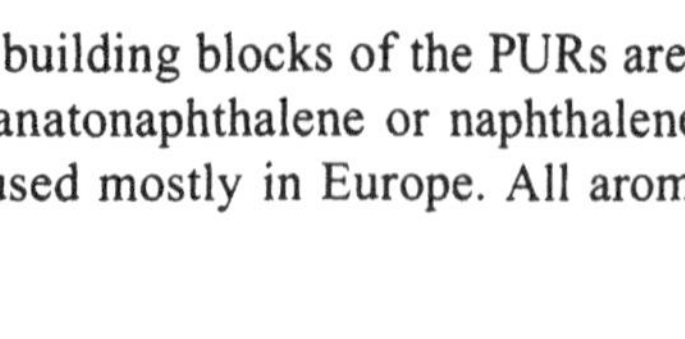

PMDI

The more specialized building blocks of the PURs are discussed in subsequent sections. 2,6–Diisocyanatonaphthalene or naphthalenediisocyanate (NDI) is a PUR building block used mostly in Europe. All aromatic isocyanate building

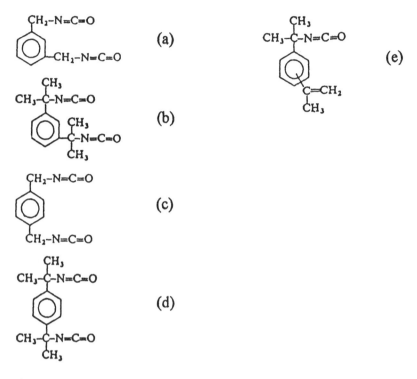

Figure 3.38 Nonaromatic isocyanates. (a) *m*-Xylenediisocyanate, Sherwin-William's MXDI; (b) *p*-Xylenediisocyanate, Takeda's XDI; (c) Tetramethyl-*m*-xylenediisocyanate, Cyanamid's TMXDI; (d) Tetramethyl-*p*-xylenediisocyanate, (e) isopropenyldimethyl-benzylisocyanate, Cyanamid's TMI.

blocks are susceptible to light-catalyzed discoloration, and hence in coating and similar applications are replaced by aliphatic isocyanates [87,88].

Nonaromatic Isocyanates. These structures in which the isocyanato functions are not attached directly to an aromatic ring (Figure 3.38). Most often they are also referred to as "aliphatic isocyanates," even though they may actually be benzylic isocyanates [85].

TMI is useful for "capping" hydroxy-terminated PURs or, for that matter, capping amino- or carboxy-terminated prepolymers. TMI also can serve as a heterodifunction monomer if the PUR structure is to be combined with a polyvinyl chain [85].

The problem of discoloration of PURs derived from benzylic isocyanates is not solved completely because of slow oxidation at the benzylic hydrogens.

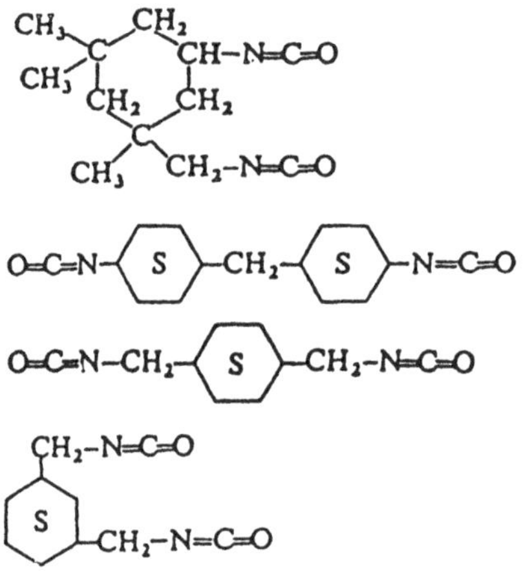

(a)

(b)

(c)

(d)

Figure 3.39   Alicyclic isocyanates. (a) Isophorone diisocyanate (or 3–isocyanato-methyl-3,5,5–trimethylcyclohexyl isocyanate), Huels' IPDI; (b) 1,4–bis(isocyanatomethyl)cyclohexane, Mobay's p-Desmodur; (c) 1,3–bis(isocyanatomethyl)cyclohexane, Takeda's H_6XDI ($3.50); (d) Methylene bis(4–cyclohexylisocyanate), Mobay's Desmodur W ($3.00), also known as H_{12}MDI, RMDI (reduced MDI), and PACM [bis-p-aminocyclohexyl methane].

Alicyclic Isocyanates.   These are shown in Figure 3.39.

Truly Aliphatic Isocyanates.   These include hexamethylenediisocyanate (HDI) derived from acrylonitrile,

$$OCN-(CH_2)_6-NCO$$

and trimethylhexamethylenediisocyanate, actually a 1:1 mixture of 2,2,4– and 2,4,4–trimethylhexamethylenediisocyanate, Nuodex's (a Huels subsidiary) TMDI:

$$OCN-CH_2C(CH_3)_2CH_2CH(CH_3)CH_2CH_2-NCO$$

$$OCN-CH_2CH(CH_3)CH_2C(CH_3)_2CH_2CH_2-NCO$$

Multiple methyl substituents in TMDI inhibit intermolecular associations of the urethane segments derived from these building blocks and hence promote flexibility in the resulting PURs. An interesting heterodifunctional isocyanate is Dow's isocyanatoethyl methacrylate (IEM)

$$[CH_2=C(CH_3)CO \cdot O-CH_2CH_2-NCO]$$

since it facilitates the grafting of vinyl oligomer or polymer side chains when the isocyanato group is attached first to an appropriate prepolymer. Also, it facilitates a build-up of an oligomer or polymer side chain attached by way of the isocyanato group when the methacrylic moiety is copolymerized first with another vinyl system. Similar possibilities are also available by means of the heterodifunctional isopropenyldimethylbenzylisocyanate (TMI) [85].

Choice of Polyol Components.

Short-Chain Diols. Examples of short-chain diols are [85]:

Ethylene glycol (EG)
Diethylene glycol (DEG)
Propylene glycol (PG)
Dipropylene glycol (DPG)
Neopentyl glycol
1,4–Butanediol, or tetramethylene glycol
1,6–Hexamethylene glycol derived from adipic acid
N-substituted diethanolamines
N-methyldiethanolamine (MDEA) in particular, and others

Short-chain diols are used as chain extenders, especially when both hydroxyl groups are primary. Secondary hydroxyl groups react more slowly with iso-cyanates but first can be capped with ethylene oxide (EO) to give the primary hydroxy-terminated derivatives.

Long-Chain Polyether Diols. Examples of long-chain polyether diols of molecular weight 10^2–10^3 are [85]:

Poly(ethylene glycol) (PEG)
Poly(propylene glycol) (PPG); in this case, since only one hydroxyl group is a primary one, PPG is usually capped with some EO in order to convert the secondary hydroxyl terminal also to a primary hydroxyl.
Poly(tetramethylene ether) glycol (PTMEG), also known as polytetrahydro-furan (PTHF).

The contribution by the polyether diols of similar molecular weight to the flexibility of the PUR end product increases in the order PEG < PPG < PTMEG. The high degree of rotational freedom in PTMEG makes it an ideal polyol for the assembly of elastic PURs such as Spandex and its worldwide demand is estimated to be about 100 million lb.

Long-Chain Polyester and Acid Anhydride Diols. These compounds include the hydroxy-terminated or EO-capped derivatives of simpler building blocks:

Poly(caprolactone),

$$HO–(CH_2)_5CO[O–(CH_2)_5–CO]_n–O–CH_2CH_2–OH$$

Poly(ethylene glycol terephthalate),

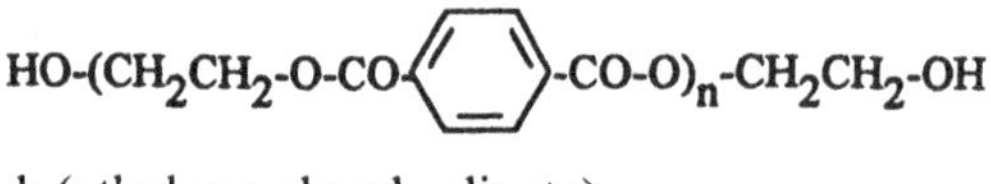

$$HO\text{-}(CH_2CH_2\text{-}O\text{-}CO\text{-}C_6H_4\text{-}CO\text{-}O)_n\text{-}CH_2CH_2\text{-}OH$$

Poly(ethylene glycol adipate),

$$HO\text{-}[CH_2\text{-}CH_2\text{-}O\text{-}CO\text{-}(CH_2)_4\text{-}CO\text{-}O]_n\text{-}CH_2CH_2\text{-}OH$$

Poly(carbonate-linked ethylene glycols) of the general structure, HO-(R-O-CO-O)n-R-OH, PPG's Duracarb 120, ($2.85), where R represents a difunctional linear, aliphatic moiety such as PPG (molecular weight 850–1500), suitable for the preparation of flexible PURs, while the analogous diol in which R represents an alicyclic moiety, PPG's Duracarb 140 ($2.93; molecular weight 600–1000) is suitable for the preparation of rigid PURs.

Polyazelaic polyanhydride, Emery's PAPA,

$$HO\text{-}CO\text{-}(CH_2)_7\text{-}CO\text{-}[O\text{-}CO\text{-}(CH_2)_7\text{-}CO]_n\text{-}OH$$

Generally, the contribution of intermolecular attractive forces operating within PUR end products increases in the order polyethers < polyesters < polycarbonates or polyanhydrides.

The polyether diols derived from polyesters seem to represent a compromise between cost and their contribution to superior elongation and tensile properties as compared to plain polyether diols; polyether diols from polyester account for about 10% of the flexible PUR production. Examples of useful products formed from hydroxy-terminated esters of terephthalic acid (TPA)—formed, in turn, by the reaction of an excess of glycols with dimethyl terephthalate (DMT) or TPA—and TDI or MDI-PMDI are flexible ski clothing, gaskets, rollers for printing presses, etc. [85].

Miscellaneous Trifunctional and Higher Alcohols.   These include:

Glycerol ($0.90/lb)
1,1,1–Trimethylolpropane (TMP; $0.76/lb)
Pentaerythritol and dipentaerythritol ($0.71/lb and $1.42/lb, respectively)
Triethanolamine ($0.35/lb) and the polyols of more complex structure; 1) Sorbitol SZM-II, powder ($0.68/lb), 2) Methyl glucoside, and 3) Sucrose polyether polyols, Dow's Voranols ($0.78/lb).

Choice of Isocyanate-Reaction Partners that Form Bonds Other than Urethane Bonds.

Primary Amines.   These react to give ureas

$$R-N=C=O + H_2N-R' \longrightarrow R-NH-CO-NH-R'$$

Commonly employed for this purpose are:

Ethylenediamine ($1.30/lb), a popular chain extender.
4,4'-Methylene dianiline (crude $1.75/lb, purified $2.25/lb), the precursor of MDI.
Toluenediamines (precursors to TDI).

An amine can be generated in situ by addition of a controlled amount of water that reacts with isocyanate to yield a spontaneously decomposed carbamic acid with a release of CO_2 and formation of foam:

$$R-N=C=O + H_2O \longrightarrow R-NH-CO-OH \longrightarrow R-NH_2 + CO_2$$
$$\text{a carbamic acid}$$

Carboxylic Acids. These react to give amides and, at the same time, carbon dioxide is released (foam formation):

$$R-N=C=O + HO\cdot CO\cdot R' \longrightarrow R-NH-CO-R' + CO_2$$
$$\text{amide}$$

It should be noted that intermolecular attractive forces between individual PUR chains increase in the order

Aliphatic–NH–CO–R < aromatic–NH–CO–R <
$$-NH-CO-NH- < -NH-CO-NH-CO-NH-$$

It is clear that an admixture of significant amounts of amines, amine-generating water, or carboxylic acids gives rise to hybrid urethane-urea or even urethane-urea-biuret products. The extent to which such hybrid macromolecules are formed will affect the thermomechanical properties of the end products.

Choice of the Degree of Crosslinking. The degree of crosslinking determines whether one produces an elastomer or a thermoset solid. A small degree of crosslinking provides "elastic memory," meaning that initially rather unorganized molecular chains tend to recover their original chaotic arrangement on removal of an external force (compression or stretching) that was applied to the material. Elastic memory causes the material to behave like a rubber or an elastomer. Crosslinking of macromolecules that creates elastic memory can be achieved in two different ways:

1. The use of a relatively small amount of polyfunctional reactants; in the case of PURs these can be either isocyanates or polyols.

2. Blocks of both highly disorganized, flexible, "soft" segments, and "hard" segments that exhibit strong intermolecular attractions for each other (by way of Coulombic forces among ionic centers, hydrogen bonding, dipole-dipole, or even dipole-induced dipole forces) can be incorporated into the polymeric chains. This is achieved by the assembly of block polymers, that is, polymeric chains that contain alternating hard and soft segments.

The most significant difference between these two modes of achieving crosslinking is that the first generates a traditional thermoset elastomer, while the second gives rise to a thermoplastic elastomer (TPE) [85].

The concept of traditional thermoset elastomers was pioneered by Goodyear's discovery in 1839 that heating natural rubber with some sulfur converted the material from one that was tacky when warm and brittle when cold into a "vulcanized rubber" that was conveniently useful over a wide temperature range. Crosslinking of the macromolecules of rubber with sulfur bonds endowed the naturally occurring material with some elastic memory and caused it to behave as we have come to expect elastomers to behave. Excessive sulfur crosslinking converts the stretchable, compressible, bouncy rubber into hard rubber such as the material found in the heads of mallets used in machine shops to pound sheet metal into desired shapes. A small dose of crosslinking prevents the macromolecules of natural rubber to crystallize at low temperatures and turn into a brittle solid and to become a tacky, sticky semifluid at elevated temperatures.

Thermoplastic elastomers constitute a relatively recent class of elastomers that offer several practical advantages when compared to the traditional, thermoset materials:

• Ease of processing because injection and extrusion molding is more efficient than curing of materials confined to a mold
• Facile recycling of scrap and recovered TPE materials
• Greater opportunity for molecular engineering, that is, designing of materials for specific applications

Historically speaking, the first TPEs belong to the PUR family, but since 1950 they were joined by block polymers assembled from hard blocks of polymeric styrene that are attached to soft blocks of polymeric ethylene and butylene, butadiene, or isoprene monomers, and in the 1960s there appeared hard blocks of polymeric aromatic polyesters joined to soft blocks of oligomeric or polymeric ethers. Still later, hard polyamide blocks derived from aromatic dicarboxylic acids were combined with soft blocks of aliphatic components and the world of TPEs continues to expand.

In the case of PUR block polymers, for example, the hard segments may consist, of a combination of aromatic diisocyanates and soft chain diols, and more linear MDI is preferred over a less linear TDI. Such a hard segment may be assembled by allowing a short-chain diol of relatively inhibited freedom of rotation to react with an excess of MDI to give first an isocyanate-terminated prepolymer PUR segment. This can then be treated with a small amount of a chain extender such as ethylenediamine, for example, to lengthen the hard segment while still retaining the isocyanate-terminated nature of the prepolymer. Finally, the latter is allowed to react with a relatively long-chain (molecular weight = 10^2–10^3) of a highly flexible diol, and thus one would obtain a block polymer of structure:

Soft polyol segment–Hard PUR segment–Soft polyol segment

The elastic textile fiber Spandex (Du Pont's Lycra) is an excellent example of a successful TPE material (U.S. Patent 2,692,873 issued in 1954, but not commercialized by Du Pont until 1962). Another example is the development of Du Pont's Adiprene PU rubber, which is formed by the reaction of 3 parts of PTMEG (an oligomer of 9–11 tetramethylene ether units) with 4.3 parts of TDI to give an isocyanate-terminated prepolymer that is subsequently treated with 0.3 part of TDI and some water to form an urea-linked end product.

A water-retaining hydrogel (W. R. Grace's Hypol) illustrates the many possibilities of molecular engineering within the realm of PURs. In this case, a crosslinked polyol is treated with TDI, for example, to give a hydrophilic semisolid that can be employed as a carrier of medicinals for the treatment of surface wounds, fragrances in personal-care products, and so on.

As for controlling the degree of crosslinking and design of new polymers with exact properties, a new generation of catalysts called "metallocenes" must be mentioned. Metallocene catalysts were discovered in the 1980s and metallocene-based polymers ranging from crystalline to elastomeric materials have been available commercially since 1991. The catalyst consists of a transition-metal atom sandwiched between ring structures to form sterically-hindered sites. Transition metals such as titanium and zirconium have been found very effective. Stereoselective catalytic sites can polymerize almost any monomeric system, beyond the traditional C_3 to C_8 α-olefins, in an exact manner [142]. Application areas of metallocenes are not obviously not limited to polyolefins and significant enhancements in polymerization fields are expected.

Catalyst Selection. The choice of catalysts determines not only the rate of PUR formation by means of the fundamental reaction between isocyanates and their partner reactants, but also the contribution of some of the more "exotic" reactions of the isocyanates. The rate of PUR formation is, of course, of critical importance to the reaction-injection molding and reinforced-reaction-injection molding technology. Among the exotic reactions of the isocyanates the following may be worthy of note.

Formation of Isocyanurates by Trimerization. If we keep in mind that when each R group contains at least one isocyanate function, it is clear that even a small degree of isocyanurate formation creates rigid centers that bind two or more PUR chains and create a focus of mechanical and thermal strength. A high degree of isocyanurate formation leads to polyisocyanurates (PIR), which can be manufactured to give flame-retardant, resilient, rigid materials.

Formation and Some Reactions of Carbodiimides.

$$2R–N{=}C{=}O \dashrightarrow R–N{=}C{=}N–R + CO_2$$
$$\text{carbodiimide}$$

The carbodiimide can react with another molecule of an isocyanate; the resulting highly strained four-membered ring intermediate undergoes further reactions that are not detailed here. Suffice it to say that these reactions lead to an oligomerization of isocyanates.

An interesting and practical illustration of this behavior is the case of MDI that starts out by being a solid (melting point 37°C) and thus is not easy to handle on a large, industrial scale. However, a small degree of carbodiimide formation depresses the melting point to give the much more convenient "liquid" MDI.

Formation of Substituted Biuret Structures. By analogy with the well-known formation of biuret by heating urea:

$$2H_2N–CO–NH_2 \dashrightarrow H_2N–CO–NH–CO–NH_2 + NH_3$$

The analogous biuret structure can be produced when moisture decomposes some isocyanate to the corresponding amine and the latter adds to two intact isocyanates:

$$R–NH_2 + 2O{=}C{=}N–R \dashrightarrow R–NH–CO–NR–CO–NH–R$$

It is clear that a biuret moiety offers greater hydrogen-bonding opportunities than the urea or carbamate moieties.

Formation of Allophanates. The reaction of the N-H moiety of one urethane function with another molecule of isocyanate gives rise to an allophanate:

$$R–NH–CO–O–R' + O{=}C{=}N–R'' \dashrightarrow R–N(CO–NH–R'')–CO–O–R'$$
$$\text{allophanate}$$

The allophanates represent a hybrid of the urea and urethane structures, and their hydrogen-bonding capability lies between those of the parents.

There are numerous catalysts employed by the PUR industry and in all practical situations they function in a poorly understood manner. Consequently, some of the more common catalysts will be listed without a pretense to clarify the mechanisms that explain their involvement.

Among the organometallic catalysts the most prominent are di-n-butyltin diacetate and dilaurate:

$$n-Bu_2Sn(O \cdot CO-R)_2$$

where R is CH_3 or n-$C_{11}H_{23}$, respectively. Numerous tertiary amines, substituted amino alcohols, and quaternary ammonium salts (with appropriate anions since these catalysts function as dipolar ion pairs (or zwitterions) have a selective catalytic effect on some of the reactions of isocyanates [85,88]. For example, N'-hydroxyethyl-N,N,N'-trimethylenediamine [Me_2N-$(CH_2)_3$-$N(CH_3)$-CH_2-CH_2-OH] promotes the cyclotrimerization of an isocyanate to the corresponding isocyanurate, while the simple amines—2–dimethylaminoethanol, dimethylaminoethanolamine, and triethylenediamine (the original Air Product's DABCO)—catalyze the formation of urethanes more so than cyclotrimerization. The latter is believed to require a tertiary ethylenediamine system with a primary hydroxyl group.

In general, the order of reactivity of active hydrogen compounds with isocyanates in uncatalyzed systems is as follows (with the most active one first): [10]

aliphatic amines > aromatic amines > primary alcohols > water > secondary alcohols > tertiary alcohols > phenols > carboxylic acids > ureas > amides > urethanes.

The self-addition reactions of isocyanates do not usually proceed as readily as reactions with active hydrogen compounds [10].

The catalytic activity of tertiary amines in the phenyl isocyanate-butanol reaction falls off as the size of the substituent groups in tertiary amines increases [140]. The base strengths (pKa) of trimethylamine, ethyldimethylamine, diethylmethylamine, triethylamine, and triethylenediamine are 9.9, 10.2, 10.4, 10.8, and 8.2, respectively, whereas the relative catalytic activities of those are 2.2, 1.6, 1.0. 0.9, and 3.3, respectively [140].

Hostettler and Cox [141] reported relative reaction rates of phenyl isocyanate with n-butanol, water, and diphenyl urea by various catalysts. Taking the rate of phenyl isocyanate with n-butanol as 1, relative rates catalyzed by N-methylmorpholine (II), triethylamine, tetramethyl-1,3–butanediamine, triethylenediamine, tributyltin acetate, and dibutyltin diacetate are 40, 86, 260, 1200, 80,000, and 600,000, respectively. The relative rate with water when uncatalyzed is 1.1, while the relative rates catalyzed by N-methylmorpholine (II), triethylamine, tetramethyl-1,3–butanediamine, triethylenediamine, tributyltin acetate, and dibutyltin diacetate are 25, 47, 100, 380, 14,000, and 100,000, respectively. The uncatalyzed relative rate with diphenylurea is 2.2, while catalyzed rates on the relative scale are 10, 4, 12, 90, 80,000, 12,000 for the above catalysts, respectively [141].

Furthermore, tertiary amines like triethylenediamine (DABCO) are also effective catalysts for isocyanate self-addition reactions, whereas organometallic catalysts are generally ineffective and tin compounds are particularly poor catalysts for these reactions [10].

Analytical Method for Determination of Phosgene in Air

As mentioned earlier, phosgene is a highly toxic chemical intermediate used primarily for industrial production of isocyanates and polycarbonates [89]. Surreptitious exposure to phosgene may also occur through decomposition of vapors of chlorinated hydrocarbons under the influence of ultraviolet light. In 1983, it was estimated that 2,358 workers were exposed to phosgene in the United States. Exposure to concentration of 3–5 ppm causes irritation of the eyes and throat, while 25 ppm is dangerous for exposures of 30–60-min duration and may result in delayed onset of pulmonary edema. The current allowable exposure level in the United States is 0.1 ppm (0.4 mg/m^3) as an 8–hr time weighted average.

The sampling and analytical method for determining phosgene in air recommended by the National Institute for Occupational Safety and Health (NIOSH) employs a midget impinger containing a solution of 0.25% 4–(4'-nitrobenzyl)-pyridine and 0.5% N-phenylbenzylamine in diethylphthalate. Phosgene in air reacts with the solution to produce a brilliant red color which can be measured spectrophotometrically at 475 nm. This method was originally described by Lamouroux and later modified and improved by other investigators. However, the technique has the following deficiencies: 1) the color formed is unstable and a 10–15% decay is reported after 8 hr of storage; 2) water vapor interferes with the measurement with an 11% reduction in color intensity observed at 73% relative humidity, and 3) the use of impingers in the field has the potential of spillage of solution, breakage of glassware, and contamination with diethylphthalate.

Phosgene generally reacts with primary and secondary amines to form substituted urea derivatives. For example, aniline has been used to convert phosgene to N,N'-diphenylurea for quantitative measurement by either gravimetry or spectrophotometry. While modern chromatographic procedures could be utilized to overcome lack of specificity and sensitivity of these gravimetric or spectrophotometric analyses, the use of a primary amine like aniline can result in a mixture of polysubstituted urea compounds with phosgene, making the chromatographic approach less reliable.

Certain secondary amine compounds are commonly used for chemical derivation of isocyanates which, like phosgene, form substituted urea compounds amenable to chromatographic determination. One of the most common of these reagents is 1-(2-pyridyl)-piperazine (PYP); PYP reacts on a one-to-one basis with isocyanate compounds yielding a stable urea derivative which can be quantified specifically and sensitively with reversed phase, high performance liquid chromatography (HPLC) and ultraviolet absorbance detection.

The use of a single derivatizing agent for chromatographic determination of both isocyanates and phosgene would be of particular benefit since these materials are usually encountered together in isocyanate manufacturing facilities. Therefore, adapting the PYP method to measure phosgene in the air is desirable. PYP reacts with phosgene as follows:

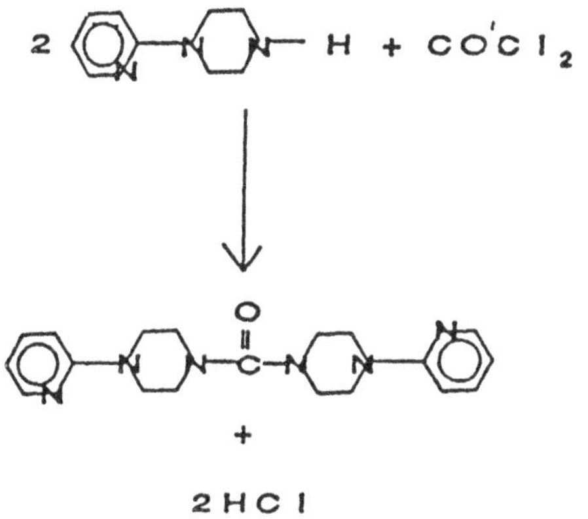

The PYP-urea derivative exhibits a unique chemical structure and is amenable to chromatographic separation and UV absorbance detection. Based on this, the PYP sampling technique has been modified for collection of phosgene on a coated solid sorbent. Phosgene is converted to the PYP urea derivative on the sorbent; subsequently solvent is desorbed and determined by HPLC.

In addition, the use of triphosgene (bis-(trichloromethyl)-carbonate) as a surrogate standard for phosgene was explored as a suitable safe alternative to handling of phosgene gas. Triphosgene has been successfully used as a substitute for phosgene in various organic syntheses. In general, reactions require a one-third equivalent of triphosgene in comparison to phosgene. Reaction products of triphosgene with compounds containing labile hydrogen are usually identical to those obtained from phosgene. The great advantage of triphosgene is that the material is a stable, crystalline solid (melting point 81–83°C), making it safer and more convenient to handle than phosgene. However, it should be noted that under certain conditions, triphosgene has been noted to release significant amounts of phosgene.

Table 3.24   Human Effects of Phosgene Exposure [90]

Doctors	Exposure variables	Exposure	Effects
Theiss &	a) Unknown	Brief	Pulmonary edema
Goldmann	b) 1 mole of phosgene	Brief	Pulmonary edema
	c) Unknown	30 min	Pulmonary edema, death
Gerritsen &	a) Unknown	Indefinite	Pulmonary edema,
Buschmann	b) Unknown	3 hr	Pulmonary edema
Spolyar et al.	Unknown (15 ppm)	< 3.5 hr	Pulmonary edema, death
Delepine	Unknown	Brief	Bronchial irritation, death
English	Unknown	8 hr	Bronchitis, reactivation of a duodenal ulcer
Steel	Unknown	Brief	Acute bronchitis and delirium

Source: [90]

Phosgene Inhalation Exposures And Human Health Effects

Phosgene is extremely toxic and its inhalation may have fatally dangerous effects on animal and human health, if the exposure time is longer than brief. However, such health effects differ from person to person and also depend on the conditions of exposure. Table 3.24 shows the human health effects of phosgene exposure [90].

B. Diisocyanates

Introduction

Organic isocynate [8,85,91] compounds have been known for a long time, but first became commercially interesting in the last decades based on the development work by Bayer. The reaction of di- and polyisocyanates with di- and polyols forms polyurethanes (PUR) with many uses. The preferred use of PUR in the automobile industry, in construction, and in refrigeration technology led to a considerable increase in the production capacity for feedstock diisocyanates.

Toluene diisocyanate (TDI), in the form of its 2,4- and 2,6-isomers, is the most significant diisocyanate. Total TDI production in the United States, Western Europe, and Japan is given in Table 3.25.

In the last few years, however, 4,4'-diphenylmethane diisocyanate (methane diphenyldiisocyanate, MDI), whose precursor 4,4'-diaminodiphenylmethane is obtained from the condensation of aniline with formaldehyde, has overtaken TDI. In 1990, capacities in Western Europe, the United States, and Japan for MDI were 600,000, 530,000, and 190,000, tons, respectively; for TDI, they were 400,000, 360,000, and 110,000 tons [91].

Table 3.25 TDI and MDI Production Capacities [91]

Year	TDI production (1000 ton)			MDI production (1000 ton)		
	1985	1987	1989	1985	1987	1989
United States	294	368	333	347	379	436
Western Europe	286	316	331	358	410	510
Japan	84	91	109	94	129	165

Another important component of polyurethanes is hexamethylene-1,6–diisocyanate (HDI, formerly HMDI) whose precursor hexamethylenediamine and its manufacture will not be discussed here.

Manufacturing Diisocyanate

Toluene diisocyanate is generally manufactured in a continuous process involving three steps [91]:

1. Nitration of toluene to dinitrotoluene
2. Hydrogenation of dinitrotoluene to toluenediamine
3. Phosgenation to toluene diisocyanate

Nitration of Toluene to Dinitrotoluene. The continuous nitration of toluene can be done under milder conditions than are necessary for benzene due to the activating effect of the methyl group. For example, the H_2O content of the nitrating acid can be as high as 23%, compared to 10% for the nitration of benzene. The mixture of mononitrated products of toluene consists of the three isomers *o*-, *p*- and *m*-nitrotoluene, whose distribution is influenced only slightly by reaction conditions. A typical composition is 63% *o*-, 33–34% *p*- and 4% *m*-nitrotoluene. The mixture can be separated by distillation or crystallization.

Nitrotoluenes are intermediates for dyes, pharmaceuticals, and perfumes, and precursors for the explosive 2,4,6–trinitrotoluene (TNT). A mixture of *o*- and *p*-nitrotoluene can be nitrated to dinitrotoluenes, the feedstocks for the manufacture of diisocyanates. The isomeric 2,4– and 2,6–dinitrotoluenes are obtained in a ratio of roughly 80:20

Hydrogenation of Dinitrotoluene to Toluenediamine. The hydrogenation of the dinitrotoluene mixture to toluenediamines is once again a standard process in aromatic synthesis. This reaction can be carried out with iron and aqueous hydrochloric acid like the reduction of nitrobenzene, but catalytic hydrogenation is preferred (e.g., in methanol with a Raney nickel catalyst at about 100°C and over 50 bars, or with palladium catalysts).

The dinitrotoluenes are reduced quantitatively in a succession of high pressure hydrogenations. The selectivity to the toluenediamines is 98–99%. In contrast to the manufacture of aniline from nitrobenzene, gas-phase hydrogenation is not used commercially due to the readily explosive decomposition of the dinitrotoluenes at the required reaction temperatures. Purification is done in a series of distillation columns.

Phosgenation to Toluene Diisocyanate. Phosgenation of the toulenediamines can be carried out in several ways. Base phosgenation (i.e., the reaction of the free primary amine with phosgene) is the most important. In the first step, the amine and phosgene are reacted at 0–50°C in a solvent such as *o*-dicholorobenzene to give a mixture of carbamyl chlorides and amine hydrochlorides. The reaction product is fed into the hot phosgenation tower where, at 170–185°C, it is reacted further with phosgene to form the diisocyanates:

The excess phosgene can be separated from HCl in a deep-freezing device and recycled to the process.

The phosgenation of the toluenediamine hydrochlorides is the second possible manufacturing process for the diisocyanates. In the Mitsubishi Chemical process, for example, the toulenediamines are dissolved in o-dichlorobenzene and converted into a salt suspension by injecting dry HCl. Phosgene is reacted with the hydrochlorides at elevated temperatures and with strong agitation to give the diisocyanates. The HCl which evolves is removed with an inert gas stream:

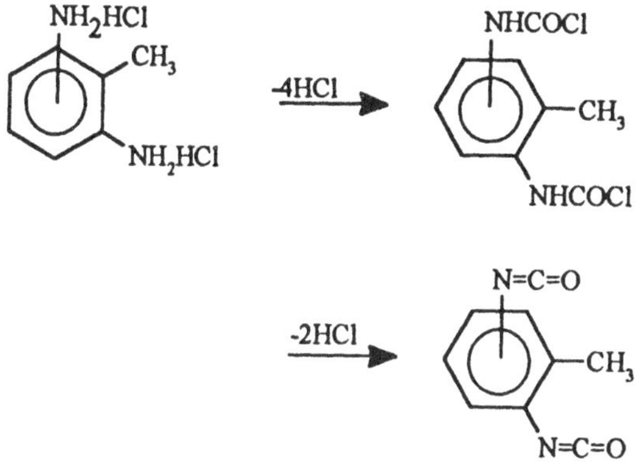

The purification is done by fractional distillation. The selectivity to toluene diisocyanates is 97% (based on diamine). In the Mitsubishi process, the overall selectivity to diisocyanates is 81% (based on toluene). In addition to pure 2,4–toluene diisocyanates, two isomeric mixtures are available commercially, with ratios of 2,4– to 2,6–isomer of 80:20 and 65:35.

New Manufacturing Routes

Because of the increased commercial interest in diisocyanates [85,91], new manufacturing routes without the costly phosgenation step (i.e, without total loss of chlorine as HCl) have been developed. Processes for the catalytic carbonylation of aromatic nitro-compounds or amines are the most likely to become commercially important.

In a process developed by Atlantic Richfield Company (ARCO), the nitrobenzene feedstock for the manufacture of 4,4'-diphenylmethyl diisocyanate (MDI) is first reacted catalytically (e.g., SeO$_2$, KOAc) with CO in the presence of ethanol to give N-phenylethyl urethane. After condensation with formaldehyde, thermolysis at 250–285°C is used to cleave ethanol and form MDI:

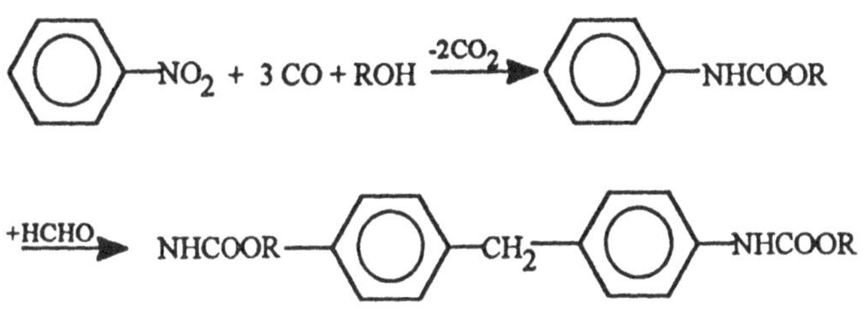

$$\xrightarrow{-2ROH} \quad O=C=N-\!\!\langle\bigcirc\rangle\!-CH_2-\!\!\langle\bigcirc\rangle\!-N=C=O$$

Mitsui Toatsu and Mitsubishi Chemical have formulated a similar carbonylation process for the conversion of dinitrotoluene to toluene diisocyanate (TDI) [91].

Another manufacturing process [91] for 4,4'-diphenylmethyl diisocyanate (MDI) was introduced by Asahi Chemical. In contrast to the ARCO route, aniline is used for the carbonylation to N-phenylethyl urethane; otherwise, the same steps are followed. The oxidative carbonylation of aniline is done in the presence of metallic palladium and an alkali iodide promoter at 150–180°C and 50–80 bar. The selectivity is more than 95% with a 95% aniline conversion:

$$\langle\bigcirc\rangle\!-NH_2 + CO + 0.5\,O_2 + C_2H_5OH$$

$$\xrightarrow{Pd} \langle\bigcirc\rangle\!-NHCOOC_2H_5 + H_2O$$

Pd function in aniline/CO/O$_2$ reaction:

$$C_6H_5NH-Pd-H + CO \longrightarrow C_6H_5NH-CO-Pd-H$$

$$+ ROH \longrightarrow C_6H_5NHCOOR + H-Pd-H$$

$$+ O_2 \longrightarrow Pd + H_2O$$

The final condensation with formaldehyde at 60–90°C and atmospheric pressure takes place in the presence of H_2SO_4 in two phases and then, after removal of the water phase, an additional treatment with, for example, trifluoroacetic acid in a homogeneous phase; this has over 95% selectivity to the diurethane at a urethane conversion of about 40%. The last step, the thermal elimination of ethanol, is done at 230–280°C and 10–30 bar in a solvent; the selectivity to MDI is over 93%. The Asahi process has not been used commercially [91].

Another way of avoiding less economical and toxic phosgene involves the use of dimethyl carbonate for the production of isocyanates and polycarbonates. In Japan, Ube have operated the first pilot plant for the highly selective gas-phase carbonylation of methanol. Further liquid-phase processes for the production of dimethyl carbonate are in operation at Daicel and Mitsui Sekka.

C.   Uses of Diisocyanates

The diisocyanates are used mainly in the manufacture of polyurethanes (PUR). These are produced by polyaddition of diisocyanates and dihydric alcohols, in particular the polyether alcohols (i.e., polyethylene glycols, polypropylene glycols, and the reaction products of propylene oxide with polyhydric alcohols). In addition, oligomeric esters from dicarboxylic acids and diols (polyester alcohols) are also used [91]:

$$nO=C=N-R'-N=C=O + (n + 1)HO-R''-OH \longrightarrow$$
$$H-OR''-O-CO-NH-R'-NH-CO-)_n-OR''-OH$$

Polyurethane can be crosslinked by adding tri- or polyhydric alcohols (e.g., glycerol, trimethylolpropane), and caused to foam by adding a small amount of water, which causes saponification of the isocyanate group to the amino group and CO_2. Polyurethanes are processed to flexible and rigid foams; they are also used in textile coatings and as elastomers (e.g., Spandex fibers). Other products using PUR are, for example, artificial leather, synthetic rubber, dyes, paints, and adhesives. However, the greatest use of PURs is for foams (Table 3.26) [91].

Environmental: Workplace Air Concentrations

Exposure to diisocyanates shall be controlled so that no employee is exposed at concentrations greater than the limits specified in Table 3.27 [92]. These limits expressed in mg/m^3 are equivalent to a vapor concentration of 5 ppb as a TWA time weighted average concentration for up to a 10–hour workshift, 40–hr workweek, and 20 ppb as a ceiling concentration for any 10–min

Table 3.26   Polyurethane Use in the USA, Western Europe, and Japan (wt%)

	USA	Western Europe		Japan
	1990	1987	1990	1990
Flexible	52	41	43	32
Rigid foams	27	25	24	17
Others:	21	34	33	51
Integral-skin and filling foams				
Paint raw materials				
Elastomers				
Thermoplastic polyurethanes				
Artificial leathers				
Total use ($\times 10^6$ ton)	1.54	1.35	1.40	0.60

Table 3.27 Workplace Air Concentrations

	TWA (mg/m^3)	Ceiling (mg/m^3)
Toluene diisocyanate (TDI)	35	140
Diphenylmethane diisocyanate (MDI)	50	200
Hexamethylene diisocyanate (HDI)	35	140
Napthalene diisocyanate (NDI)	40	170
Isophorone diisocyanate (IPDI)	45	180
Dicyclohexylmethane 4,4'-diisocyanate (hydrogenated MDI)	55	210

If other diisocyanates are used, employers should observe environmental and safety limits equivalent to a ceiling concentration of 20 ppb and a TWA concentration of 5 ppb.

sampling period. The mg equivalents for selected diisocyanates are given in Table 3.27.

D. Polycarbonates

Introduction

The characteristic properties of polycarbonate (PC) include good transparency associated with great toughness, heat resistance up to about 150°C with standard polycarbonate, high strength, good aging resistance, and high electrical resistance. Even at the beginning of 1990s these properties have ensured continuous growth for PC in world market [8,85,93–95]. The average annual growth in worldwide consumption in the past three to four years has been about 4%, although in the United States and European markets it has significantly weakened; over the past three years in Europe, for example, it has sunk to a growth rate of about 1% per year. Estimated worldwide consumption of PC, including that used for blending, was 660,000 tons in 1992.

While at the end of the 1980s there was still a shortage of PC resulting from a previous period of frenetic growth, the picture at the beginning of the 1990s was totally changed. The preliminary decisions of producers for removing bottlenecks from production lines and the almost simultaneous construction of new PC production lines resulted in a level of plant loading of some 75% in 1992. A world-wide capacity of 830,000 tons in 1992 now contrasts with a consumption of only 660,000 tons. Simultaneously there was an erosion of the market price of PC in industrial region, especially in Europe and Japan. This, in combination with the high cost of investment in PC plant and that for bisphenol A supply, as well as a flatter growth curve, clearly demonstrates the increased risks of reinvestment.

In 1993 a further 50,000 tons was added to Japan's PC capacity. Some 70% of world capacity is provided by the two largest PC producers, GE Plastics and Bayer-Miles. The producers in the United States are GE Plastics, Bayer-Miles, and Dow; in Europe they are Bayer, GE-Plastics, Dow, and Enichem, and in Japan Teijin, Mitsubishi Gas Chemical, Mitsubushi Kasei, Idemitsu, and GE-Plastics/Mitsui Petrochemical. Smaller quantities of PC come from Sam Yang Kasei/Korea, Polycarbonatas do Brasil, and Shanghai Zhang Lian Chemical (PRC). From the consumption and regional capacity figures, it can be deduced that there is a significant PC over-capacity in the industrial regions—United States, Europe, and Japan—which has been relieved somewhat by the demand and powerful growth in the large Southeast Asia region [94].

Polycarbonate Manufacturing

The production of the most prominent polycarbonate polymer based on bisphenol A is the second largest consumer of phosgene, while poly(alkylene carbonates) are obtained from epoxides and CO_2.

The production of the most common "polycarbonate resin" pellets ($1.93) involves the reaction of bisphenol A (BPA) (polycarbonate grade $0.86, epoxy grade $0.82, and phosgene $0.55) [106]. The reaction is carried out in a convenient solvent such as methylene chloride, and this process offers an excellent illustration of phase-transfer catalysis [85,94].

Phase-Transfer Catalysis. Since the efficiency of the reaction requires that bisphenol A be used in the form of water-soluble phenolate anions, while phosgene must be dissolved in a chemically inert and hence water-insoluble solvent, the desired reaction would have to depend on the diffusion rate of the two reactants to the interface between the immiscible solvents. The area of the interface can be increased by vigorous stirring of the two-phase reaction mixture, but a more efficient way to accelerate the process is to induce one of the reactants to migrate into a phase that is not particularly receptive to it. In this example, the sodium phenolate ions are in equilibrium with a phase-transfer catalyst such as tetra-*n*-butylammonium chloride, and while one of the products of the equilibrium (sodium chloride) remains in the aqueous phase, the other products of the equilibrium (the BPA anion-tetra-*n*-butylammonium cation ion pairs) are of sufficiently covalent character to migrate into the nonaqueous phase where they encounter phosgene and the reaction takes place.

The disadvantage of the preceding process is the insolubility of the polymer as its molecular weight increases [85,94]. An alternative process that avoids this limitation involves the reaction of molten BPA (melting point 155–157°C), with a convenient simple organic carbonate that produces a volatile byproduct. Diphenyl carbonate has been employed for that purpose because phenol is sufficiently volatile (boiling point 182°C at atmospheric pressure) and provides a good leaving group. Another advantage of this process is the absence of traces of HCl from the end product. Diphenyl carbonate can be prepared from phenol by means of phosgene or by transesterification with dimethyl carbonate; similarly, one can use di-n-butyl carbonate. The use of organic carbonates in place of phosgene allows a better control of the molecular weight distribution of the resulting PC.

Polycarbonate is an engineering-quality thermoplastic and forms useful blends with PBT, ABS, PS, and other polymers. Its structural uses include the manufacture of components of typewriters and other office equipment, including computers and scratch-resistant eyeglasses (a two billion dollar market). Until recently, a special high-quality PC required for the manufacture of compact discs and other electronic and telecommunication devices had to be imported from Bayer in West Germany. Mobay is now manufacturing such material in the United States. According to an announcement by Mobay in 1987, the coextrusion of PC-PVDC films is now possible to give excellent moisture barrier properties in plastic films suitable for packaging moisture-sensitive products.

PC Production No Longer Dependent on Phosgene?

Various processes are technically available for PC manufacture, but the one used almost exclusively at present is *phase boundary interfacial polymerization* [94]. With this method, PC can be made very economically from phosgene and bisphenol A, and its properties profile can be varied widely. Thus, molecular weight, structural uniformity and the PC structure itself can be modified and tailored to the needs of the application and the processing method.

Many new products made from monomers of the bisphenol type were studied during the development of PC [106–109], but were rated unsuitable on grounds of transparency, toughness, melt flow, hydrolytic stability, and production cost. With the discovery of trimethyl-cyclohexanone-bisphenol-polycarbonate (TMC-BP-PC) and the possibility of blending it with BPA-PC, it became possible to develop and introduce to the market a new spectrum of PCs, with heat-deflection temperature range of 160–205°C, depending on the bisphenol and TMC content. These high-performance PCs exhibit hitherto unattained combinations of transparency, impact strength, good flow properties, and high-heat resistance.

Catalyzed esterification, in the melt of diphenyl carbonate with bis-phenol-A, liberates phenol and is a well-known process. Developed by Bayer, this process was used from 1958 to the beginning of the 1970s for the large-scale production of PC, and also, under license, by Mitsubishi Gas Chemical and Teijin in Japan. The product did not achieve the recognized quality standards of today for thermal stability and optical properties.

Further development of the technology of this process, which also involved the manufacture of diphenyl carbonate by an alternative process, led in 1993 to the construction of a new, large scale production (25,000 ton/yr) in Japan by GE Plastics/Mitsui Petrochemical. The manufacturing sequence is:

$$\text{methanol} + CO/O_2 \longrightarrow \text{dimethyl carbonate} \longrightarrow \text{diphenyl carbonate} \longrightarrow \text{PC}$$

and thus uses no phosgene to make the polymer. The properties of the PC made by this process are reputed to be good.

Product Development and Applications

Growth rates in the various market sectors and the changing boundary conditions, such as more economical manufacture and higher quality, affect product development and applications to a great extent. The principal area of application for PC is still the electrical sector in which it has replaced classical materials like ceramics, metals, and thermosets, because of an advantageous combination of properties such as electrical insulation, impact strength, stiffness, heat resistance, and transparency. This process is essentially complete even if isolated substitution events take place [94].

If the applications of PC in Europe are analyzed by market sector, then those in the electrical-engineering sector in 1992 had still only achieved a 32% share of the total. In the light-housings/covers area the development of easy-flow grades triggered a substitution move. The required combination of good flow and good mechanical properties can be achieved by various technologies. Incorporation of aliphatic "soft block" segments results in co(polyestercarbonates) with considerably higher flow ability than standard PC of the same molecular weight. Conversely, a PC of significantly higher molecular weight with the same flow properties can be used; this provides a bigger reserve of impact strength. Typical applications are thin-wall articles with large flow-lengths which, because processing conditions are favorable, can be molded economically.

Little Movement in Sheet Market. In 1992, the European market for sheet PC was about 43,000 tons. In view of the current economic climate for construction, no increase in the consumption of PC in the sheet market is to be seen, though in the long-term additional low-level growth is predicted. Higher growth in thin-walled sheet during 1992 further widened the lead over solid sheet. Corrugated PC sheets (0.8–1.0 mm) recently came on to the European market; up to now, this has been the market for PVC and glass-reinforced unsaturated polyester (UP-GF). In the United States and Asia, PC corrugated sheets have already taken a considerable share of the market [94].

The single-layer coextrusion process based on the use of low-volatility UV absorbers has been adopted for thin-walled sheet and profile manufacture, both in Europe and worldwide. Coextrusion coating using PC is important also from the point of view of recycling, because single-polymer systems increasingly are being required.

CD Market Still Dynamic. The compact disc market continued its dynamic development in 1993 and, over the next few years, will continue to grow considerably faster than the PC market in general. The use of PC for CD-related applications in Europe during 1992 was some 16,000 tons. At the end of the 1990s, when CD substitution for old-style records will be completed, growth in this application will fall back to the general rate for the music sector. Growth rates depend on further developments in the various CD-technologies (CD-ROM, WORM, CD-photo, CD-interactive, CD-video, mini-disk, and so on) that are expected in future. Demands for better quality (purity), combined with optimized processability will set the course of PC product development for this market [94].

PC Returnable Bottle Coming Soon. The market for bottles, and especially for PC water bottles, is among the largest growth areas. This market was first developed in the United States for water bottles (five gallon), and these are now increasingly being used in Asia. In Asia, quality water is generally available; however, in the absence of drinking water, there is a need for very light, unbreakable containers for transporting drinking water over long distances. Polycarbonate offers the right combination of nontoxicity, toughness, transparency, and stiffness. The world market in 1995/96 for this application is estimated to be around 20,000 ton/yr.

The returnable PC bottle is ready for a large-scale launch onto the market. Extensive trials on the replacement of glass by PC milk bottles are in progress in every European country, so a marketing decision is to be expected during 1994.

An increasing demand for PC is expected in the medical field. Radiation is being used increasingly alongside steam sterilization for sterilizing PC parts, particularly in Japan and in the United States. This method uses hard radiation, so new stabilizers had to be found for PC to ensure adequate color stability. Suitable grades of PC have been introduced [94].

New Applications Accessible to TMC-BP-Co-PC

Growth in the use of transparent thermoplastics with continuous-use temperatures well above those for standard polycarbonates is being seen in many areas. The development of trimethyl-cyclohexanone-bisphenol-polycarbonate (TMC-BP-PC) provided, for the first time, the combined benefits of good heat resistance, toughness, transparency, light stability, and flow ability. As their melt viscosities are considerably lower, these Co-PCs are more easily processable than aromatic polyester carbonates with a similar heat deflection temperature. It is clear from Table 3.28 that the processability of TMC-PC copolycarbonates and their corresponding homo-PCs is better than that of other amorphous thermoplastics. Depending on the composition, stepwise adjustments to the heat resistance for a given flow ability and impact strength, can be made by changes in composition. These mixtures have outstanding light transmission and are therefore particularly suitable for applications de-

Table 3.28 Comparision of bisphenol-TMC polycarbonate, bisphenol-TMC copolycarbonates and other amorphous thermoplastics [94]

Type of thermoplastics	Share TMC-BP (mol%)	T_g °C	Impact strength[a] (kJ/m^2)	Melt viscos-ity[b] (Pa)	UV resist-ance[c]
Polycarbonate (from 100 mol% TMC-BP)	100	239	8 n.b. 2 150	400	+
Polyetherimide	0	219	110	360	–
Polycarbonate from 45 mol% BPA/55 mol-% TMC-BP	55	205	n.b.	120	
Polyarylate (from 1/1 BPA/isoterephthalate)	0	190	n.b.	300	–
Polysulfone (from BPA and 4,4-dichlorphenylsulfone)	0	189	n.b.	270	–

[a] measured according to ISO 180: n.b. = no break.
[b] measured at 395°C and shear rate 10 s^{-1}.
[c] + = good, – = poor.
T_g: glass transition temperature.

manding transparency. Their 90% light transmission at 1 mm thickness approximates that of standard PC.

Application areas for the new Co-PCs include electrical engineering/electronics, domestic appliances, lighting, automotive engineering, and medical technology. In automotive engineering the headlamp lens will be offered by headlamp manufacturers as the plastics alternative to main headlamps and fog lamps. The best compromise on optical data, mechanical strength, and heat deflection resistance for their application is shown by PC.

The properties and application potential of aromatic PCs are still far from exhausted. By control of molecular weight, modification of the polymer chain, optimization of purity, and development of new formulations, it will be possible to achieve variable use in existing markets or in newly developed segments of the market. Thus, PC is one of the most important polymeric construction materials in the plastics industry, with solid growth potential for the long future.

Outlook of Polycarbonate

Polycarbonate is the toughest transparent material available. Its resistance to impact is remarkable, so much so that some jurisdictions require express notification of its use in glazing or removable panels because firemen and rescue squads might have difficulty in breaking through in the event of an emergency. Polycarbonate sheet is made by extrusion process, and maximum sheet thickness is limited to 0.5 in. However, heavier thicknesses, principally used for protective glazing applications, are produced by lamination of several thicknesses. The sheet is supplied either flat or on reels (in thickness up to 0.250 in.). Maximum width is 8 ft [95].

Colorless clear polycarbonate has a distinct bluish tint which is added to mask its natural straw color. Several medium-density tints are available as well as a number of translucent colors. Over the years the weatherability of polycarbonate has been the subject of much discussion and conflicting claims. Efforts are continuously being made to improve or extend its retention of appearance after prolonged exposure outdoors. The use of coatings holds some promise in this regard. Nonetheless, polycarbonate remains the material of choice when impact resistance is required, for example, to withstand attack by vandals or for safety guards and other industrial plant applications that must meet OSHA requirements [86,95].

Abrasion-Resistant Polycarbonate Sheet. Abrasion-resistant polycarbonate sheet retains the extremely high impact properties and also provides resistance to abrasion. The surface coating also gives some added protection against weathering.

Weather-Resistant-Grade Polycarbonate. Weather-resistance-grade poly-carbonate is material with a surface coating which, it is claimed, provides improved resistance against weathering.

Bullet-Resistant Polycarbonate. Bullet-resistant polycarbonate is listed by UL as resistant to super-power small arms in thickness of 1 in. or greater. In 0.125 in. thickness, polycarbonate is rated as burglar-resistant.

Double-Skinned Sheet. Double-skinned or hollow-core polycarbonate sheet is available in two thicknesses, 0.220 in. and 0.625 in., in widths of 36 and 48 in. for thin material and 47.25 in. wide for thick material. Standard lengths range from 8 to 16 ft. Double-skinned sheet that is 0.220 in. thick is used primarily as a thermal barrier when light transmission is also required. The thicker sheet has more substantial stiffness and is used as primary glazing.

The principal reasons for considering plastics such as polycarbonate in architectural applications are: transparency, formability, shatter resistance, and low weight. Obviously, if transparency alone were required, glass would be the prime candidate for the application in question. But any combination of requirements, along with transparency, shifts the emphasis to a plastic material. Sometimes plastics are the material of choice even when light transmission is not involved, but transparency and translucency remain the strongest reasons for the use of polycarbonate in architecture. Being able to enclose or shield a space from the elements without sacrificing natural daylight (or controlling the light or transparency) is a primary incentive to use plastic materials.

E.   Chlorinated Isocyanurics

Backgrounds

The starting point for the production of chlorinated isocyanurates is isocyanuric acid [8,85]. The most important member of this group is potassium dichloro-cyanurate, made by chlorinating isocyanuric acid and neutralizing with potassium hydroxide. Isocyanuric acid may be made from phosgene and ammonia or from urea [85].

$$COCl_2 + NH_3 \longrightarrow NH_2COCl + HCl$$

$$3NH_2COCl \longrightarrow 3HCl +$$

or $$3NH_2CONH_2 \longrightarrow 3NH_3 +$$

Isocyanuric acid

In the countries where chlorine is expensive, such as Japan, isocyanuric acid is made by pyrolysis of urea, yielding ammonia as a byproduct.

These products have a great advantage of being transportable in concentrated, solid form, instead of in solution as is the case with liquid chlorine bleaches. They are used in household bleach formulations, as swimming pool disinfectants, commercial bleaches and scouring compounds, and in dishwashing preparations.

Recent Developments

Polycyanurate resins are a success story that began by accident and suffered a bad stumble along the way [96]. But perseverance by chemists at several companies worldwide recovered the situation, and polycyanurates are emerging today as a broad family of commercial products.

Polycyanurates take their name from the propagation of their three-dimensional, densely crosslinked structures through three-way cyanuric acid (2,4,6–triazinetriol) linkages. In actual production, they are formed by trimerization of cyanate ester monomers (Table 3.29) of the form Ar-O-CN, where Ar is an aromatic group (Figure 3.40).

Today, the largest application of polycyanurates is in circuit boards, where they have begun to edge out epoxides. Their transparency to microwave and radar energy makes them useful for the conical radome nose cones that house radar antennas of military and weather reconnaissance planes. Microcrack resistance gives them a role in assembling communications satellites. Impact resistance fits them out for aircraft structures and engine pistons. They also find use in such friction materials as brake linings, grinding wheels, and in high-performance adhesives and coatings.

Difunctional cyanate ester monomers were discovered 30 years ago by Ernst Grigat, an organic chemist now retired from Bayer of Levetkusen, Germany, who was also the first to succeed in making organic cyanate esters at all. Bayer licensed dicyanate/polycyanurate technology to Mitsubishi Gas Chemical Co., Tokyo, and to the then Celanese Corp., New York City. Mitsubishi improved the resin by copolymerization of bisphenol A dicyanate with the bis-(maleimide) of 4,4'-methylenedianiline.

One routine achievement in industrial polycyanurate processing is to obtain conversion of 90–97% of the cyanate groups. In laboratories, conversions as

Table 3.29 Seven Cyanurate Monomers that are Commercially Available

Monomer structure and precursor	Trade name and supplier	T_g (C)[a]
 Bisphenol A	AroCy B Ciba-Geigy BT-2000 Mitsubishi Gas Chemical	289
 Tetramethylbisphenol F	AroCy M Ciba-Geigy	252
 Hexafluorobisphenol A	AroCy F Ciba-Geigy	270
 Bisphenol E	AroCy L-10 Ciba-Geigy	258
 Bisphenol M	RT-366 Ciba-Geigy	192
 Phenolic novolac resin	Primeset PT AlliedSignal REX-371 Ciba-Geigy	270– >350
 Dicyclopentadiene bisphenol	XU-71787 Dow Chemical	244

[a] Glass transition temperature of resin.
Source: Cieba-Geigy.

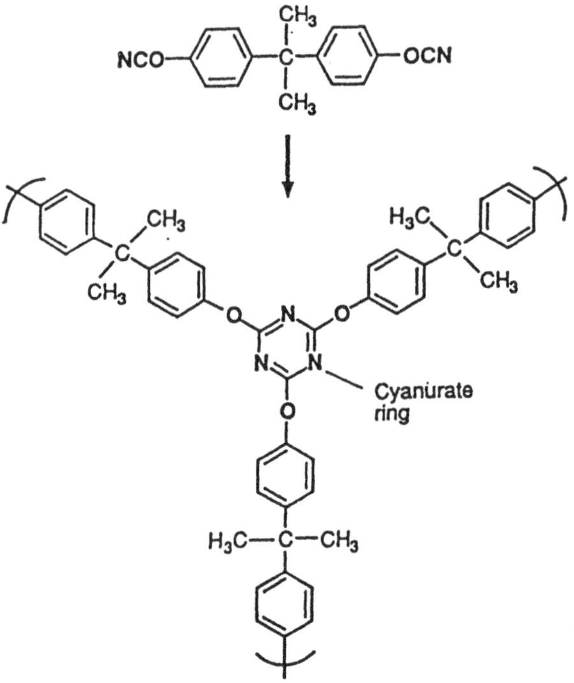

Figure 3.40 Cyanurate monomers polymerize with cyanurate ring formation.

high as 99% have been detected, and the limits of instruments may have missed some cases of 100% conversion. For more details, see [96].

F.   Herbicides and Pesticides

Herbicides

Chemicals used as herbicides vary widely in their properties [84–87,97–100,104,105]. Many of the early ones were byproducts of the chemical industry or compounds of very low value. Examples are arsenic trioxide (a smelter waste), iron sulfate (a byproduct of the steel industry), and waste oils of low value.

In contrast to these chemicals, the newer organic compounds that have come into use within the past two decades are higher cost materials, but they are also much more toxic. Examples are dalapon, which competes with pantoate and inhibits enzymatic synthesis of pantothenic acid (D,L-N-[2,4–dihydroxy-3,3–dimethylbutyryl]-β-alanine), and the substituted ureas, s-triazines

Table 3.30 A Chemical Classification of Herbicides [97]

Inorganic herbicides	
Acids	H_2SO_4, HCl, H_3PO_4
Salts	$CuSO_4$, $FeSO_4$, $Cu(NO_3)_2$, NaCl, KCl, $NaClO_3$, NH_4CNS, $Na_2B_4O_7 \cdot 10H_2O$, Na_2CrO_4, AMS
Organic herbicides	
Substituted phenols	DNC, DNBP, DNAP, PCP
Chlorophenoxy compounds	2,4-D, 2,4,5-T, MCPA, 3,4-DA, 4-CPA, 2-(2,4-DP), 2-(MCPP), 2-(2,4,5-TP) (silvex), 2-(3,4-DP), 2-(4-CPP), 4-(2,4-DB), 4-(MCPB), 4-(2,4,5-TB), 4-(3,4-DB), 4-(4-CPB), 2,4-DES (sesone), 2,4,5-TES, MCPES, 2,4-DEB
Chloro- and methyl-substituted acetic and propionic acids	TCA, 2,4-DPA (dalapon), 2,2,3-TPA, 2,3-DBA, erbon
Amide, thioamide, and amidine herbicides	
Amides	CDAA, CDEA, NPA, MH
Ureas	DCU, fenuron, monuron, diuron, neburon
Carbamates	IPC, CIPC, BCPC, CEPC, CPPC, barbane
Thiocarbamates	EPTC, Avadex
Dithiocarbamates	CDEC, SMDC (Vapam)
Triazoles	ATA (amitrol)
s-Triazines	Simazine, chlorazine, propazine, trietazine, ipazine, atrazine, simetone, prometone, atratone
Benzoic acids	2,3,6-TBA, 2,3,5,6-TBA, PBA, amiben, nitroben
Chlorinated benzenes	DCB; TCB
Miscellaneous herbicides	NPA, endothal, PMA, KOCN, HCA, IPX, OCH, EXD, MAA, EBEP, DIPA, DMA, AMS, fenac, diquat, acrolein

$(C_3H_3N_3)$, and carbamates, which inhibit the release of oxygen in the process of photosynthesis.

It is not necessary to understand the exact mode of action involved in the functioning of a given herbicide in order to use it intelligently. Far more important from the practical standpoint is the knowledge of the uptake and distribution of the chemical in the plant, in other words, the physiology of action. Table 3.30 presents a chemical classification that groups herbicides into groups of similar chemical properties. The scheme presented in Table 3.31 is based on the common methods of application and each category is related to a mechanism of action. Table 3.32 follows a classification of chemicals that also is divided broadly into inorganic and organic [98].

Table 3.31 A Classification of Herbicides [97]

Condition of application	Application to foliage		Application to soil
	Contact	Translocated	Residual
Selective herbicides			
Pre-planting	Na arsenite, H₂SO₄, DNC, DNBP, DNAP, PCP, medium or light weed oil	2,4-D, MCPA, dalapon, amitrol, amitrol-T, amiben, MH	IPC, CIPC, sesone, Sesin, Natrin, TCA, EPTC, CEPC, CPPC, Avadex, fenac, Mylone, CDEC, CDAA, amazine, endothal
Pre-emergence	H₂SO₄, DNC, DNBP, DNAP, PCP, light weed oil	2,4-D, MCPA, dalapon, amitrol, amitrol-T, amiben, MH, 2,4-DB	IPC, CIPC, CEPC, CPPC, EPTC, DNBP, DNAP, monuron, diuron, simazine, atrazine, CDEC, CDAA, Biuret, fenac, Mylone, Calcium cyanamid, SMDC, neburon, 2,4-D, Falone, endothal, NPA
Post-emergence	Water-soluble salts of DNC, DNAP, DNBP, Na monochloroacetate, H₂SO₄ PMAS, KOCN, endothal, MAA, Karsil, Dicryl, Solan	2,4-D, 2,4,5-T, MCPA, 4-CPA, 2-(2,4-DP), 2-(MCPP), silvex, 2-(4-CPP), 4-(2,4-DB), 4-(MCPB), 4-(2,4,5-TB), 4-(4-CPB), 2,3,6-TBA, Carbyne	IPC, CIPC, CEPC, CPPC, EPTC, sesone, Natrin, fenuron, monuron, diuron, neburon, simazine, atrazine, 2,3,6-TBA, pelleted IBC, CIPC, endothal, NPA, dalapon, Biuret, fenac, Falone
Post-emergence selective by directed spray	DNC, DNAP, DNBP, diquat, weed oil	2,4-D, MCPA, 4-(2,4-DB), 2-(2,4-DP), dalapon, amiben	

Turf weed control, spray or pellets	KOCN, Cacodylic acid, DAC-893	2,4-D, MCPA, 2,4,5-T, silvex	PMA, MAA, neburon, Zytron, Na arsenite, lead arsenate, chlordane, DMA, monuron, Ca arsenate
Nonselective herbicides Miscellaneous weeds (no crop)	Na arsenite, NaClO$_2$, DN and OCH in oils, aromatic oils	Acid-arsenical, NaClO$_2$, AMS, amitrol, dalapon, 2,4-D, erbon, amitrol-T, Garlon	CaCN$_2$CH$_2$Br, SMDC, EPTC, Mylone, CS$_2$, chloropicrin, borates, NaClO$_2$, As$_2$O$_3$, 2,3,6-TBA, pelleted fenuron, monuron, diruon, neburon, simazine, atrazine, erbon, TCA, Urox, Urab, CBMM, CBFM, CBDM, CBM, BMM, BDM, HCA
Chemical fallow	Na arsenite, weed oils, diquat, DNC, DNBP, DNAP	2,4-D, dalapon, amitrol	Monuron, diuron, simizine, erbon
Aquatic weed control in irrigation and drainage systems, lakes	Aromatic solvents, chlorinated benzene, acrolein, dichlone, CuSO$_4$, Na arsenite	Silvex, 2,4-D	Fenuron, monuron, diuron, neburon, chorinated benzene
Cut surface treatment on trees		AMS, 2,4-D amine formulation in cuts, 2,4,5-T, 2,4-D, and 2,4,5-T	

Table 3.32   Organic and Inorganic Herbicides [98]

A. *Inorganic Chemicals*[1]

Ammonium sulphamate (AMS)	$(NH_4)O.SO_2.NH_2$
Ammonium sulphate	$(NH_4)_2SO_4$
Ammonium thiocyanate	NH_4CNS
Calcium cyanamide	$CaCN_2$
Cupric sulphate	$CuSO_4$
Cupric nitrate	$Cu(NO_3)_2$
Ferrous sulphate	$FeSO_4$
Magnesium sulphate/potassium chloride (Kainit)	$MgSO_4/KCl$
Potassium cyanate	$KCNO$
Sodium arsenite	$Na_3AsO_3(+NaAsO_2+Na_4As_2O_5)$
Sodium tetraborate	$Na_2B_4O_7$
Sodium chlorate	$NaClO_3$
Sodium chloride	$NaCl$
Sodium nitrate	$NaNO_3$
Sulphuric acid	H_2SO_4

Table 3.32 (continued)

B. *Organic Chemicals*

1. *No nitrogen*

$$OCH_2COOH$$

PHENOXYACETIC ACIDS

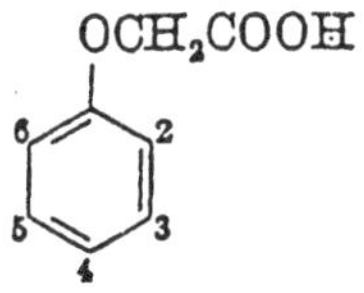

Common Name	Specific Groups
4 CPA*†	4-chlorophenoxyacetic acid
MCPA*†, MCP	4-chloro-2-methylphenoxyacetic acid
2,4-D*†	2,4-dichlorophenoxyacetic acid
3,4-DA	3,4-dichlorophenoxyacetic acid
2,4,5-T*†	2,4,5-trichlorophenoxyacetic acid

$$CH_3$$
$$O.CH.COOH$$

α-PHENOXYPROPIONIC ACIDS

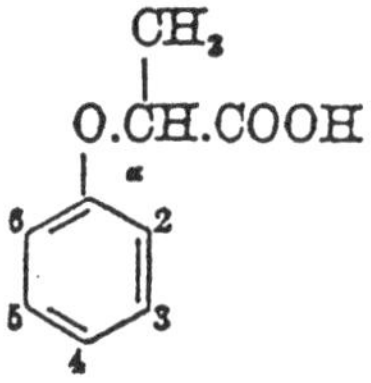

Common Name	Specific Groups
2-(4-CPP)	α-(4-chlorophenoxy)propionic acid
Dichlorprop*,2-(2,4-DP)	α-(2,4-dichlorophenoxy)propionic acid
2-(3,4-DP)	α-(3,4-dichlorophenoxy)propionic acid
Mecoprop*, 2-(MCPP)†	(±)-2-(4-chloro-2-methylphenoxy)propionic acid
Fenoprop*, silvex†, 2,4,5-TP	2-(2,4,5-trichlorophenoxy)propionic acid

[1] An asterisk (*) signifies a .common name approved by the British Standards Institution, while a dagger (†) signifies one approved by the Weed Society of America.

Table 3.32 (continued)

$$\overset{\gamma}{O}CH_2\overset{\beta}{C}H_2\overset{\alpha}{C}H_2COOH$$

γ-PHENOXYBUTYRIC ACIDS

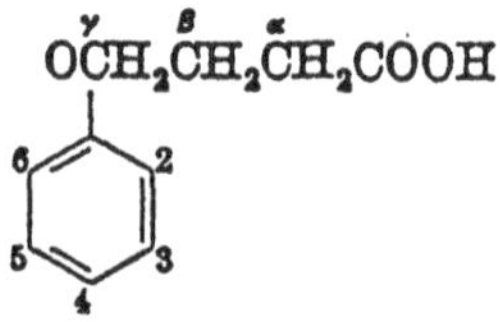

Common Name	Specific Groups
4-(4-CPB)	γ-(4-chlorophenoxy)butyric acid
MCPB*, 4-(MCPB)†	γ-(4-chloro-2-methylphenoxy)butyric acid
4-(2,4-DB)†, 2,4-DB	γ-(2,4-dichlorophenoxy)butyric acid
4-(3,4-DB)	γ-(3,4-dichlorophenoxy)butyric acid
4-(2,4,5-TB)†, 2,4,5-TB	γ-(2,4,5-trichlorophenoxy)butyric acid

$$\overset{\beta}{O}CH_2\overset{\alpha}{C}H_2(X)$$

β-PHENOXYETHYL (x)

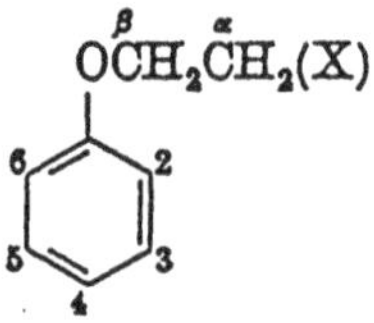

Common Name	Specific Groups (Group X)
2,4-DEB†	2,4-dichlorophenoxyethyl benzoate
2,4-DEP†	tris-(2,4-dichlorophenoxyethyl) phosphite
2,4-DES*, sesone†, SES	2,4-dichlorophenoxyethyl hydrogen sulphate
MCPES	sodium 4-chloro-2-methylphenoxyethyl sulphate
2,4,5-TES†	sodium 2,4,5-trichlorophenoxyethyl sulphate

CH_2COOH

PHENYLACETIC ACIDS

$COOH$

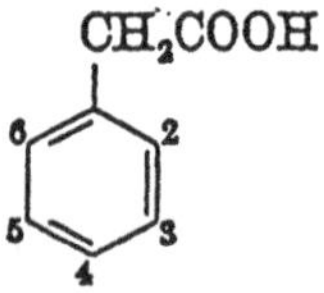

BENZOIC ACIDS

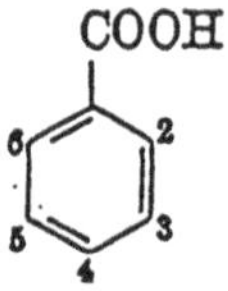

Common Name	Specific Groups
Fenac†	2,3,6-trichlorophenylacetic acid
2,3,6-TBA†	2,3,6-trichlorobenzoic acid
2,3,5,6-TBA	2,3,5,6-tetrachlorobenzoic acid
Amiben†	3,amino-2,5-dichlorobenzoic acid

Table 3.32 (continued)

HALOGENATED ALIPHATIC ACIDS

Common Name	Chemical Name and Formula
—	Sodium-monochloroacetate; $ClCH_2COONa$
TCA†	trichloroacetic acid; $Cl_3C.COOH$
Dalapon*†	2,2-dichloropropionic acid $CH_3C.Cl_2COONa$
2,2,3-TPA	2,2,3-trichloropropionic acid; $CH_2Cl.CCl_2.COOH$

2. Containing Nitrogen

$$R_1\!\!\diagdown\!\!N.\overset{\displaystyle O}{\overset{\displaystyle \|}{C}}\!\!-\!\!R_3$$
$$R_2\!\!\diagup$$

AMIDES

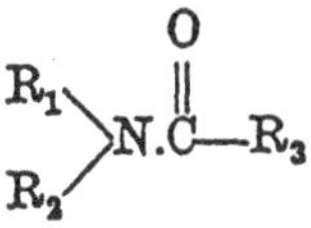

	Specific Groups		
Common Name	R_1	R_2	R_3
CDAA†, Randox	$-CH_2CH:CH_2$	$-CH_2CH:CH_2$	$-CH_2Cl$
Diphenamid†	$-CH_3$	$-CH_3$	$-CH$-diphenyl
CDEA†	$CH_3.CH_2-$	$CH_3.CH_2-$	$-CH_2Cl$
FW 734	3,4-dichloro-phenyl	H	CH_3CH_2-
Karsil	3,4-dichloro-phenyl	H	$\overset{\displaystyle CH_3}{\underset{\displaystyle \mid}{-CH.CH_2CH_2.CH_3}}$
Solan	3-chloro-4-methylphenyl	H	$\overset{\displaystyle CH_3}{\underset{\displaystyle \mid}{-CH.CH_2CH_2.CH_3}}$
Dicryl	3,4-dichloro-phenyl	H	$\overset{\displaystyle CH_3}{\underset{\displaystyle \mid}{-C:CH_2}}$
Naptalam* NPA†, (N-1-naphthyl phthalamic acid)	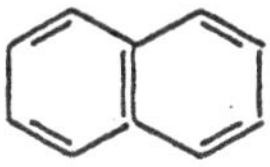 (naphthyl)	H	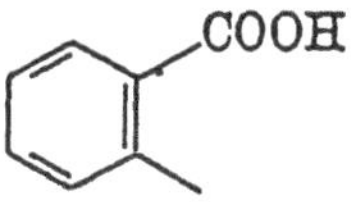 (benzene ring with COOH and CH₃)

MH†
Maleic hydrazide

$$O=\overset{\displaystyle H\quad\; H}{\underset{}{\overset{\displaystyle C=C}{C}}}\!\!\diagup\!\!\diagdown\overset{}{\underset{}{C=O}}$$

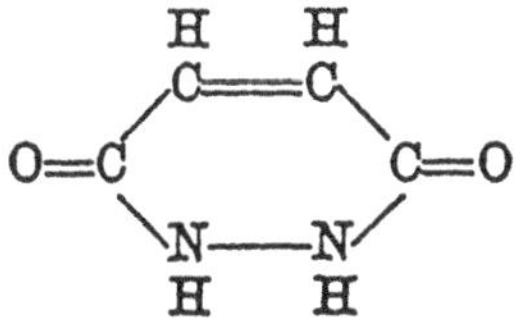

1,2-dihydropyridazine-3,6-dione

Table 3.32 (continued)

UREAS

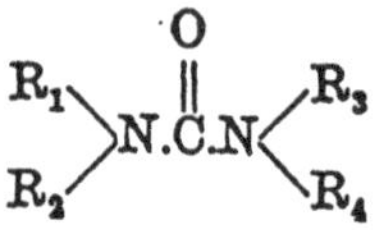

$$R_1 \diagdown \underset{R_2 \diagup}{N.\overset{\overset{O}{\|}}{C}.N} \diagup R_3 \diagdown R_4$$

Common Name	Specific Groups			
	R_1	R_2	R_3	R_4
DCU†	$Cl_3C.CH(OH)$	H	H	$Cl_3C.CH(OH)$
OMU	cyclo-octyl	H	$-CH_3$	$-CH_3$
Fenuron*† } PDU	phenyl	H	$-CH_3$	$-CH_3$
Monuron*† } CMU	4-chloro-phenyl	H	$-CH_3$	$-CH_3$
Diuron*†	3,4-dichloro-phenyl	H	$-CH_3$	$-CH_3$
Neburon†	3,4-dichloro-phenyl	H	$-CH_2\overset{\overset{CH_2.CH_3}{\|}}{CH_2}$	$-CH_3$
Linuron	3,4-dichloro-phenyl	H	$-CH_3$	$-O-CH_3$

CARBAMATES (URETHANES)

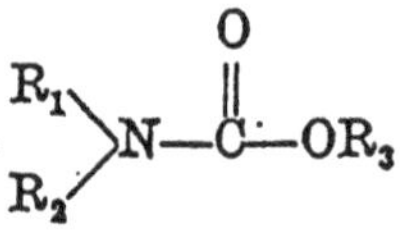

$$R_1 \diagdown \underset{R_2 \diagup}{N} - \overset{\overset{O}{\|}}{C} - OR_3$$

Common Name	Specific Groups		
	R_1	R_2	R_3
Propham* } IPC†	phenyl	H	$-CH \diagup^{CH_3}_{\diagdown CH_3}$
CEPC	3-chlorophenyl	H	$-CH_2-CH_2Cl$
Chloro-propham* } CIPC†	3-chlorophenyl	H	$-CH \diagup^{CH_3}_{\diagdown CH_3}$
BiPC	3-chlorophenyl	H	$-CH \diagup^{CH_3}_{\diagdown C\equiv CH}$
Barban*†	3-chlorophenyl	H	$-CH_2C:C.CH_2 \overset{\underset{\|}{Cl}}{}$

Table 3.32 (continued)

THIOLCARBAMATES

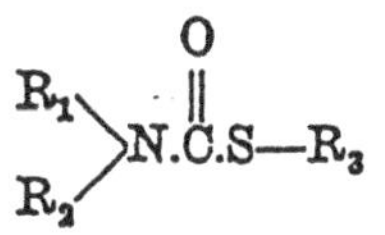

$$\begin{matrix} R_1 \\ \\ R_2 \end{matrix} \!\!> N.\overset{\displaystyle O}{\overset{\displaystyle \|}{C}}.S\!-\!R_3$$

	Specific Groups		
Common Name	R_1	R_2	R_3
EPTC†	$CH_3CH_2CH_2$	$CH_3CH_2CH_2$	CH_2CH_2
Diallate* ⎫ Avadex ⎬ DATC ⎭	$HC\!\!\begin{matrix} \diagup CH_3 \\ \diagdown CH_3 \end{matrix}$	$HC\!\!\begin{matrix} \diagup CH_3 \\ \diagdown CH_3 \end{matrix}$	$\begin{matrix} CHCl \\ \| \\ -CH_2.CCl \end{matrix}$
Tillam	$CH_3CH_2CH_2CH_2-$	$CH_3.CH_2-$	$CH_3CH_2CH_2-$

DITHIOCARBAMATES

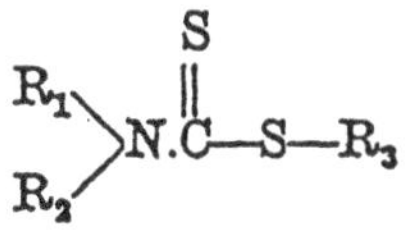

$$\begin{matrix} R_1 \\ \\ R_2 \end{matrix} \!\!> N.\overset{\displaystyle S}{\overset{\displaystyle \|}{C}}\!-\!S\!-\!R_3$$

	Specific Groups		
Common Name	R_1	R_2	R_3
CDEC	CH_3CH_2	CH_3CH_2	$CH_2CCl : CH_2$
SMDC, Metham	$-CH_3$	H	Na

Table 3.32 (continued)

TRIAZINES

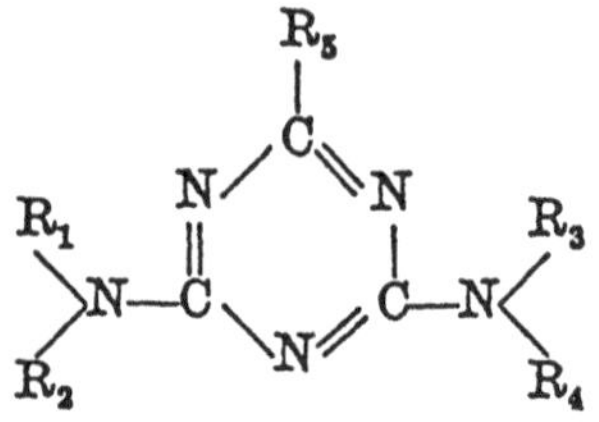

Common Name	Specific Groups				
	R_1	R_2	R_3	R_4	R_5
Simazine*†	H	$-C_2H_5$	H	$-C_2H_5$	Cl
Propazine*†	H	$-CH{<}^{CH_3}_{CH_3}$	H	$-CH{<}^{CH_3}_{CH_3}$	Cl
Chlorazine†	$-C_2H_5$	$-C_2H_5$	$-C_2H_5$	$-C_2H_5$	Cl
Trietazine*†	H	$-C_2H_5$	$-C_2H_5$	$-C_2H_5$	Cl
Atrazine*†	H	$-CH{<}^{CH_3}_{CH_3}$	H	$-C_2H_5$	Cl
Ipazine,*† Isodiazine	H	$-CH{<}^{CH_3}_{CH_3}$	$-C_2H_5$	$-C_2H_5$	Cl
Simeton*†	H	$-C_2H_5$	H	$-C_2H_5$	$-OCH_3$
Prometon*†	H	$-CH{<}^{CH_3}_{CH_3}$	H	$-CH{<}^{CH_3}_{CH_3}$	$-OCH_3$
Atraton*†	H	$-CH{<}^{CH_3}_{CH_3}$	H	$-C_2H_5$	$-OCH_3$
Ametryne*	H	$-CH{<}^{CH_3}_{CH_3}$	H	$-C_2H_5$	$-S.CH_3$
Prometryne*	H	$-CH{<}^{CH_3}_{CH_3}$	H	$-CH{<}^{CH_3}_{CH_3}$	$-S.CH_3$
Simetryne*	H	$-C_2H_5$	H	$-C_2H_5$	$-S.CH_3$

Table 3.32 (continued)

SUBSTITUTED PHENOLS

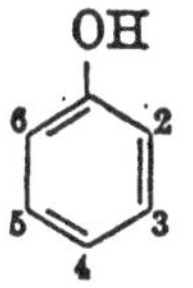

$$OH$$

Common Name	Specific Groups
DNOC* DNC†, Sinox	2-methyl-4,6-dinitrophenol
Dinoseb*, DNBP†	4,6-dinitro-2-s-butylphenol
PCP†	pentachlorophenol
Dinosam* DNAP	4,6-dinitro-2-s-pentylphenol

BIPYRIDYLIUM QUATERNARY SALTS

Common Name	Chemical Formula
Diquat*† (dibromide)	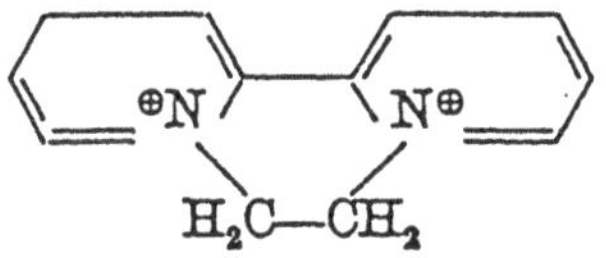 $\overset{\oplus}{N}$ $\overset{\oplus}{N}$ H_2C—CH_2 $2Br^-$
Paraquat*† (dimethyl-sulphate)	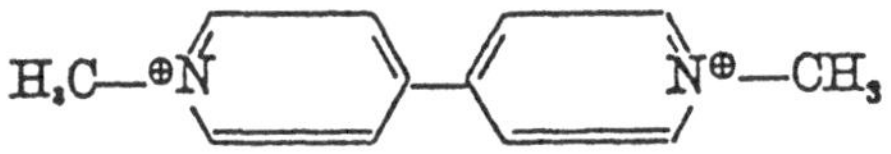 H_3C—$\overset{\oplus}{N}$ $\overset{\oplus}{N}$—CH_3 $2CH_3SO_4^-$

TOLUIDINES

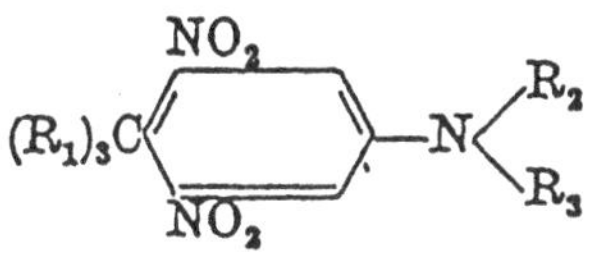

$$(R_1)_3C \underset{NO_2}{\overset{NO_2}{\diamond}} -N\overset{R_2}{\underset{R_3}{<}}$$

Common Name	R$_1$	R$_2$	R$_3$
		Specific Groups	
Dipropalin†	—H	n-propyl	n-propyl
Trifluralin†	—F	n-propyl	n-propyl

A*

Table 3.32 (continued)

MISCELLANEOUS

Common Name	Chemical Name	Formula
Amitrole†, AT	3-amino-1,2,4-triazole	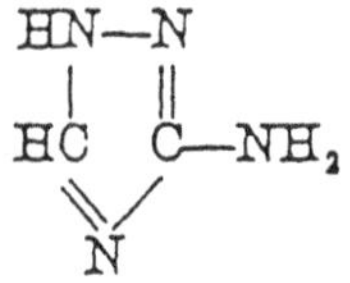
Endothal*†	7-oxabicyclo [2,2,1] heptane-2,3-dicarboxylic acid	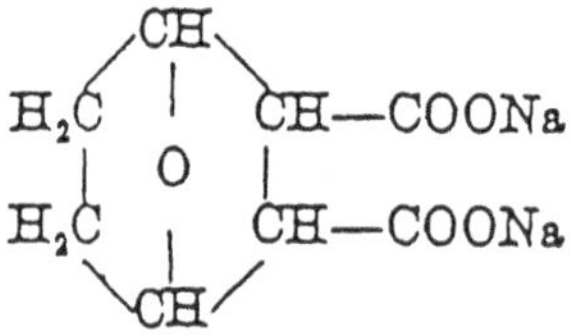
Casoron	2,6-dichloro-benzonitrile	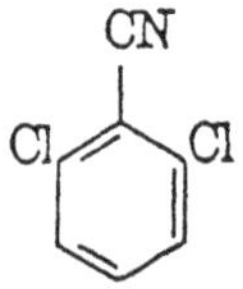
Dichlone	2,3-dichloro-1,4-naphthaquinone	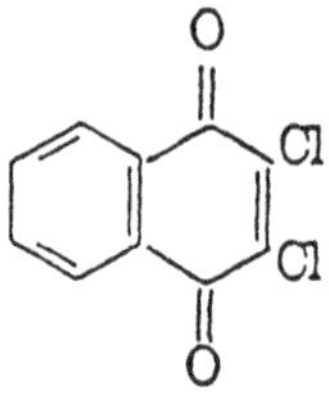
DMTT	3,5-dimethyltetrahydro-1,3,5,2H-thiadiazine-2-thione	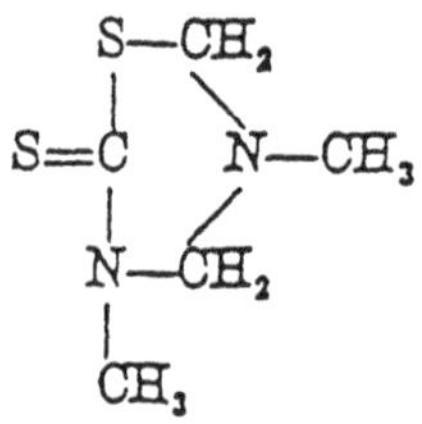
Oils		

Pesticides

Economics. Advocates of pesticides point with pride to the high quality, variety, and volume of food produced with the aid of pesticides. For example, concurrent with the postwar pesticide era, the yield of potatoes doubled. Although new varieties played a role in this increase, credit must go in a major way to control of the potato leafhopper, potato flea beetle, and the fungus disease. In the United States the control of the northern corn rootworm on corn and the development of effective herbicides revolutionized corn production with respect to both profit and volume. The reduced need for cultivation made it possible for much larger and efficient production units with greatly reduced manpower. Pest control in cotton, tobacco, citrus, and deciduous fruits contributed to increased production and decreased cost per unit. In agriculture, generally, the United States estimates indicate a $1.00 to $10.00 return for every $1.00 invested in pesticides, but the return differs greatly in different situations.

Economic benefits as a function of the World Health Organization's (WHO) efforts to eradicate malaria also can be cited as phenomenal when one considers the number of ill days converted to working days through this program.

Weighted against the positive economic effects are some facts that make the generalizations of economic benefits less dramatic. Nova Scotia orchardists did not have a significant problem with spider mites until DDT was used for codling moth control; the red-banded leafroller was not a problem to apple growers in eastern North American until modern pesticides were used; fall panicum was not a problem in corn until the herbicide atrazine became used widely, and modern pesticides, as used during the past 25 years, have created as many problems as they have solved in citrus and cotton. In the United States the pesticide portion of the production cost of agricultural crops rose from 1% in 1955 to 4.6% in 1968.

Use of Pesticides. The use of pesticides by consumers in the United States is estimated to range between some 660 million and 1.2 billion lb/yr. More than 60% of the pesticides are represented by herbicides, and the rest are insecticides, fungicides, rodenticides, etc. [85,86,98–103].

Weed control is essential to mechanized agriculture, and the importance of insecticides can be illustrated by the grasshopper plague in 1985, the greatest in the United States since the late 1930s. the damage caused by grasshoppers to food, forage, and forest crops averages about $390 million/yr, but in 1985 the USDA was forced to treat nearly 14 million acres infested by an estimated 55 million adult grasshoppers. This time, an integrated pest-control management (IPM) program was used, which combines chemical and biological defenses: carbaryl, acephate, and malathion (a member of the thiophosphate family of insecticides) were used in combination with a grasshopper parasite *Nosema locustae*. In April, 1988, the locust plague that was sweeping across Africa was likely to cause greater famine than that experienced in recent history, and the appropriate United Nations agencies were even considering the use of highly maligned chlordane to control the disaster.

Insecticides and fungicides must be used to prevent postharvest damage to fruit estimated at about 20% of the crop. Nematodes cause an estimated annual crop loss valued at $4 billion in the United States.

The traditional route to carbaryl and other methylcarbamate pesticides involves the reaction of methyl isocyanate (MIC) with a given substance like phenol: [85]

$$CH_3-NH_2 + Cl_2CO \longrightarrow CH_3-N=C=O + 2HCl$$

methyl isocyanate

$$CH_3-N=C=O + Q-O-H \longrightarrow CH_3-NH-C(=O)-O-Q$$

Environmental Concerns. The subject of pesticides in the environment is a large and controversial one. Its magnitude derives from the extent of its two major components [8,90,92,103]. First the number of different chemicals being used runs in the hundreds and each of these is available in many formulations. Each chemical has its own set of chemical and biological properties which may be modified by the formulation dispensed. The second, even larger and less well-understood component is the environment. Included in this are all the biological, physical, and chemical components that make up our surroundings and the interactions among these dictated by principles that we can grasp only in a superficial way.

As an example, some 14 million Americans are routinely exposed to agricultural weed killers in their drinking water, charges a report, "Tap Water Blues," released by the Environmental Working Group (EWG) and Physicians for Social Responsibility (PSR) at a press conference in October 1994 in Washington, D.C. The contamination is particularly acute from April through August in the Midwest. Each spring farmers apply to their fields about 150 million lb of five herbicides analyzed in the report: alachlor, atrazine, cyanazine, metolachlor, and simazine [101].

V. PURE HYDROGEN

In chemical industries, most pure hydrogen applications require amounts that are small compared to those involved in other synthesis gas derivatives. Sometimes, the amount is too small to consider steam reforming of methane. Obviously, there are other sources for pure hydrogen especially in small amounts. Depending upon the size of the business and the supportive infrastructure around the industry, the source for pure hydrogen will be determined.

The most common source of pure hydrogen in small amounts is as a byproduct from caustic-chlorine electrolysis plants. A typical 100-ton/day chlorine plant would produce around one millon scfd of pure hydrogen, that is far in excess of the needs of a typical small consumer of hydrogen such as a fats and oils hydrogenation unit.

Hydrogen is quite costly to transport and therefore, generating it in small amounts frequently makes a lot of sense. Three sources have become increasingly important. They are:

Catalytic decomposition of ammonia:

$$2NH_3 \longrightarrow 3H_2 + N_2$$

Steam reforming of methanol:

$$CH_3OH + H_2O \longrightarrow 3H_2 + CO_2$$

Electrolysis of water:

$$H_2O \longrightarrow H_2 + \frac{1}{2}O_2$$

Hydrogen from ammonia is obtained at 75 mol% purity and further purification has to be carried out by low-temperature fractionation. Hydrogen from the steam reforming of methanol contains CO_2 with very small amounts of CH_4 and CO as virtually the only impurities. These impurities can be removed by any of the conventional washing processes and CO can be converted to methane. All these processes are significantly inferior to steam reforming of methane, in terms of production cost. But, these methods are used in situations where the cost of hydrogen has a slight influence in economics.

The uses of pure hydrogen in the organic industry can be classified into the following major categories:

1. Double-bond hydrogenation:

$$\diagdown C = C \diagup + H_2 \longrightarrow \diagdown C - C \diagup$$
$$\underset{H \quad H}{| \quad |}$$

2. Reduction of nitro-groups to amines:

$$-NO_2 + 3H_2 \longrightarrow NH_2 + 2H_2O$$

3. Reduction of aldehydes to alcohols:

$$-CHO + H_2 \longrightarrow -CH_2OH$$

4. Reduction of carboxylic acids to alcohols:

$$-COOH + 2H_2 \longrightarrow -CH_2OH + H_2O$$

5. Hydrogenation of nitriles to amines:

$$-CN + 2H_2 \longrightarrow -CH_2NH_2$$

6. Hydrotreating of heavier feedstocks including petroleum crude, synthetic crudes, and coal liquids.

A. Production of Hydrogen

In this section, several of commercial hydrogen production processes based on recovery of hydrogen from syngas mixture or other petrochemical offgas mixture are briefly summarized. Other commercial hydrogen production processes using natural gas as feedstock are covered in Chapter 4.

Air Products and Chemicals, Inc. developed a process called Gemini Hyco that recovers and purifies carbon monoxide (CO) for use as a chemical feedstock, or separates hydrogen and CO into individual high purity products from synthesis gas streams using an adsorption process. Figure 3.41 shows a schematic of this process [61].

Syngas is produced by steam reforming of methane. The syngas is fed to the Hyco adsorption separation system. First, the gas enters the CO adsorption beds where the CO is adsorbed. Then CO is desorbed from the adsorbent under a vacuum cycle and compressed to the desired pressure. The remaining

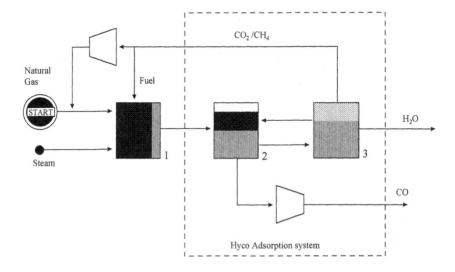

Figure 3.41   GeminiHyco process. *Source*: [121].

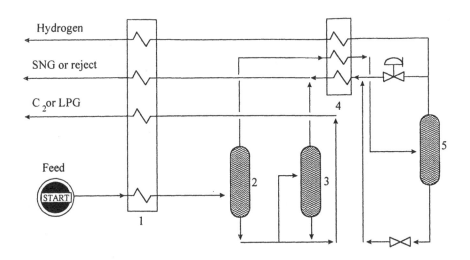

Figure 3.42 Cryofining process. *Source*: [121].

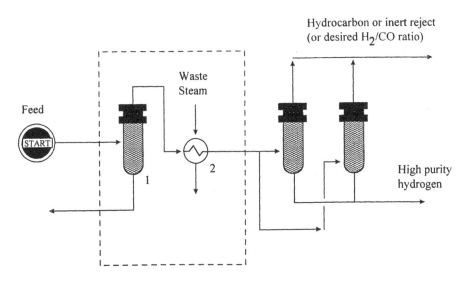

Figure 3.43 MEDAL process. *Source*: [121].

syngas stream is processed further using a pressure swing adsorption (PSA) system to yield a purified hydrogen product. The hydrogen product purity can be as high as 99.999%. Unrecovered syngas can be used as fuel and/or recycled to the reformer feed stream to enhance the CO production.

When integrated with a steam reformer, the Hyco process results in a CO product cost savings of up to 15% over conventional systems of CO_2 solvent removal plus cryogenics. This process eliminates the need for a separate CO_2 removal system. The purity of CO is greater than 99.5% and the high purity CO can be produced in a single stage. The product CO is suitable for use in the production of isocyanates, acetic acid, oxo alcohols, and specialty chemicals that are covered in this book.

Another process by Air Products and Chemicals, Inc. called Cryofining process [61], recovers high purity hydrogen of up to 98% purity and liquid hydrocarbon byproducts from refining fuel gas. Figure 3.42 show a schematic of cryofining process. The process involves cryogenic separation of offgases and purges containing 30–70% H_2 and 15–30% hydrocarbon liquids. Gases at 400 psi or higher are chilled to approach hydrate temperature to condense H_2O and then dried to < 1 ppm. At Stage 1, the inlet gas is condensed against returning product streams, and condensed hydrocarbon fractions are separated at Stages 2 and 3. Refrigeration is obtained at Stage 4 by throttling condensed hydrocarbons and exchanging against feed. This process is favored by moderate impurity levels (30–70%), by reasonably high purity requirements (90%-98%), and by concurrent need for hydrocarbon liquid recovery.

The MEDAL process is versatile, and recovers and purifies hydrogen from various sources [61]. The feed for this process may come from either petroleum refining sources or petrochemical sources. The former includes hydroprocessing purges, offgases from catalytic reformer, steam reformer streams, and fuel gas streams. The latter include ethylene plant streams, dehydrogenation plant streams, ammonia plant purge, methanol plant purges, benzene plant purges, cryogenic unit feed, and waste gas streams. The hydrogen recovery ranges from 80 to 95+% and the hydrogen purity can be 99+%. Purification is based on polyaramide or polyimide hollow-fiber membrane separators. Polyaramide accepts ammonia, alcohols, aromatics, acids, and olefins. A schematic of the process is shown in Figure 3.43. The process typically consists of a coalescing filter, a preheat exchanger, and membrane separators. This process is widely used in commercial applications.

B. Fats, Oils, and Waxes

Background

Fats, oils, and waxes belong to a class of compounds called lipids. These compounds are insoluble in water, but soluble in organic solvents. Lipids are one of the three major classes of food nutrients, with the other two being proteins and carbohydrates. Lipids differ from proteins and carbohydrates, in that they produce twice the energy per unit weight when combusted to CO_2 and H_2O. Lipids provide over 40% of the calories in the U.S. diet [107].

Oils and fats are distinguished by their melting points and their sources. At ambient temperatures, oils are in a liquid form and fats are in a solid or semisolid form. Oils are usually extracted from plants and marine animals, while fats are usually of animal origin. Fats and oils are primarily glycerol esters of fatty acids. These esters can be considered to be formed from the combination of alcohols and fatty acids with the loss of water. Since glycerol is a trialcohol, one molecule can combine with three fatty acid molecules to form a triglyceride [107]. Fats and oils obtained from animal or vegetable sources, are predominantly triglycerides with traces of mono- and diglycerides, free fatty acids and other minor constituents [107].

Waxes are different from fats and oils in that they are mono- or dihydric alcohols, and are glycerides of ethylene glycol instead of glycerol. Waxes may originate from either animal or plant material. The wax alcohols may be aliphatic or alicyclic [107].

The common commercial fats and oils are primarily mixed triglycerides of five common fatty acids: palmitic, stearic, oleic, linoleic, and linolenic acids. All are unbranched carboxylic acids and have 18 carbon atoms except for palmitic which has 16 carbon atoms.

The difference between stearic, oleic, linoleic, and linolenic acids is the number of double bonds (i.e., the degree of unsaturation) in the chain: none for stearic, one for oleic, two for linoleic, and three for linolenic. Unsaturated

$$CH_3(CH_2)_{14}CO_2H \qquad \text{Palmitic Acid}$$

$$CH_3(CH_2)_{16}CO_2H \qquad \text{Steric Acid}$$

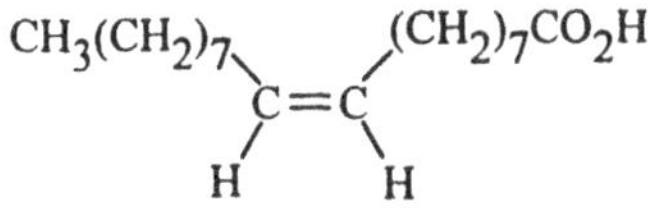

Oleic Acid

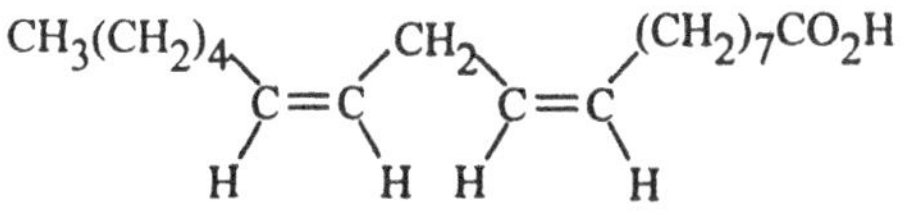

Linoleic Acid

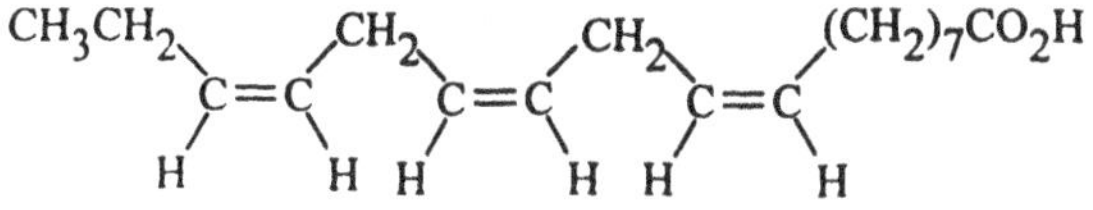

Linolenic Acid

acids usually are in the *cis* configuration, even though *trans* is thermodynamically more stable. Multiple double bonds are always separated by methylene groups, such as -CH=CH-**CH₂**–CH=CH- although conjugated double bonds are thermodynamically preferred. When conjugated unsaturated bonds occur, they are either a mixture of *cis-trans* or all *trans* [107].

Physical properties of fatty acids show a regular characteristic pattern. Melting points increase with chain length of saturated acids, but decrease with degree of unsaturation and chain branching. Each molecule of the triglyceride contains three fatty acids that do not have to be the same. As a result, fatty acids do not have an exact melting point, but rather a temperature range. The first molecules to solidify are saturated and can be removed by filtration. This separation is called winterization [107].

Production

The hydrogenation reaction of fats and oils is undertaken to reduce the amount of unsaturation and to improve chemical stability. Hydrogenation is a simple reaction, where many hydrogenation catalysts such as Ni, Pd, Pt, Cu, Ag, Au/Rh, Ir, Ru, and Os have been used with varying temperatures and pressures. However, selective or partial hydrogenation of a certain type of unsaturation is more difficult. No catalyst is entirely selective, so selectivity is counterbalanced with the desired properties of the final product. Another problem with hydrogenation catalysts is that they may also cause isomerization of an unsaturated bond from the *cis*- to a *trans*-configuration, or cause the unsaturated bonds to migrate up and down the fatty acid chain. All these reactions alter the properties of fats and oils and must be considered for the final product. A comparison of activity, double-bond migration, *cis-trans* isomerization, and selectivity for five noble metals is presented:

Activity:	Pd > Rh > Pt > > > Ir > Ru
Double-bond migration:	Pd > Rh ᴖ Ru > Ir > Pt
cis-trans isomerization:	Pd > Rh > Ru > Ir > Pt
Selectivity:	Pd > Rh > Pt > Ru > > Ir

Notice the similar sequences for all the metal, but especially for double-bond migration and *cis-trans* isomerization, inferring similar reactions. Common reduction conditions are 400–465 K and 50–500 KPa hydrogen pressures [113].

Use

Vegetable oils go to a number of end uses, edible and industrial. Some of the edible uses of vegetable oils are: shortening, margarine, salad oils, frying oils, hard butters, and surfactants. Further processing such as hydrogenation is used to produce hard fats or to enhance oxidative stability [107–114].

Fatty acids are formed from the hydrolysis of the triglyceride with glycerol as a byproduct as shown:

$$\begin{array}{lll} CH_2O_2R & CH_2OH & RO_2H \\ | & | & \\ CHO_2R' \xrightarrow{H_2O} & CHOH & + R'O_2H \\ | & | & \\ CH_2O_2R'' & CH_2OH & R''O_2H \end{array}$$

The fatty acids are then separated into saturated and unsaturated forms. The fatty acids then can be hydrogenated to form fatty alcohols. The same catalyst used to reduce the unsaturation can be used to form the fatty alcohols. Similar temperatures are used with an increased hydrogen pressure (2.5 Mpa). Fatty alcohols have the same use as higher alcohols and are discussed in the higher alcohols section.

Outlook

The fats, oils, and waxes industry will continue to prosper. An emphasis on more selective hydrogenation catalysts and development of a *trans-* to *cis-* configuration catalyst are expected to dominate the industry.

C. Tetrahydrofuran

Background

Tetrahydrofuran (THF) is a planar 5–membered ring containing 4 carbons and one oxygen atom. Its boiling point is 65–66˚C, its specific gravity is 0.985, and its molecular weight is 80.17 g/mole [118]. It is a colorless polar organic solvent that is flammable and is an irritant so care must be used in handling. It is a high-performance solvent that is used as a raw material of elastic fiber [115]. It also can be directly polymerized which accounts for approximately 70% of the THF consumption [116]. Its 1990 production capacity was 160,000 tons [108].

Production

Tetrahydrofuran can be produced in six different ways:

1. Furfural route
2. Reppe process
3. Maleic anhydride route
4. 1,4-Dichlorobutene route
5. Direct synthesis
6. Mitsubishi Kasei Corporation (MKC) process

The furfural route is the oldest industrial route to THF. It is so named because furfural is the starting material. The reaction scheme is as follows.

$$\text{Furfural} \xrightarrow[\text{Cat.}]{H_2O} \text{Furan} + CO_2 + H_2 \xrightarrow[\text{Cat.}]{H_2} \text{THF}$$

Furfural—found in corn cobs, oat hulls, rice hulls, cotton seed hulls, and extracted from residue from sugar cane and sugar beets—undergoes hydrolysis to form furan and CO_2 and H_2 as byproducts. This step is carried out at 400°C in the presence of $ZnCrO_2$ and $MnCrO_2$ catalyst. The furan is then hydrogenated to form THF. This has been accomplished with either Ni or Pd catalyst [115,117].

The Reppe process is based on acetylene and formaldehyde feed; originally, it was developed in Germany during World War II to form butadiene. The last step of the Reppe process was eliminated and the intermediate product was used to produce THF. The process is:

$$C_2H_2 + HCHO \xrightarrow{\text{Cat.}} \underset{\substack{\text{1,4-dihydroxy-}\\\text{2-butyne}}}{HOCH_2C\equiv CCH_2OH} \xrightarrow[\text{Cat.}]{H_2} \underset{\substack{\text{1,4-dihydroxy-}\\\text{butane}}}{HOCH_2CH_2CH_2CH_2OH}$$

$$HOCH_2CH_2CH_2CH_2OH \xrightarrow{H_3PO_4} \underset{\text{THF}}{\bigcirc} + H_2O$$

The feed materials, acetylene and formaldehyde are reacted in the presence of CuC_2 or SiO_2 catalyst at 90–100°C at 5 atm to form 1,4–dihydroxy-3–butyne(butynediol), which is then hydrogenated in the presence of Ni or Pd at 100°C to form 1,4–dihydroxy-butane. The 1,4–dihydroxy-butane in the presence of H_3PO_4 at 100–130°C cyclizes to form THF and water as a byproduct [117].

The maleic-anhydride route is based on the direct hydrogenation of maleic anhydride with H_2O formed as a byproduct:

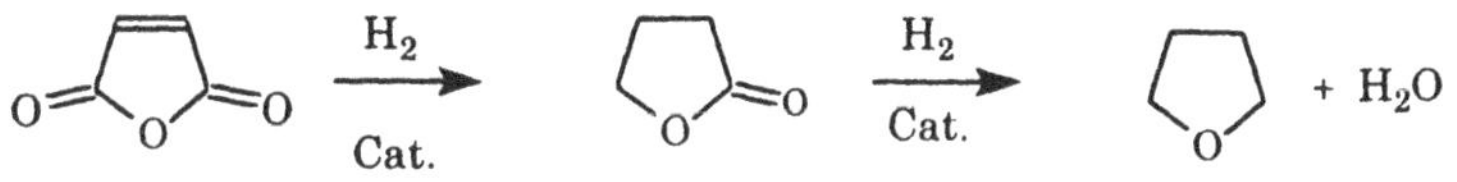

$$O = \underset{O}{\diamond} = O \xrightarrow[\text{Cat.}]{H_2} \underset{O}{\diamond} = O \xrightarrow[\text{Cat.}]{H_2} \underset{O}{\bigcirc} + H_2O$$

Maleic Anhydride   THF

This can be accomplished in one direct step or two steps with little difference between the two. The catalyst used is a Ni-precious metal with high pressures required [117].

The 1,4-dichlorobutene route starts with 1,3–butadiene. The mechanism is:

$$H_2C{=}CHCH{=}CH_2 \xrightarrow{Cl_2} ClCH_2CH{=}CHCH_2Cl \xrightarrow{NaOH} HOCH_2CH{=}CHCH_2OH$$

1,3-Butadiene   + NaCl

                      1,4-dichloro-2-Butene   1,4-dihydroxy-2-Butene

$$HOCH_2CH{=}CHCH_2OH \xrightarrow[\text{Cat.}]{H_2} HOCH_2CH_2CH_2CH_2OH$$

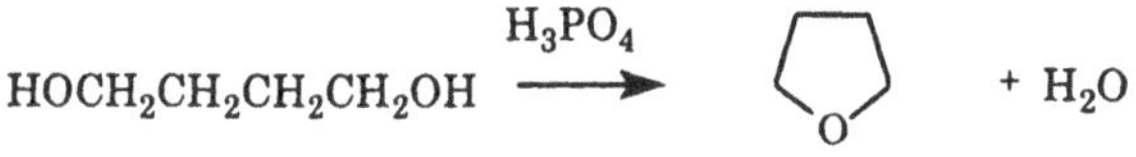

$$HOCH_2CH_2CH_2CH_2OH \xrightarrow{H_3PO_4} \underset{O}{\bigcirc} + H_2O$$

THF

The 1,3–butadiene is chlorinated with Cl_2 to form 1,4–dichloro-2–butene and 2–chlorobutene as a byproduct. The 2–chlorobutene is hydrolyzed with NaOH to form 2–butene-1,4–diol and NaCl. The 2–butene-1,4–diol is hydrogenated to form 1,4–butanediol which, in the presence of H_3PO_4 at 100–130°C, cyclizes to form THF and water as a byproduct [117].

The direct synthesis starting materials are ethylene and ethylene oxide. They are reacted directly to form THF using $Ru(CO)_3[(C_6H_5)_3]_2$ as a catalyst and HI as a promoter.

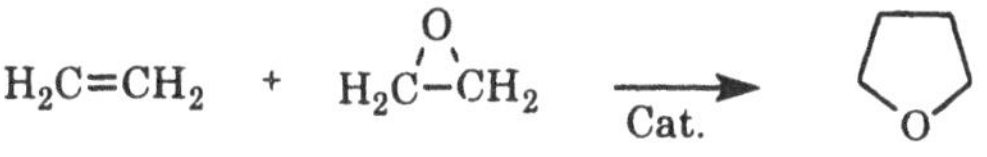

$$H_2C{=}CH_2 + H_2\overset{O}{\overset{/ \backslash}{C}}CH_2 \xrightarrow{\text{Cat.}} \underset{O}{\bigcirc}$$

THF

Low conversions have hindered this process [117].

The Mitsubishi Kasei Corporation (MKC) developed an alternative method to produce THF. The MKC method involves the reaction of butadiene and acetic acid. The scheme is:

$$CH_2=CHCH=CH_2 + CH_3\overset{O}{\overset{\|}{C}}OH \xrightarrow[\text{Cat.}]{O_2} CH_3\overset{O}{\overset{\|}{C}}OCH_2CH=CHCH_2O\overset{O}{\overset{\|}{C}}CH_3$$

1,3-Butadiene   Acetic Acid   1,4-Diacetoxybutene-2

$$CH_3\overset{O}{\overset{\|}{C}}OCH_2CH=CHCH_2O\overset{O}{\overset{\|}{C}}CH_3 \xrightarrow[\text{Cat.}]{H_2} CH_3\overset{O}{\overset{\|}{C}}OCH_2CH_2CH_2CH_2O\overset{O}{\overset{\|}{C}}CH_3$$

1,4-Diacetoxybutene-2   1,4-Diacetoxybutane

$$CH_3\overset{O}{\overset{\|}{C}}OCH_2CH_2CH_2CH_2O\overset{O}{\overset{\|}{C}}CH_3 \xrightarrow[\text{Cat.}]{H_2O} \left[\text{O} \right] \;+\; 2\; CH_3\overset{O}{\overset{\|}{C}}OH$$

1,4-Diacetoxybutane   THF   Acetic Acid

1,3–butadiene is reacted with acetic acid to form 1,4–diacetoxybutene-2. This is accomplished with a homogeneous Wacker-type Pd catalyst. Because of the mild reaction conditions (70°C and 70 Kg/cm^2) the hydrogenation is selective. The 1,4–diacetoxybutene-2 is then hydrogenated to form 1,4–diacetoxybutane using noble metal catalyst at 60°C and 50 Kg/cm^2. Excess water is added to the 1,4–diacetoxybutane and then is combined with an ion-exchange catalyst to form THF and acetic acid. The THF is purified with distillation. The acetic acid solution is distilled and then recycled back to the process [115].

Environmentally Friendly THF Process

DuPont's new tetrahydrofuran plant, which is slated to start up in the second quarter of 1996 in Asturias, Spain, is based on a new process that is claimed to be environmentally friendly [133]. The process is harnessed with a circulating fluidized-bed (CFB) reactor for the partial oxidation of *n*-butane. The catalyst used for this stage is vanadium phosphorus oxide (VPO). To enhance the catalytic life in the CFB reactor, attrition-resistant catalyst particles have been developed. A highly selective catalyst for hydrogenation of maleic acid to THF also has been developed. This THF conversion reaction is carried out in a bubble column reactor under relatively mild conditions. This process has several significances:

1. It minimizes the byproduct formation.
2. It has significant environmental improvements over the conventional process.
3. It is the first commercial use of the CFB for catalytic processing.

The overall chemistry of the DuPont process is as follows:

$$CH_3(CH_2)_2CH_3 \xrightarrow{\text{VPO}} HO_2CCH = CHCO_2H \xrightarrow[H_2]{\text{Pd/Re/C}} \langle \!\!\! \bigcirc \!\!\! \rangle$$

The process yields butanediol as a coproduct. A schematic of the DuPont process is shown in Figure 3.44 [133].

Use

Tetrahydrofuran is an excellent solvent for high molecular weight polymers such as poly(vinyl chloride), rubbers, and many others; it is also used as the solvent for the formation of Grignard reagents, an important industrial reactant. Tetrahydrofuran is also used as a plasticizer in thermoplastic elastomers, such as polyurethanes and polyesters [117].

Tetrahydrofuran also can undergo cationic polymerization, which accounts for the majority of its usage. Since the THF ring contains an oxygen atom

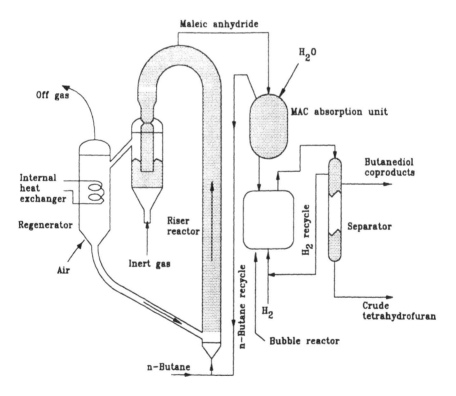

Figure 3.44   DuPont THF process. *Source*: [133].

with two sets of paired electrons, THF is a nucleophilic monomer. The reaction is slightly exothermic, -3350 J/mol [117].

Tetrahydrofuran polymerization proceeds with cationic ring-opening mechanism. The propagating species is a tertiary oxonium ion associated with negatively charged counter ion. Stabilizing counters are those that contain very electronegative ions, such as PF_6^-, AsF_6^-, SbF_6^-, $SbCl_6^-$, BF_4^-, $SO_3CF_3^-$, and ClO_4^- [117].

The polymerization scheme is as follows:

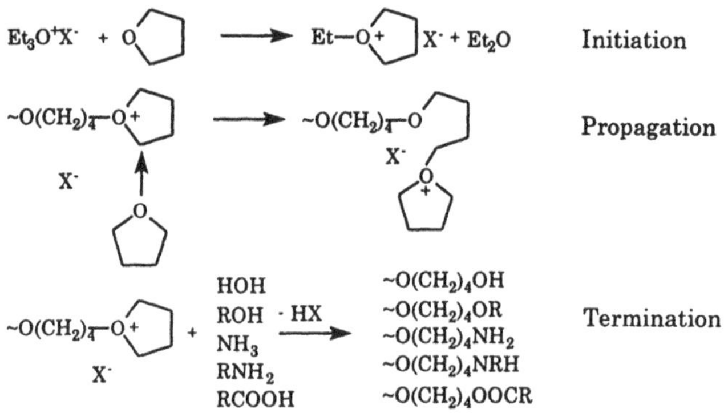

Tetrahydrofuran polymerization is initiated by a nucleophilic attack of the free electrons on the oxygen atom in THF by tertiary oxonium salts, carbonium salts, and superacid esters. The polymer is formed in the propagation step by a S_N2 mechanism. The termination step uses common nucleophiles such as water, alcohols, amines, and carboxylic acids.

Outlook

The THF market is expected to continue to expand as the production of THF gets cheaper and more uses are found for THF.

D. Sorbitol

Background

Sorbitol or D-Glucitol is a white odorless nontoxic powder, with a sweet taste. It contains 0.5 mol of water of crystallization and melts at 97–98°C. Anhydrous sorbitol melts at 110°C. It is very hygroscopic, slightly soluble in methanol and ethanol, and insoluble in ether and most nonpolar solvents [119]. The occurrence of sorbitol in nature is not widespread, but it is found in some fruits [120].

Production

Sorbitol is a hexahydric alcohol obtained from the hydrogenation of fructose-free glucose, usually obtained from corn [8]. The reaction is:

$$
\begin{array}{ccc}
\underset{|}{\text{CHO}} & & \underset{|}{\text{CH}_2\text{OH}} \\
\text{H}-\overset{|}{\underset{|}{\text{C}}}-\text{OH} & & \text{H}-\overset{|}{\underset{|}{\text{C}}}-\text{OH} \\
\text{HO}-\overset{|}{\underset{|}{\text{C}}}-\text{H} & \xrightarrow[\text{Cat.}]{\text{H}_2} & \text{HO}-\overset{|}{\underset{|}{\text{C}}}-\text{H} \\
\text{H}-\overset{|}{\underset{|}{\text{C}}}-\text{OH} & & \text{H}-\overset{|}{\underset{|}{\text{C}}}-\text{OH} \\
\text{H}-\overset{|}{\underset{|}{\text{C}}}-\text{OH} & & \text{H}-\overset{|}{\underset{|}{\text{C}}}-\text{OH} \\
\text{CH}_2\text{OH} & & \text{CH}_2\text{OH}
\end{array}
$$

Glucose   Sorbitol

The conditions of the reaction are 140–150°C and 4040–5050 kPa. A nickel catalyst has been used but Pd catalyst is the preferred catalyst [8,119,120].

Uses

Industrially, sorbitol is used as a humectant in cosmetic and pharmaceutical preparations, and in the printing, resin, and glue industries. It is also a raw material for the complex production of ascorbic acid (Vitamin C) which involves many biological and chemical transformations [118,119]. Sorbitol is also used as a sugar substitute for diabetics. It is used in diabetic foods, such as chocolate, because of its slow rate of intestinal absorption versus sugars such as glucose [120].

Propoxylated derivatives of sorbitol are used as the polyols in the formation of rigid polyurethane foams. Sorbitol can be dehydrated and esterified with stearic acid to give sorbitol mono- or tristearate. The ester then can be ethoxylated with about 20 ethylene oxide units to give an ethoxylated surfactant. Hydrolysis of sorbitol results with glycerin, which is used in the pharmaceutical and personal care products field [108,109].

Outlook

The sorbitol market is mature and will remain steady. Growth may come from the polymer industry with the use of sorbitol derivatives as plasticizers.

E. Hydrogen Peroxide

Background

Hydrogen peroxide (H_2O_2) is a colorless liquid that readily decomposes to form H_2O and O_2 without stabilizers. Hydrogen peroxide is a popular oxidizing agent for a wide range of compounds, such as aldehydes, ketones, phenols, and amines [123]. Its highly reactive nature and innocuous byproduct, water, explains its extensive use in industrial applications. Some of the industrial applications are bleaching operations, treatment of municipal wastes, polymerization initiator, and the manufacture of organic peroxides, epoxides, and glycols [123–125]. The advantage of H_2O_2 over other oxidizing agents is its high selectivity. Despite hydrogen peroxide's advantages of selective reactions, few byproducts and nonhazardous byproduct, it is still more expensive than O_2 when compared on a price per pound basis [123].

Hydrogen peroxide is available commercially in a variety of concentrations, with the most common aqueous solutions being 35, 50, and 70% by weight H_2O_2. The 1990 world capacity for H_2O_2 production was 850,000 tons, based on 100% H_2O_2 [108].

Production

There are four processes that are used for the production of H_2O_2:

1. Oxidation of isopropanol
2. Electrochemical oxidation of sulfuric acid or ammonium sulfate
3. Anthraquinone auto oxidation process
4. H_2 and O_2 direct reaction

The oxidation of isopropanol process is used only in a few plants in Russia and the electrochemical processes are currently utilized only in a few plants in Western Europe because of the high operating costs due to the large use of electricity. Because of their minor importance in H_2O_2 production [126], they will not be discussed further. The anthraquinone process accounts for 95% of the H_2O_2 production. The direct reaction of H_2 and O_2 is a recent discovery, so it is not yet fully utilized [126]. The reaction cycle is:

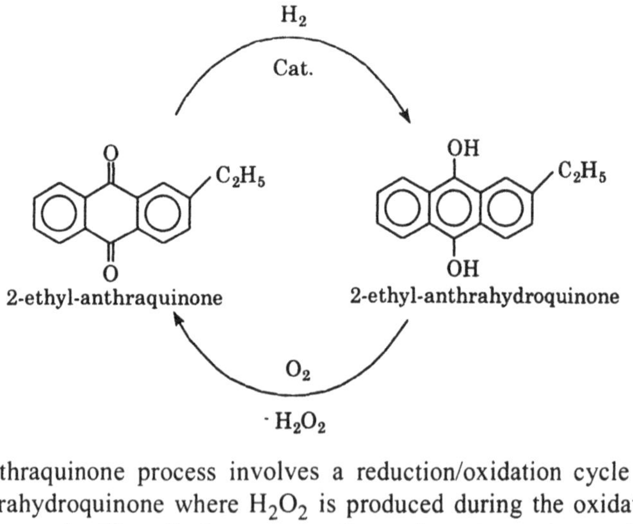

The anthraquinone process involves a reduction/oxidation cycle with 2–alkyl-anthrahydroquinone where H_2O_2 is produced during the oxidation portion of the cycle. The alkyl group on the anthraquinone is usually ethyl, although *t*-butyl, *t*-amyl, and *sec*-amyl have been used [108].

The first step is the catalytic hydrogenation (reduction) of the 2–alkyl-anthraquinone to 2–alkyl-anthrahydroquinone. Pd is the preferred reduction catalyst on gauze carriers or in suspension, although Ni and other hydrogenation catalysts can be used [108,109,126–129]. This reaction step is operated at temperatures around 40°C and at pressures up to 5 bar. Only 50% conversion is achieved to limit the amount of side reactions [108].

The second step involves the oxidation of 2–alkyl-anthrahydroquinone with air to form 2–alkyl-anthraquinone and H_2O_2. This step is operated at 30–80°C and up to 5 bar. The 2–alkyl-anthraquinone working solution then is filtered to remove the hydrogenation catalyst [108].

The H_2O_2 formed during the oxidation step is extracted from the working solution using H_2O. The extraction process is 98% efficient and results with H_2O_2 solutions of 15–35% by weight. Organic compounds in the solution must be removed and stabilizers such as diphosphates are added before further H_2O_2 purification. H_2O_2 is purified to its final concentration by distillation. The working solution is dried and freed of byproducts using active aluminum oxide before being recycled to the reactor [108].

The solvent mixture used is complex because of its many functions. The solvent mixture must dissolve both the quinone and hydroquinone compounds, have low solubility in H_2O, and be a nonparticipant in the reaction. A mixture of aromatic compounds such as naphthalene or trimethylbenzene are used to dissolve the quinone. Polar compounds such as diisobutylcabinol or methyl-cyclohexanol-acetate are used to dissolve the hydroquinone [108].

The long sought direct reaction of H_2 with O_2 only became available within the last ten years [124]. The process is:

$$H_2 + O_2 \xrightarrow[\text{Cat.}]{} H_2O_2$$

The process involves bubbling H_2 and O_2 through an aqueous medium containing a proprietary 2–5 wt% Pd catalyst that is loaded on activated carbon. The solution contains aqueous HCl or H_3PO_4 to improve yield, and some organic solvents. Operating conditions range between 0 and 25°C, and 500 and 2500 psi. Up to 13 wt% H_2O_2 concentration have been achieved with this process [124].

Presently, the investment cost is about half as much as the anthraquinone process for the production of H_2O_2, but operating costs are higher for the direct reaction; however, improvements are being made and should make the direct reaction process an economic alternative in the near future [130].

Uses

Hydrogen peroxide is replacing chlorine as a bleaching agent used in the paper industry; it is also used as an oxidizing agent for the treatment of municipal waste [131].

In industrial reactions, H_2O_2 is used to form peracids. Strong aliphatic acid solutions, mainly acetic acid, can form organic peracids when H_2O_2 is added. Peracids are also used as oxidizing agents and can be used for a wide range of reaction conditions. Peracids react with olefins to form epoxides or diols; both are used as plasiticizers in the polymer industry [108,125].

Hydrogen peroxide also can react with olefins to form organic peroxides. Organic peroxides are used as free-radical polymerization catalysts and vulcanization agents [130]. Hydrogen peroxide also reacts with phenol to form p- and o-dihydroxybenzene in the presence of Co^{2+} and Fe^{2+} catalyst used in photography. Hydrogen peroxide also will react with amines in the presence of acid at mild conditions to form nitrides. Dimethylsufoxide can be formed from dimethylsulfide with H_2O_2 in the presence of an acid [108,109]. Hydrogen peroxide also reacts with ethene and propene over titanium silicate microporous catalysts, producing ethylene oxide and propylene oxide, respectively. These reactions are commercially important.

Outlook

The outlook for H_2O_2 is excellent. Its market is currently increasing and is expected to continue to do so. The growth of the H_2O_2 market is mainly from the replacement of Cl_2 with H_2O_2 for bleaching in the paper industry. As H_2O_2 production becomes cheaper, it will better be able to compete with O_2 as an oxidizing agent.

F. Higher Alcohols

Background

Higher alcohols in the range of C_6 to C_{18} have many industrial applications. Commercial interest includes the whole group of primary and secondary, branched and unbranched, and even- and odd-numbered alcohols. Higher tertiary alcohols are not industrially significant [109]. The C_6 to C_{11} alcohols are called "plasticizer" alcohols and the C_{12} to C_{18} are called surface-active or "detergent" alcohols because of their respective major end use [109]. Alcohols > C_8 are often called fatty alcohols because that was their first original primary source. The world production capacity is currently 3 million ton/yr, of which 88% are synthetic [108,109].

Production

Higher alcohols are produced from three routes that involve H_2. These three routes differ in feed stocks and the nature of the process, so alcohols can be produced for specific process needs. The three routes are hydrogenation of fatty acids or esters, hydroformylation, and the Alfol process [108,109].

Fatty acids or their methyl esters obtained from the saponification or methanolysis of triglycerides, can be hydrogenated to produce alcohols and water or methanol, as shown:

$$ROOH \xrightarrow[\text{Cat.}]{H_2} ROH + H_2O$$

$$ROOCH_3 \xrightarrow[\text{Cat.}]{H_2} ROH + CH_3OH$$

The fatty acids or their methyl esters are hydrogenated at about 200 bar and 50–350°C in the presence of Cu-Cr oxide, Ni, and Pd catalyst [108].

Hydroformylation is the reaction of an olefin with CO and H_2 to form an aldehyde with one more carbon atom than the original olefin. Both normal and isoaldehydes can be formed. The ratio can be controlled through the catalyst choice. The aldehydes then can be hydrogenated to form normal and isoalcohols. This is a good process to achieve odd number alcohols [132].

$$RHC=CH_2 + CO + H_2 \longrightarrow + \begin{array}{l} RCH_2CH_2CHO \longrightarrow RCH_2CH_2CH_2OH \\ RCH_2CH(CHO)CH_3 \longrightarrow RCH_2CH(CH_2OH)CH_3 \end{array}$$

This reaction is carried out homogeneously with $Co_2(CO)_8$, $Rh(CO)_8$, or $Ru(CO)_8$, but $-NR_2$ or $-PR_2$ groups can be used instead of CO [108,109,134].

Higher alcohols also can be formed from ethylene to form a primary alcohol with an even number of carbon atoms. This is called the Alfol process. The first two steps involve the catalyst regeneration. First aluminum powder and triethyl aluminum are hydrogenated to form diethyl aluminum hydride at 110–140°C and 50–200 bar. H_2 is used in this step:

$$Al + 2Al(C_2H_5)_3 + 1.5H_2 \longrightarrow 3AlH(C_2H_5)_2$$
triethylaluminum   diethylaluminumhydride

Triethyl aluminum is obtained from the reaction with ethylene at 100°C and 25 bar. These two steps are done to increase the amount of triethylaluminum. This step is:

$$AlH(C_2H_5)_2 + H_2C=CH_2 \longrightarrow Al(C_2H_5)_3$$
diethylaluminumhydride   triethylaluminum

The next step is the strongly exothermic chain growth or insertion reaction. It takes place in a flow reactor at 120°C and an ethylene pressure of 100–140 bar.

$$Al(C_2H_5)_3 + 3nH_2C=CH_2 \longrightarrow Al[(CH_2CH_2)_nC_2H_5]_3$$

The trialkyl aluminum then is oxidized at 50–100°C with extremely dry air to the corresponding alkoxides:

$$Al[(CH_2CH_2)_nC_2H_5]_3 + 1.5O_2 \longrightarrow Al[O(CH_2CH_2)_nC_2H_5]_3$$

The final step is the saponification of the aluminum alkoxides with water to form the alcohols and aluminum hydroxide:

$$Al[O(CH_2CH_2)_nC_2H_5]_3 + H_2O \longrightarrow 3CH_3(CH_2CH_2)_nCH_2OH + Al(OH)_3$$

The selectivity to alcohols reaches 85–90% with small amounts of ester, ethers, acids, and aldehydes produced as byproducts. The alcohols are recovered by separating the water and organic phase and purified by distilling the organic phase [108].

Use

The higher alcohols are used as plasticizers and solvents in the polymer industry. They also have several other important industrial uses. They can be sulfated using H_2SO_4 to give a long-chain sulfate:

$$ROH \xrightarrow{\ H_2SO_4\ } RSO_4$$

These long-chain sulfates are used as anionic detergents [109]. Higher alcohols can react with ammonia and methyl amines:

$$ROH + NH_3/H_3CNH_2 \longrightarrow RNH_2/R_2NH/R_3N$$

to form primary, secondary, and tertiary amines [109]. These are used as nonionic detergents. Higher alcohols can undergo dehydration to form α-olefins, which are used to produce longer chained molecules [109]. Alcohols can react with H_2S to form mercaptans. Higher alcohols also can react with aldehydes to form long-chain esters which are used as plasticizers [118].

Outlook

The higher alcohols business is going to continue to grow. Much of the future work is going to go toward improving hydroformylation process and finding an economical heterogeneous process, as well as expanding the many industrial uses of higher alcohols.

REFERENCES

1. Kirschner, J. I., Growth of top 50 chemicals, in *Chemical and Engineering News*, American Chemical Society, April 8: 17 (1995).

2. Kroschwitz, J. I. and Howe-Grant, M. (eds.), *Kirk-Othmer Encyclopedia of Chemical Technology*, 4th Edition, Wiley and Sons, New York, Vol. 2, 1992, pp. 638–689.

3. Kent, J. A., (Ed.), *Riegel's Handbook of Industrial Chemistry*, 9th Edition, Van Nostrand Reinhold Co., New York, 1990.

4. Kroschwitz, J. I. and Howe-Grant, M. (Eds.), *Kirk-Othmer Encyclopedia of Chemical Technology*, 4th Edition, Wiley and Sons, New York, Vol. 10, 1993, pp. 433–514..

5. Lowrison, G. C., *Fertilizer Technology*, John Wiley & Sons, Inc., New York, 1989.

6. Grayson, D. and Eckroth, D. (Eds.), *Kirk-Othmer Encyclopedia of Chemical Technology*, 3rd Edition, Vol. 23, 1983. Urea.

7. Cotton, F. A. and Wilkinson, G., *Advanced Inorganic Chemistry*, 5th Edition, John Wiley & Sons, New York, 1988.

8. Hahn, A. V., *The Petrochemical Industry: Market and Economics*, McGraw-Hill Book Co., New York, 1970.

9. Petrochemical processes, *Hydrocarbon Processing*, March, 1993, pp. 70–139..

10. Saunders, K. J., *Organic Polymer Chemistry*, 2nd Ed., Chapman and Hall Ltd., London, 1988.

11. Buchner, W., Schliebs, R., Winter, G., and Buchel, K. H., *Industrial Organic Chemistry*, VCH Publishers, New York, 1989.

12. Busby, J. A., Knaption, A. G., and Budd, A. E. R., Catalytic processes in nitric acid manufacture, *Fert. Soc. Proc.*, No. 169, (1978).

13. Kroschwitz, J. I. and Howe-Grant, M. (Eds.), *Kirk-Othmer Encyclopedia of Chemical Technology*, 4th Edition, Wiley and Sons, New York, Vol. 13, 1993, pp. 560–561.

14. Lee, S., *Methanol Synthesis Technology*, CRC Press, Boca Raton, FL, 1990.

15. Ozturk, S. and Shah, Y., Comparison of gas and liquid methanol synthesis process, *The Chemical Engineering Journal*, 37: 177–192 (1988).

16. Tjandra, S., Anthony, R., and Akgerman, A., Low H_2/CO ration synthesis gas conversion to methanol in a trickle bed reactor, *Ind. & Eng. Chem. Res.* 32: 2602–2607 (1993).

17. Westerterp, K., Bodewes, T. Vrijland, M., and Kuczynski, M., Two new methanol converters, *Hydrocarbon Processing*, Nov.: 69 (1988).

18. Wagialla, K. and Elnashaie, S., Fluidized-bed reactor for methanol synthesis. A theoretical investigation, *Ind. Eng. Res.*, 30: 2298–2308 (1991).

19. Lee, S., Research support for liquid phase methanol synthesis process development, *Electric Power Research Report*. AP-4429, pp. 1–312, Palo Alto, CA, February 1986.

20. Palekar, V., Jung, H., Tierney, I. and Wenda, I., Slurry phase synthesis of methanol with a potassium methoxide/copper chromite catalytic system, *Appl. Catal. A: General*, 102: 13–34 (1993).

21. Chanchlanii, K., Hudgins, R. and Silveston, P., Methanol synthesis from H_2, CO, and CO_2 over Cu/ZnO catalysts, *J. Catal.*, 136: 59–75 (1992).

22. Walker, A., Lambert, R., and Nix, R., Methanol synthesis over catalysts derived from CeCu$_2$, *J. Catal.*, 138: 694–713 (1992).

23. Andriamasinoro, D., Kieffer, R., and Kiennemann, A., Preparation of stabilized copper-rare earth oxide catalysts for the synthesis of methanol from syngas, *Appl. Catal. A: General*, 106: 201–212 (1993).

24. Sizek, G., Curry-Hyde, H., and Wainwright, M., Methanol synthesis over copper ZnO promoted copper surfaces, *Appl. Catal. A: General*, 115: 15–28 (1994).

25. Lee, S., Parameswaran, V. R., and Wender, I., The roles of carbon dioxide in methanol synthesis, *Fuel Sci. Technol. Internat.*, 7(8): 1021–1057 (1989).

26. Rogerson, P. L., Imperial Chemical Industries' low pressure methanol plant, *Chem. Eng. Prog. Symp. Ser.*, 66(98): 28 (1970).

27. Refining processes, *Hydrocarbon Processing*, Nov.: 71–130 (1979).

28. Hansen, J. B. and Joensen, F., in *Natural Gas Conversion Symposium Proceedings* (Holman, A., Ed.), Elsevier, Amsterdam, 1991. pp. 457–467.

29. Prescott, J. T., *Chemical Engineering*, April 5: 60 (1971).

30. Calkins, W., Chemicals from methanol, *Catal. Rev.-Sci. Eng.*, 26(3&4): 347–358 (1984).

31. Wender, I., Chemicals from methanol, *Catal. Rev.-Sci. Eng.*, 26(3&4): 303–321 (1984).

32. Juran, K. and Porcelli, R., Convert methanol to ethanol, *Hydrocarbon Processing*, Oct.: 85 (1985).

33. Klissurski, D., Presheva, Y., and Sbsdjieva, N., Multicomponent oxide catalyst for the oxidation of methanol to formaldehyde, *Appl. Catal.*, 77: 55–66 (1991).

34. Winkelman, J., Sijbring, H., and Beenackers, A., Modeling and simulation of industrial formaldehyde absorbers, *Chem. Engin. Sci.*, 47(13/14): 3785–3792 (1992).

35. Deng, J. and Wu, J., Formaldehyde production by catalytic dehydrogenation of methanol in inorganic membrane reactors, *Appl. Catal. A: General*, 109: 63–76 (1994).

36. Baldwin, T. and Burch, R., Partial oxidation of methane to formaldehyde on chlorine promoted catalysts, *Appl. Catal.*, 75: 153–178 (1991).

37. Parmaliana, A., Frusteri, F., Mezzapica, A., A basic approach to evaluate methane partial oxidation catalysts, *J. Catal.* 143: 262–274 (1993).

38. Monti, D., Reller, A., and Baiker, A., Methanol oxidation on K$_2$SO$_4$–promoted vanadium pentoxide: activity, reducibility, and structure of catalysts, *J. Catal.* 93: 360–367 (1985).

39. Chang, F., Chen, J., and Guo, J., The ethylation kinetics of formaldehyde in a three phase slurry reactor, *Chem. Engin. Sci.*, 47: 3793–3800 (1992).

40. Ai, M., Formation of acrolien by the reaction of formaldehyde with ethanol, *Appl. Catal.*, 77: 123–132 (1991).

41. Boniface, A., BASF integrated MMA production, *The Chemical Engineer*, June 28: 15 (1990).

42. Sofianos, A., Conversion of synthesis gas to dimethyl ether over bifunctional catalytic systems, *Ind. Eng. Chem. Res.*, 30: 2372–2378 (1991).

43. Lee, S. and Gogate, M. R., A single-stage, liquid-phase DME synthesis from syngas: I. Dual catalytic activity and process feasability, *Fuel Sci. Technol. Internat.*, 9(6): 653–679 (1991).

44. Lee, S. and Gogate, M. R., A single-stage, liquid-phase DME Synthesis from syngas: II. Comparison of per-pass conversion, reactor productivity and hydrogenation extent, *Fuel Sci. Technol. Internat.*, 9(7): 889–911 (1991).

45. Lee, S. and Gogate, M. R., A single-stage, liquid-phase DME synthesis from syngas: III. Dual catalyst crystal growth, deactiviation, and activity conservation studies, *Fuel Sci. Technol. Internat.*, 9(8): 949–976 (1991).

46. Cavacanti, F., Stakheev, A., and Sacther, W., Direct synthesis of methanol, dimethyl ether, and paraffins from syngas over Pd/Zeolite Y catalysts, *J. Catal.*, 134: 226–241 (1992).

47. Haggin, J., Dimethyl ether from syngas in one step, *Chemical and Engineering News*, July 22: 20 (1991).

48. Anderson, R. B., *The Fischer Tropsch Synthesis*, Academic Press, New York, 1984.

49. Mills, A., Status and future opportunities for conversion of synthesis gas to liquid fuels, *Fuel*, 73: 8 (1994).

50. Piel, W., *High Value Structures for Oxygenates in Cleaner Burning Gasoline*, Presented at 11th Annual International Pittsburgh Coal Conference, Pittsburgh, PA., Sept. 12–16, 1994.

51. Rhodes, A., U.S. refiners choose a variety of routes to MTBE, *Oil & Gas J.*, Sept. 7: 36 (1992).

52. Bitar, L., Hazbun, E., and Piel, W., MTBE production and economics, *Hydrocarbon Processing*, Oct.: 63 (1984).

53. Herwig, J., Schleppinghoff, B., and Schulwitz, S., New low energy process for MTBE and TAME, *Hydrocarbon Processing*, June: 86 (1984).

54. Chu, P. and Kuhl, H., Preparation of MTBE over zeolite catalysts, *Ind. Eng. Chem. Res.*, 26: 365–369 (1987).

55. Al-Jarallah, A. and Lee, A., Economics of new MTBE design, *Hydrocarbon Processing*, July: 51 (1988).

56. Nicolaides, C., Stotijn, C., van der Veen E., and Visser, M., Conversion of methanol and isobutanol to MTBE, *Appl. Catal. A: General*, 103: 223–232 (1993).

57. Ladisch, M. Hendrickson, R., Brewer, M., and Westgate, P. Catalyst-induced yield and enhancement in a tubular reactor, *Ind. Eng. Chem. Res.*, 32: 1888–1894 (1993).

58. Jayadeokar, S., and Sharma, M., Absorption of isobutylene in aqueous ethanol mixed alcohols: cation exchange resins as catalysts, *Chem. Eng. Sci.*, 47(13/14): 3777–3784 (1992).

59. Fite, C., Iborra, M., Tejero, J., and Cunhill, F., Kinetics of the liquid phase synthesis of ETBE, *Ind. Eng. Chem. Res.*, 33: 581–591 (1994).

60. Cunhill, F., Vila, M., Izquierdo, J., and Cunhill, J., *Sci. Technol. Internat.*, 9(7): 889–911 (1991).

61. Refining processes, *Hydrocarbon Processing*, Nov.: 85–148 (1994)

62. Dean, J. A., *Lange's Handbook of Chemistry*, McGraw-Hill, New York, 1985.

63. Ignatius, J., Jarelin, H., and Lindquist, P., Use of TAME and heavier ethers to improve gasoline properties, *Hydrocarbon Processing*, Feb.: 51–53 (1995).

64. -2Rock, T., TAME: Technology merits, *Hydrocarbon Processing*, May: 86–88 (1992).

65. Brockwell, H. L., Sarathy, P. R., and Trott, R., *Hydrocarbon Processing*, 70: 133 (1991).

66. Fox, J. M., The different catalytic routes for methane vaporization, *Catal. Rev. Sci. Eng.*, 35(2): 169 (1992)

67. O'Hara, J. and Bela, A., Fischer-Trospch plant design criteria, *Chem. Eng. Prog.*, 73: 141–143, (1977).

68. Fuijimoto, K. and Asami, K., Two-stage reaction system for synthesis gas conversion to gasoline, *Ind. Eng. Chem. Proc. Res. Dev.*, 25: 262–267 (1986).

69. Kuo, J., Conversion of methanol to gasoline components, U.S. Patent 3,931,349 (1976).

70. Daviduk, J., Method for producing gasoline from methanol, U.S. Patent 3,998,899 (1976).

71. Chen, N. and Yan, T., M2 forming—A process for aromatization of light hydrocarbons, *Ind. Eng. Chem. Des. Dev.*, 25: 151–155 (1986).

72. Le Van Mao, R. and McLaughlin, G. P. Conversion of light alcohols to hydrocarbons over ZSM-5 zeolite and asbestos-derived zeolite catalysts, *Energy and Fuels* 3: 620–624 (1989).

73. Chang, C. D., Hydrocarbons from methanol, *Catal. Rev. Sci. Eng.*, 25(1): 1–118 (1983).

74. Givens, E., Manufacture of Light Olefins, U.S. Patent 4,079,095 (1978).

75. Butter, S. and Kaeding, W., Olefins synthesis from methanol, *J. Catal.*, 61: 155 (1980).

76. Gelsthorpe, M. R. and Theocharis, C. R., Modified aluminophosphate molecular seives: Preparation and characterization, *Catalysis Today* 2(5): 613–620 (1988)

77. Hatch, L. F., *Higher Oxo Alcohols*, Wiley and Sons, New York, 1957.

78. Ruwet, M. Ceckiewicz, S., and Delmon, B., Pure and Mo-doped BiPO$_4$, promoted by O$_2$ as a new catalyst for butyraldehyde production, *Ind. Eng., Chem. Res.*, 26: 10 (1987).

79. Cybulski, A., Liquid-phase methanol synthesis: Catalysts, mechanism, kinetics, chemical equilibria, vapor-liquid equilibria, and modeling—A review, *Catal. Rev.-Sci. Eng.* 36(4): 557–615 (1994).

80. Srinivas, G. and Chuang, S., An in-situ infrared study of the formation of n- and iso-butyraldehyde from propylene hydroformulation on Rh/SiO$_2$ and sulfided Rh/SiO$_2$, *J. Catal.*, 144: 131–147 (1993).

81. Raizada, V., Tripathi, V., and Lal, D., Kinetic studies on dehydrogenation of butanol to butyraldehyde using zinc oxide as catalyst, *J. Chem. Tech. Biotechnol.*, 56: 265–270 (1993).

82. Massoudi, R., Kim, J., King, R., and King, A., Homogeneous catalysis of the reppe reaction with iron pentacarbonyl: The production of propionaldehyde and 1–propanol from ethyene, *J. Am. Chem. Soc.*, 109: 7428–7433 (1987).

83. Chuang, S. and Pien, S., Synthesis of aldehydes from synthesis gas over Na-promoted Mn-Ni catalysts, *J. Catal.*, 128: 569–573 (1993).

84. Brown, H. S., Lessons from Bhopal, *Chemical and Engineering News*, Oct. 10 (1994).

85. Szmant, H. H., *Organic Building Blocks of the Chemical Industry*, John Wiley and Sons, New York, 1989.

86. Lewis, Sr., Richard J., *Sax's Dangerous Properties of Industrial Materials*, 9th Ed., Van Nostrand Reinhold, New York, 1995.

87. *Preliminary Studies For H. M. Factory, Gretna and Study For An Installation of Phosgene Manufacture*, Ministry of Munitions, Department of Explosive Supply, London.

88. Malatesta, L. and Bonati, F., *Isocyanide Complexes of Metals*, Wiley and Sons, New York, 1969.

89. Rando, R. J., Poovey, H. G., and Chang, S., H., *J. Liq. Chrom.*, 16(15): 3291 (1993).

90. *Occupational Exposure to Phosgene*, U.S. Department of Health, Education, and Welfare, Cincinnati, OH, 1976.

91. Weissermel, K., *Polycarbonates Recent Development*, Noyes Data Co., Park Ridge, NJ, 1970.

92. *Occupational Exposure to Diisocyanates*, U.S. Department of Health, Education, and Welfare, Cincinnati, OH, 1976.

93. Johnson, K., *Polycarbonates Recent Development*, Noyes Data Co., Park Ridge, NJ, 1970.

94. Leverkusen, W. U., Polycarbonate (PC), *Kunststoffe*, 83: 760 (1993).

95. Montella, R., *Plastics in Architecture: A Guide to Acrylic and Polycarbonate*, Marcel Dekker, New York, 1985.

96. Stinson, S. C., Polycyanurates find applications: Their chemistry remains puzzling, *Chemical and Engineering News*, Sept. 12: 31 (1994).

97. Crafts, A. S., *The Chemistry and Mode of Action of Herbicides*, Interscience Publ., New York, 1961.

98. Audus, L. J., *The Physiology and Biochemistry of Herbicides*, Academic Press, London, 1964.

99. Frear, D. E. H., *Chemistry of Insecticides, Fungicides and Herbicides*, D. Van Nostrand Co., New York, 1948.

100. Strum, C. V., *A Bitter Fog: Herbicides and Human Rights*, Sierra Club Books, San Francisco, 1983.

101. McEwen, F. L., *The Use and Significance of Pesticides in the Environment*, John Wiley & Sons, New York, 1979.

102. Tisdale, W. H. and Williams, J. W. Responsibilities of the chemist in the development of insecticides, *Agricultural Control Chemicals*, presented at the 115th National Meeting of the American Chemical Society, San Francisco, March 28 to April 1, 1949, and the 116th National Meeting of the American Chemical Society in Atlantic City, Sept. 18–23, 1949.

103. Lee, R. E., Jr., *Air Pollution from Pesticides and Agricultural Processes*, CRC Press, Boca Raton, FL, 1976.

104. FMC to build plant for new herbicide products, *Chemical and Engineering News*, Nov. 7: 8 (1994).

105. Zurer, P., Report says herbicides foul drinking water, *Chemical and Engineering News*, Oct. 24: 8 (1994).

106. Wu., J. S., Shen, S. C., and Chang, F. C. J., *J. Appl. Sci.*, 50: 1379 (1993).

107. Kent, J. (Ed.), *Riegel's Handbook of Industrial Chemistry*, 8th Edition, Van Nostrand Reinhold Co., New York, 1983.

108. Weissermk, K. and Arpe, H., *Industrial Organic Chemistry*, 2nd Edition, (Linley, C., trans.), Weinheim, New York, 1993.

109. Szmant, H., *Organic Building Blocks of the Chemical Industry*, Wiley and Sons, New York, 1989.

110. Johnson, R. and Fritz, E., (Eds.), *Fatty Acids in Industry*, Marcel Dekker, Inc., New York, 1989.

111. Markley, K., (Ed.), *Fatty Acids: Their Chemistry, Properties, Production, and Uses. Part 5*, Interscience Publ., New York, 1968.

112. Pattison, E. (Ed.), *Fatty Acids and Their Industrial Applications*, Marcel Dekker, Inc., New York, 1968.

113. Grau, K. R., Cassano, A., and Baltanas, M., Catalyst and network modeling in vegetable oil hydrogenation processes, *Catal. Rev.-Sci. Eng.*, 30(1): 1–48 (1988).

114. Ralston, A., *Fatty Acids and Their Derivatives*, Wiley and Sons, New York, 1948.

115. Mitsubishi Kasei Corp., 1,4–Butanediol/tetrahydrofuran production technology, *CHEMTECH*, 18: 759–763 (1988).

116. Brownstein, A., 1,4–Butanediol and tetrahydrofuran: Exemplary small volume commodities, *CHEMTECH*, 21: 506–510 (1991).

117. Dreyfuss, P. *Poly(Tetra-Hydrofuran)*, Gordon and Breach Science Publishers, New York, 1987.

118. Pine, S., *Organic Chemistry*, 5th Edition, McGraw-Hill Book Company, New York, 1987.

119. Phillips, M., Catalytic hydrogenation of glucose to sorbitol using a highly active catalyst, *Brit. Chem. Eng.*, 8: 767–769 (1963).

120. Shallenberger, R. and Birch, G., *Sugar Chemistry*, The AVI Publishing Company, Inc., Westport, CT, 1975.

121. Gas processes, *Hydrocarbon Processing*, April: 99–150 (1996).

122. Duvnjak, Z., Turcotte, G., and Duan, Z., Production and consumption of sorbitol and fructose by saccharomyces cerevisiate ATCC 36859, *J. CHEMTECH. Biotechnol.*, 52: 527–537 (1991).

123. Ling, P., Cha, J., Fagg, B., Use of manganese (II)-polyol complexes to accelerate high pH peroxide oxidation reactions, *Ind. Eng. Chem. Fund.*, 23: 29–33 (1984).

124. Choudhury, J., H_2O_2 Business is zooming, *Chem. Eng.*, June 20: 32–37 (1988).

125. Wallace, J., *H_2O_2 in Organic Chemistry*, DuPont, Wilmington, DE, 1962.

126. Arden, M., *Oxygen Elementary Forms and H_2O_2*, W. A. Benjamin Publ., New York, 1965.

127. Berglin, T. and Schoon, N., Selectivity aspects of hydrogenation stage of the anthroquinone process for hydrogen peroxide production, *Ind. Eng. Process Des. Dev.*, 22: 150–153 (1983).

128. Knarr, R., Velasco, M., Lynn, S., and Tobias, C., The electrochemical reduction of 2–ethyl anthroquinone, *J. Electrochem. Soc.*, 139(4): 948–954 (1992).

129. Santacesaria, E., Wilkson, P., Babini, P., and Cara, S., Hydrogenation of 2–ethyltetrahydroanthroquinone in the presence of palladium catalyst, *Ind. Eng. Res.*, 27: 780–784 (1988).

130. Technical note, A DuPont process may lead to H_2O_2 satellite plant, *Chem Week*, Dec. 9: 20 (1987).

131. Peaff, G., Water treatment companies embrace expanding international markets, *Chemical and Engineering News*, 72(46): 15–23 (1994).

132. Hatch, L., *Higher Oxo Alcohols*, Enjay Company, Inc., New York, 1957.

133. Haggin, J., Innovations in catalysis create environmentally friendly THF process, *Chemical and Engineering News*, 73: 14 (1995).

134. Petrochemical processes, *Hydrocarbon Processing*, March: 87–115 (1995).

135. Tabak, S. A. and Yurchak, S., *Catal. Today*, 6(3): 307 (1990).

136. Yurchak, S. and Wong, S. S., *Hydrocarbon Asia 1991*, July/August: 38, (1991).

137. U.S. Department of Energy Working Group on Research Needs for Advanced Coal Gasification Techniques (COGARN) (S.S. Penner, Chairman), *Coal Gasification: Direct Application and Synthesis of Chemicals and Fuels*, DOE Contract No. DE-AC01–85 ER 30086, DOE Report DE/ER-0326, U.S. Department of Energy, Springfield, VA, June 1987.

138. Lee, S., Gogate, M. R., Fullerton, K. L., and Kulik, C. J., U.S. Patent No. 5,459,166 (Oct. 17, 1995).

139. Dybkjaer, I., Development in ammonia production technology—Historical review, *Nitrogen 91 International Conference Preprints*, Copenhagen, Denmark, June 4–6, 1991.

140. O'Mant, D. M. and Twitchett, H. J., Catalytic activity of t-amines in the phenyl isocyanate-butanol reaction, cited in Johnson, P. C. in *Advances in Polyurethane Technology*, (Busit, J. M., and Gudgeon, H., Eds.), Wiley and Sons, New York, 1968, p. 10.

141. Hostettler, F. and Cox, E. F., *Ind. Eng. Chem.*, 52: 609 (1960).

142. Thayer, A. M., Metallocene catalysts initiate new era in polymer sysnthesis, *Chemical and Engineering News*, Sept. 11: 15–20, (1995).

4

Methane Derivatives Directly from Methane

Methane accounts for approximately 85 percent of the composition of natural gas with heavier hydrocarbons, nitrogen, and, in some regions, helium accounting for the other 15 percent [1]. Purification of methane is carried out at ambient or low temperature absorption (5–10 thousand ppm and 1–2 thousand ppm, respectively) and low-temperature fractionation (100 ppm) [2]. Impurities in the methane, such as heavier hydrocarbons, promote undesirable side reaction. Methane is also produced in an increasing number of organic waste-disposal plants [3]. Methane is used as feedstock to produce many chemicals, including hydrogen cyanide, carbon disulfide, and chlorinated methanes.

I. HYDROGEN CYANIDE

Like the hydrogen halides, hydrogen cyanide (HCN) is a covalent, molecular substance, but is a weak acid on aqueous solution (pK = 9.0). The colorless gas is extremely toxic, though much less so than H_2S. Hydrogen cyanide condenses at 25.6°C to a liquid with a very high dielectric constant (107 at 25°C) [183]. There are two main processes for the production of hydrogen cyanide from methane. The Andrussow process (originally developed by BASF) is the most preferred method to produce hydrogen cyanide as a major product from hydrocarbons [4]. Methane, ammonia, and air are reacted to produce hydrogen cyanide and water.

$$CH_4 + NH_3 + 1.5O_2 \xrightarrow{\ [cat.]\ } HCN + 3H_2O \qquad \Delta H = -473 \text{ kJ/mol}$$

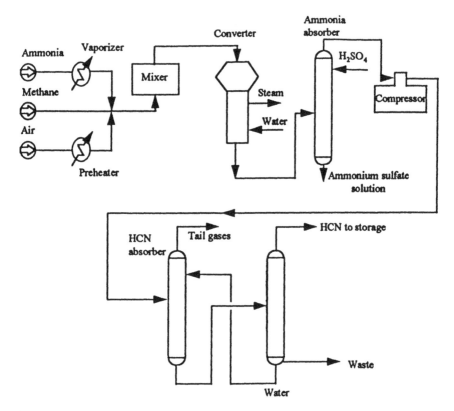

Figure 4.1   Hydrogen cyanide via the Andrussow process. *Source*: [7].

The reaction takes place over a platinum or platinum-rhodium catalyst at temperatures of 1000–1500°C and atmospheric pressure [2,4–6].

Figure 4.1 is a flow diagram of the Andrussow process [7]. To avoid the decomposition of methane and ammonia, the ratio of reactants must be carefully controlled. The products are cooled where care is taken to avoid the formation of "azulmic acids," polymers formed by the reaction between hydrogen cyanide, ammonia, and water. The products go to a scrubbing tower where unconverted ammonia is absorbed in sulfuric acid. The product is then absorbed in water, stripped, and distilled to produce greater than 99% HCN [8]. Yields are 70 and 60% for methane and ammonia, respectively.

Methane must be essentially pure to prevent the formation of soot and poisoning the catalysts. The removal of higher hydrocarbons and desulfurization of the methane is necessary [9]. This process has been adapted by

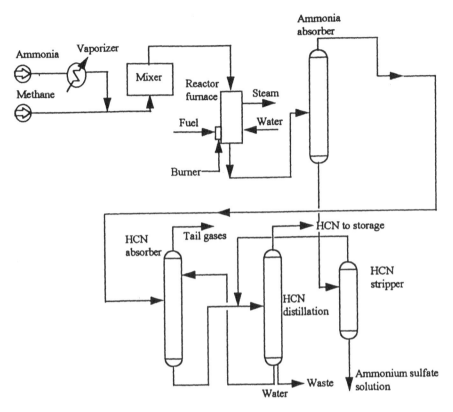

Figure 4.2 Hydrogen cyanide via the Degussa process. *Source*: adapted from [14].

American Cyanamide, DuPont, Goodrich, ICI, Mitsubishi, Monsanto, Monte-catini, Nippon Soda, and Röhm & Haas [10].

The Degussa BMA (Blausäure-Methan-Ammoniak, or hydrocyanic acid-methane-ammonia) process also is used in the production of hydrogen cyanide from methane. The difference between the Andrussow process and the Degussa process is that the latter does not use air in the synthesis of hydrogen cyanide. The reaction is as follows:

$$CH_4 + NH_3 \xrightarrow{\text{[cat.]}} HCN + 3H_2 \qquad \Delta H = 251 \text{ kJ/mol}$$

Platinum, ruthenium, or aluminum-lined ceramic tubes are used as a catalyst [11]. The reaction takes place at 1200–1400°C [12,13].

A fuel burning combustion chamber precedes the reactor to provide the endothermic heat of reaction [14] (Figure 4.2). Ammonia then is removed and hydrogen cyanide is absorbed, stripped, and distilled. The yield of HCN is around 83% (based on NH_3) which makes this process slightly cheaper than Andrussow process [15]. Aside from the higher yield, hydrogen also can be recovered economically. However, the long residence time makes the process unsuitable for large scale production [16]. The reactor effluent gases are more concentrated than in the Andrussow process and the remainder is mainly hydrogen instead of nitrogen. Yields are about 83% (based on ammonia), which presents a substantial cost saving over the Andrussow process [15]. This process is currently being used by Degussa (Germany, Belgium, United States) and Lonza (Switzerland) [17].

The Shawinigan process (also called the Flouhmic process) was developed to use hydrocarbons (originally propane as well as methane) to produce hydrogen cyanide [18]. It is used mainly in regions where methane is not available. The reaction of propane and ammonia is: [19]

$$C_3H_8 + 3NH_3 \longrightarrow 3HCN + 7H_2 \qquad \Delta H = +630 \text{ kJ/mol}$$

This process takes place in an electrically-heated bed of coke particles at 1300–1600°C [20]. In order for the process to be economical, cheap and abundant power is required [21]. The advantages of the Shawinigan process are: 1) hydrogen can be recycled to the ammonia feed unit, 2) there is no need for ammonia absorption facilities, and 3) the absence of water vapor and low concentration of ammonia suppresses undesirable polymerization [22]. This process is currently used in a commercial plant in South Africa [23].

Other processes involved in the formation of hydrogen cyanide include:

1. indirect reaction between carbon monoxide and methanol (DuPont) [24]

$$CO + CH_3OH \longrightarrow HCOOCH_3 \xrightarrow{NH_3} HCONH_2 + CH_3OH$$
$$\qquad\qquad\qquad\quad \text{methyl formate} \qquad\qquad \text{formamide}$$

$$HCONH_2 \longrightarrow HCN + H_2O$$

2. acidification of sodium cyanide [25]

$$Na + NH_3 \longrightarrow NaNH_2 + 0.5H_2$$

$$NaNH_2 + C(coke) \longrightarrow NaCN \xrightarrow{H_2SO_4} HCN$$

3. as a byproduct in the ammoxidation of propylene to acrylonitrile (Sohio Process) [26]

$$CH_3CH=CH_2 + NH_3 + 1.5O_2 \longrightarrow NC-CH=CH_2 + 3H_2O$$

acrylonitrile

The Sohio process provides strong competition against the synthesis of hydrogen cyanide from methane and ammonia [27] especially since acrylonitrile is in high demands [28]. The amount of hydrogen cyanide is 10–24 wt% relative to acrylonitrile [29]. However, improved catalyst is reducing the amount of hydrogen cyanide produced [30]. This process accounts for about 25% of the hydrogen cyanide used in the world [31]. The DuPont process and the acidification of sodium cyanide are now considered obsolete [32].

Hydrogen cyanide is a precursor for many chemicals. The United States, Western Europe, and Japan currently produce about 1.1 million tons per year [33] About 40% of the hydrogen cyanide produced is used in the production of building blocks of Nylon 6/6, 30% for the production of methyl methacrylate (MMA), 10% for cyanogen chloride, and the other 20% to produce inorganic cyanides, chelating agents, and other chemicals [34]. Many of these chemicals will be discussed in this chapter. Hydrogen cyanide cost about $0.50/lb in 1989 [35].

A. Methacrylates

The most common process used to produce methyl methacrylate ($0.71/lb) is the acetone cyanohydrin process [36]. Acetone is reacted with hydrogen cyanide and methanol to produce methyl methacrylate:

$$(CH_3)_2C=O + HCN + CH_3OH \longrightarrow CH_2=C(CH_3)CO \cdot OCH_3$$

This actual mechanism consists of three reactions as shown in Figure 4.3. The first reaction combines acetone and hydrogen cyanide to form acetone

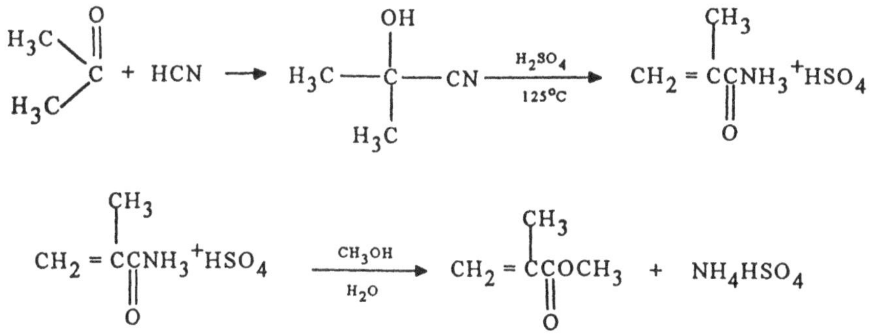

Figure 4.3   Mechanism of acetone cyanohydrin process.

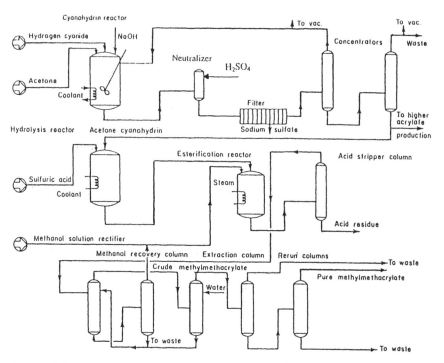

Figure 4.4   Methyl methacrylate process. *Source*: [38].

cyanohydrin. Then acetone cyanohydrin and sulfuric acid react at 125°C to produce methacrylamide sulfate. Finally, the methacrylamide sulfate is reacted with methanol and water to form methyl methacrylate. The yield of methyl methacrylate is about 85% [37].

The process flowsheet is given in Figure 4.4 [38]. Acetone and hydrogen cyanide are reacted in a strong base. Sulfuric acid is used to neutralize excess alkali to prevent the decomposition of the cyanohydrin; then sodium sulfate is filtered out and the unreacted HCN and acetone are recycled by distillation. The cyanohydrin is hydrolyzed with sulfuric acid to form methacrylamide sulfate. The methacrylamide sulfate reacts with methanol to form methyl methacrylate and ammonium bisulfate. Distillation removes methyl methacrylate and unreacted methanol which is recycled. Water extraction is used to remove any excess methanol and the monomer is purified to 99.8% in a rerun tower. This is the only process used in United States to produce methyl methacrylate [39].

Another process used in Japan for making methyl methacrylates is the Escambia process. Methyl methacrylate is made by nitric acid oxidation of isobutylene to methacrylic acid and esterification with methanol [40].

$$CH_2=C(CH_3)_2 \xrightarrow{(O)} CH_2=C(CH_3)COOH \xrightarrow{CH_3OH} CH_2=\underset{\underset{CH_3}{|}}{C}-\overset{\overset{O}{\|}}{C}-OCH_3$$

In 1991, the production of methyl methacrylate was 1.84 million tons per year [41]. The United States, Western Europe, and Japan produced 0.66, 0.54, and 0.46 million tons per year, respectively [42]. The cost for methyl methacrylate in 1991 was $0.62/lb [43]. Most acrylics start with methyl methacrylate monomer (MMA). Methyl methacrylate is used in the production of poly(methyl methacrylate) and in copolymers to improve the impact resistance of other vinyl polymers [44]. Poly(methyl methacrylate) is a colorless transparent plastic with a higher softening point, better impact strength, and better weatherability than polystyrene [45].

Acrylic monomers are polymerized by a free-radical process with the aid of peroxides. Methyl methacrylate is usually copolymerized with other acrylates (e.g., methyl or ethyl acrylate) for molding and extrusion compounds. Methyl methacrylate is used in a variety of products such as Plexiglas, lighting fixtures, plumbing and bathroom fixtures, surface-coating resins, and in automotive industries for control knobs, instrument covers, and taillights [46].

B. Cyanogen Chloride

Cyanogen chloride is produced by reacting hydrogen cyanide with chlorine:

$$HCN + Cl_2 \longrightarrow ClCN + HCl \qquad \Delta H = -89 \text{ kJ/mol}$$

The importance of cyanogen chloride is that it introduces cyano groups onto electron-rich substrates [47].

Trimerization of cyanogen chloride produces cyanuric chloride.

Cyanuric chloride ($\Delta H = -233$ kJ/mol)

Cyanuric chloride has three chlorine substituents and a stable aromatic triazine ring. Different substituents can be introduced onto the triazine because the chlorines have different reactivities. Cyanuric chloride is used in pesticides, herbicides, dyes, detergent brighteners, and so on [48]. Cyanuric chloride also is used in the manufacture of hindered-phenol triazines as antioxidants for polyolefin resins. Cyanuric chloride production is more than 100,000 ton/yr [49]. American Cyanamid, Nilok, and Geigy are three U.S. producers of cyanuric chloride [50].

Hydrolysis of cyanuric chloride yields cyanuric acid ($1.16/lb):

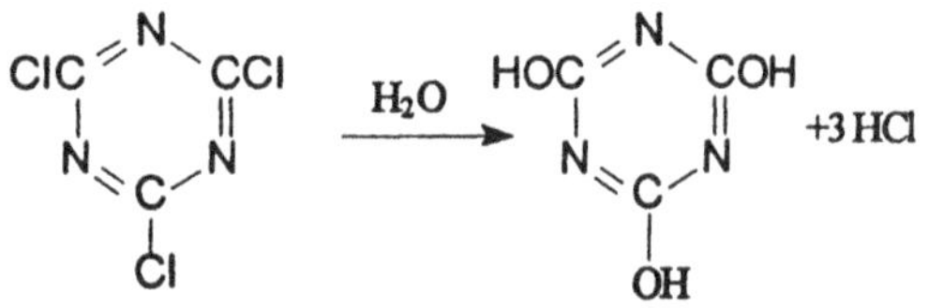

Cyanuric acid can stabilize cyanuric chloride to use as a swimming-pool chlorinating agent. Cyanuric acid also can remove nitrogen oxide pollutants.

The three chlorine groups in cyanuric chloride have different reactivities which make possible the stepwise introduction of different substituents. Atrazine, a very important herbicide, is made by reacting cyanuric chloride with monoethyl- and monoisopropylamine:

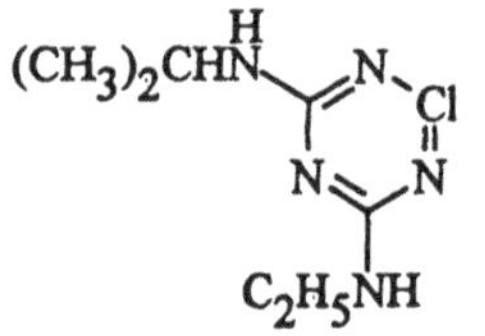

Atrazine has the general formula

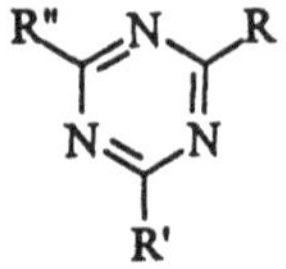

where R, R′, and R″ can be Cl itself or a substituted amino group, a methoxy radical, or a thiol group.

Triallyl cyanurate is made from cyanuric chloride and allyl alcohol. Triallyl cyanurate is used as a comonomer to impart higher temperature resistance to polyesters. It also is used as an organic catalyst for polymer grafting reactions in the solid phase, including polypropylene-g-maleic anhydride (a polypropylene-graft copolymer) [184].

Cyanuric acid amide (melamine, $0.50/lb) can be synthesized by reaction between cyanuric chloride and ammonia [51]. Another method is the trimerization of dicyandiamide. However, these reactions have been replaced with the cyclization of urea with the loss of carbon dioxide and ammonia [52]. The worldwide production capacity of melamine in 1990 was about 550,000 ton/yr [53]. Melamine is used to form melamine resins to be used as thermosetting resins, glues, and adhesives.

Thermal condensation of cyanogen chloride with acetonitrile produces malononitrile ($7.85/lb)

$$N\equiv CCl + CH_3-C\equiv N \longrightarrow N\equiv C-CH_2-C\equiv N + HCl$$

Malononitrile is used for the synthesis of fine chemicals and heterocyclic compounds [54].

C. Chelating Agents (Sequestrants)

Chelating agents are ionic complexes that can attach to a metal ion at two or more points. The simplest chelating agent known is ethylenediamine tetracetic acid (EDTA) [55]; it is formed by reaction between hydrogen cyanide, formaldehyde, and ethylenediamine (EDA) ($1.38/lb) in sodium hydroxide.

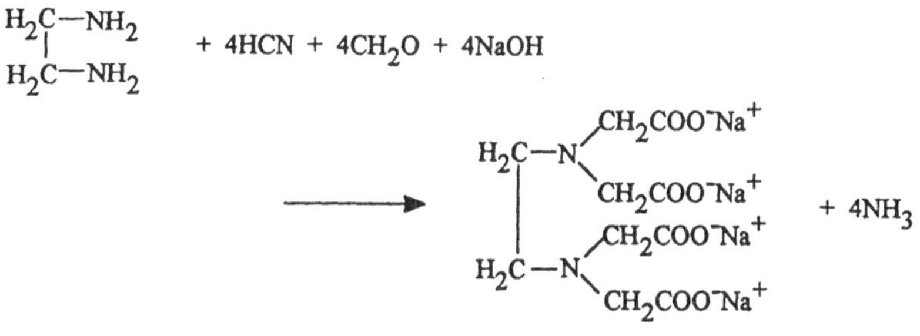

Another method for producing EDTA is by using sodium cyanide with formaldehyde followed by alkaline hydrolysis of the intermediate nitrile [56].

Chelation occurs when the electron-donating nitrogen and the acetic anion attaches to the metal ion (e.g., Cu^{2+}), to form a cyclic ring (Figure 4.5).

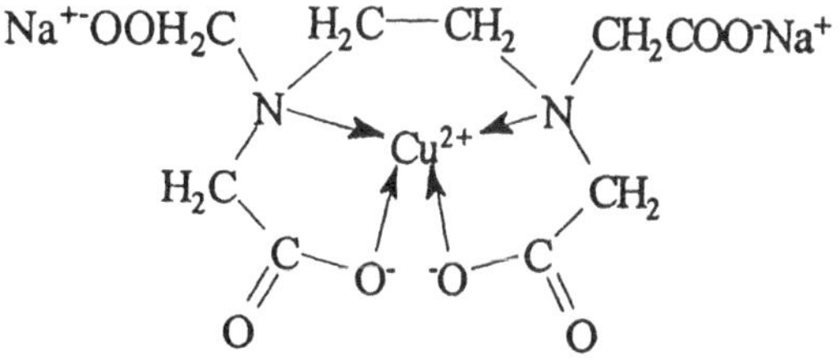

Figure 4.5 Chelation.

EDTA is used to prevent pesticides from precipitation in hard-water solutions, to soften water in soap and detergent, and to prevent the catalytic effects of metal ions present in water in the cold rubber SBR process [57]. EDTA attached to iron catalyzes the direct hydroxylation of phenolic compounds [58]; it also is the best way to supply iron-deficient citrus trees [59]. Other chelating agents include nitrilotriacetic acid (NTA) made from ammonia and diethylenetriamine pentacetic acid from diethylenetriamine [60].

Chelates are used in textile wet processing, paper processing, and water treatment [61]. They prevent the interference of the metal ions that could cause interference in a product's color and undesirable reactions, and promote stability of other compounds [62].

EDTA costs $0.43/lb in a sodium salt solution; 30% of the 80 million pounds of ethylenediamine produced in the United States is used to produce chelates [63].

D. Lactic Acid

The first commercial method of making lactic acid was by fermentation of carbohydrates. Lactic acid ($1.12/lb) can be synthesized by reacting acetaldehyde with hydrogen cyanide.

$$CH_3CHO + HCN \longrightarrow CH_3CHOHCN \xrightarrow{H_3O^+} CH_3CHOHCOOH$$

The process was developed by Monsanto to use the hydrogen cyanide byproduct of acrylonitrile from the Sohio process [64]. However, synthetic lactic acid is not as competitive as that from the fermentation of starch or sugar [65].

Ethyl lactate is produced by reacting acetaldehyde, hydrogen cyanide, and ethyl alcohol.

$$CH_3CHO + HCN \longrightarrow CH_3CHOHCN \xrightarrow{C_2H_5OH} CH_3CHOHCOOC_2H_5$$

Ethyl lactate is used as a high-boiling solvent for cellulose ether [66]. Phenyl mercuric lactate is obtained by reacting phenyl mercuric acetate with lactic acid. It is used in slime control of paper stock [67].

Worldwide production of lactic acid is estimated at 40,000 ton/yr with only 33% produced synthetically [68]. Lactic acid is used in the food industry as a flavoring acid, baking agent, pickling agent, and to keep products fresh and tender [69,70,71]. Lactic acid also is used in antiperspirants, leather tanning, as a component of permanent floor-polish formulations, as a pH control agent, and in the textile industry [72,73]. Having two reactive groups (-OH, -COOH), lactic acid can homopolymerize; poly(lactic acid) has been investigated as a material for surgical sutures. Due to the bifunctional reactive groups as well as the food-friendly nature, lactic acid is becoming more important for novel material development for biodegradability and bioavailability.

Lactic acid can be found in everyday products like cheese spreads, dry egg powder, olives, poultry, salad dressing mix, and wine [74].

E. Sodium Cyanide

Sodium cyanide ($0.65/lb) is produced by absorbing hydrogen cyanide in a sodium hydroxide solution:

$$HCN + NaOH \longrightarrow NaCN + H_2O$$

Anhydrous hydrogen cyanide and a 50% solution of sodium hydroxide are fed into a reactor. The reactants are heated in a crystallizer to remove water and form the sodium cyanide crystals. The sodium cyanide is filtered, then dried on a conveyor. Mechanical devices (e.g., cyclones) are used to produce wither briquettes or granular products. The feed must be essentially pure to obtain high quality, 99% sodium cyanide [75]. Figure 4.6 is a flow diagram of the sodium cyanide process [76].

In 1989, the amount of sodium cyanide used in the world was 340,000 tons [77]. Sodium cyanide is used in electroplating, case hardening, gold and silver extraction, synthesis of iron blues, and synthesis of a large number of chemicals [78]. Electroplating is considered the largest use of sodium cyanide; however, tighter restriction on cyanide discharge forced a considerable decline in this usage [79].

Case hardening of steel occurs when sodium cyanide oxidizes in the presence of a trace of iron or nickel oxide. The cyanide oxidizes first to cyanate and then to carbonate in the following manner:

$$2NaCN + O_2 \longrightarrow 2NaCNO$$

$$2NaCNO + 1.5O_2 \longrightarrow Na_2CO_3 + N_2 + CO_2$$

The active carbon and nitrogen then are absorbed into the steel surface [80]. Extraction is carried out by dissolving the metals in an aqueous sodium cyanide. Most metals, except lead and platinum, will dissolve in minute quantities in aqueous sodium cyanide. The metal then is recovered by precipitation with zinc dust, electrodeposit, or absorption on carbon [81].

Sodium cyanide also is used in chemical synthesis. As previously mentioned, sodium cyanide is used to produce the sequestrant EDTA [82]. Chloroacetic acid reacts with sodium cyanide in the presence of soda ash to yield cyanoacetic acid. Sodium cyanide reacting with α-chloromethylnaphthalene followed by hydrolysis yields α-naphthaleneacetamide and the corresponding carboxylic acid; these are two of several common plant hormones, growth acceler-

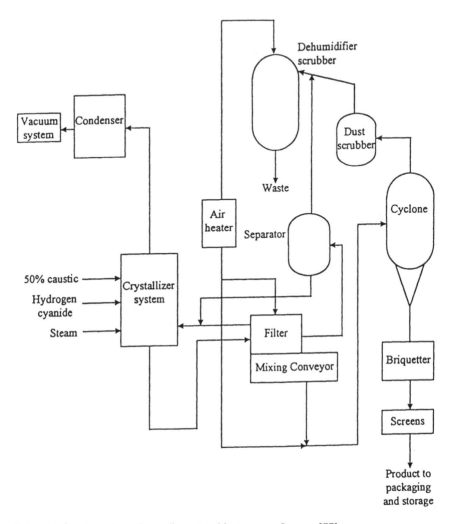

Figure 4.6   Flowsheet for sodium cyanide process. *Source*: [77].

ators, or auxins used in horticulture [83]. Sodium cyanide also is used in the production of oxymetazoline hydrochloride ($136.50/lb), a popular nasal decongestant retailed as a 0.5% solution under a variety of tradenames [84].

F. Ferrocyanides

Ferrocyanides are compounds containing iron(II) bonded to six cyano groups in the form of $[Fe(CN)_6]^{4-}$ ion [85]. Ferric ferrocyanide pigments, also known as "Prussian blue," are one of the most popular ferrocyanides. Ferric ferrocyanide is produced by absorbing hydrogen cyanide in a slurry of ferrous hydroxide in sodium (or potassium) hydroxide to form sodium (or potassium) ferrocyanide:

$$5H_2O + 6HCN + FeO + 4NaOH \longrightarrow Na_4[Fe(CN)_6] \cdot 10H_2O$$

Sodium ferrocyanide reacts with ammonium and ferrous sulfates to give a white precipitate of ferrous ferrocyanide: [86]

$$2NH_4^+ + Fe^{++} + Na_4[Fe(CN)_6] \longrightarrow Fe(NH_4)_2[Fe(CN)_6] + 4Na^+$$

$$Fe(NH_4)_2[Fe(CN)_6] + 4Na^+ \longrightarrow FeNH_4[Fe(CN)_6]$$

Oxidizing the ferrous ferrocyanide with chlorates or dichromates yields the blue ferric ferrocyanide [87]. The particle size can be controlled by initial concentration and temperature (20–60°C) [88]. Dark ferrocyanates have particle sizes of 0.01–0.05 μm. Bright blue ferrocyanates have particle sizes of 0.05–0.2 μm. Ferric ferrocyanide is used for surface coating, printing ink, and colored paper [89].

A variety of ferrocyanide complexes can be formed when one or more of the cations of the alkali or alkali earth salt is replaced by a complex cation [90].

G. Cyanoacetic Acid

Cyanoacetic acid is produced by the reaction of sodium cyanide and sodium chloroacetate. The product is neutralized to give cyanoacetic acid [91].

$$ClCH_2COONa + HCN \xrightarrow{\text{NaOH}} NCCH_2COONa \longrightarrow NCCH_2COOH$$
$$\text{cyanoacetic–acid}$$

Cyanoacetic acid is used to produce diethyl malonate [92].

$$NCCH_2COOH \xrightarrow[H^+]{C_2H_5OH} CH_2(COOC_2H_5)_2$$

Table 4.1 Alkyl Groups in Barbiturates and Thiobarbiturates [92]

R	R'	Q'	Name	Physiological effect
Et-	Et-	O	Barbital	Hypnotic
Et-	Ph-	O	Phenobarbital	Sedative, anticonvulsant
Et-	2-pentyl	O	Pentobarbital	Anesthetic, anticonvulsant
Et-	iso-Am	O	Amobarbital	Hypnotic
Allyl-	2-pentyl	O	Secobarbital	Preanesthetic
Et-	iso-Am	S	Thiopental	Hypnotic, anesthetic

This cyanoacetic ester is used to manufacture most barbiturates and thio-barbiturates, a family of sedatives. The hydrogens on the center carbon are very reactive to alkyl halides. Barbiturates and thiobarbiturates then are made from the alkyl intermediates by reaction with urea and thiourea, respectively. The reaction is:

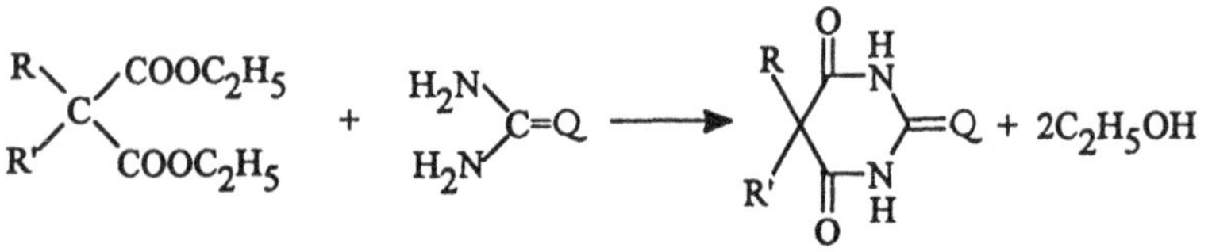

$$
\begin{array}{c}
R \\
 \\
R'
\end{array}
C
\begin{array}{c}
COOC_2H_5 \\
 \\
COOC_2H_5
\end{array}
+
\begin{array}{c}
H_2N \\
 \\
H_2N
\end{array}
C{=}Q
\longrightarrow
{=}Q + 2C_2H_5OH \qquad (18)
$$

Depending on the alkyl groups (R and R') and whether urea or thiourea is used, the product is given in Table 4.1 [93].

Diethyl malonate also is used to produce phenylbutazone, another chemical used in the pharmaceutical industry as an antirheumatic [94].

Cyanoacetic acid also is an intermediate in the production of synthetic caffeine and certain polymethine dyes [95]. Cyanoacetic esters reacting with formaldehyde are used to produce cyanoacrylate adhesives such as "miracle glue" and "crazy glue" [96].

H. Orthoformic Esters

Orthoformic esters are produced by reacting hydrogen cyanide, hydrogen chloride, and an alcohol:

$$HCN + HCl + 3ROH \longrightarrow HC(OR)_3 + NH_4Cl$$

Ethyl orthoformate is an intermediate for acrylic fiber dyes and photographic sensitizers of the polymethine family.

I. *t*-Butylamine

Most tertiary butylamine (and other tertiary alkyl amines) is produced using hydrogen cyanide in the Ritter reaction. Isobutylene [97] or methanol [98] is reacted with hydrogen cyanide under acidic conditions to generate *t*-butylamine ($1.46/lb):

$$H_3C \diagdown C=CH_2 + HCN \xrightarrow[H_2O]{Acid} (CH_3)_3CNHCH \overset{O}{\overset{\|}{}}$$

$$\xrightarrow[H_2O]{Acid} CH_3 \underset{\underset{CH_3}{|}}{\overset{\overset{CH_3}{|}}{C}}NH_2 + HCOOH$$

The Ritter reaction proceeds by a nucleophilic addition of hydrogen cyanide followed by the hydrolysis of the intermediate formamide. Higher tertiary alkyl amines are produced using the same method with higher alkenes and alcohol. This is considered the most practical way to produce tertiary amines [99].

Tertiary amines have unusually low viscosities, are more soluble in petroleum solvents, and are more selective and more stable than nontertiary amines [100]. These amines are corrosion inhibitors and are intermediates in the preparation of materials coating, plastics, textile, paper. and leathers [101]. They are used as precursors of agricultural chemicals, like herbicides and insecticides, pharmaceutical and therapeutic agents, and in fuel oil additives such as aviation jet fuels, gasoline, and in lubricants [102].

Tertiary butylamine is used to make accelerators and hardeners for epoxy resins [103] and Celiprolol [104], a chemical used in the treatment of hypertension and anginal pectoris. Tertiary butylamine also is used to make products of isobutylene by way of Friedel-Crafts reaction [105].

The only other successful process to produce tertiary butylamine is olefin amination. Isobutylene is reacted with ammonia using a zeolite catalyst at temperatures of 200–350°C and pressures as high as 18 MPa [106]. This process has a low conversion which makes recycling unconverted feed economical [107].

Tertiary butylamine had a world wide production of 1.4 million tons in 1988 [108]. Over 57% was produced using the Ritter reaction. Sterling Chemicals, Nitto Chemical Industry, and Sumitomo Chemicals use the Ritter process [109]. BASF is the only company producing tertiary butylamine with the olefins amination process.

Table 4.2 Half-Lives of Free Radical Initiators

Initiator	Half-life at									
	50°C	60°C	70°C	85°C	100°C	115°C	130°C	145°C	155°C	175°C
Azobisisobutyronitrile	74 hr		4.8 hr		7.2 min					
Benzoyl peroxide			7.3 hr	1.4 hr	19.2 min					
Acetyl peroxide	158 hr		8.1 hr	1.1 hr						
Lauryl peroxide	47.7 hr	12.8 hr	3.5 hr	31 min						
t-Butyl peracetate				88 hr	12.5 hr	1.9 hr	18 min			
Cumyl peroxide						13 hr	1.7 hr	16.8 min		
t-Butyl peroxide					218 hr	34 hr	6.4 hr	1.38 hr		
t-Butyl hydroperoxide					388 hr				44.9 hr	4.81 hr

Values are for benzene or toluene solutions of the initiators.
Source: [185].

J. Dimethyl Hydantion

Reacting hydrogen cyanide with acetone and ammonium carbonate produce dimethyl hydantion:

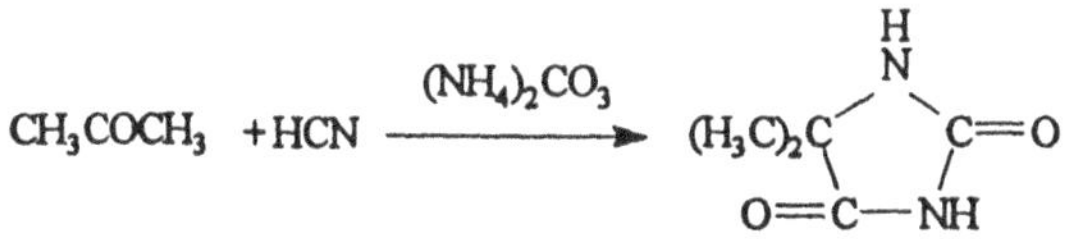

$$CH_3COCH_3 \ +HCN \ \xrightarrow{(NH_4)_2CO_3} \ (H_3C)_2C \underset{O=C-NH}{\overset{\overset{\displaystyle H}{\underset{\displaystyle N}{}}}{\diagdown}} C=O$$

The chlorinated derivative of dimethyl hydantion was the first solid bleach and was used in swimming pool chlorination [110]. Reacting dimethyl hydantion with formaldehyde produces mono- and dimethylol dimethylhydantion which are used as preservatives in cosmetics [111]. Monomethyloldimethylhydantion is converted into a resin used in the formulation of hair lacquers upon heating [112]. Dimethyl hydantion also is used in the synthesis of several α-amino acids, including phenylalanine found in aspartame [113].

K. Azobisisobutyronitrile

Azobisisobutyronitrile (AIBN) is produced by reacting acetone, hydrogen

$$CH_3COCH_3 \ \xrightarrow[HCN]{H_2N-NH_2} \ \cdots \ \xrightarrow[oxidation]{mild} \ N\equiv C-\underset{CH_3}{\overset{CH_3}{C}}-N=N-\underset{CH_3}{\overset{CH_3}{C}}-C\equiv N \quad (22)$$

AIBN

cyanide, and hydrazine, followed by mild oxidation:

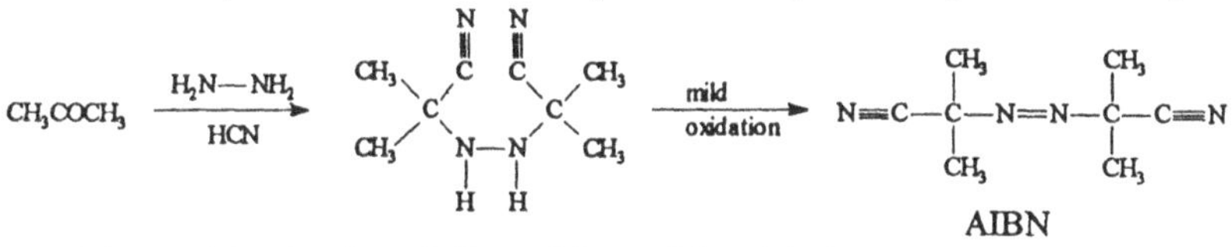

Azobisisobutyronitrile is used as a catalyst in vinyl polymerization and as a blowing agent for making foam rubber [114]. Azobisisobutyronitrile also is used as a free radical initiator (FRI) in graft copolymerization.

Azonitriles have generally been considered to be "cleaner" initiators in the sense of being devoid of chain transfer; however, recent research work indicates this is not true. Other initiators/chain-transfer agents include: *t*-butyl peroxide, cumyl peroxide, lauryl peroxide, benzoyl peroxide, *t*-butyl hydroperoxide, cumyl hydroperoxide, dicumyl peroxide, and persulfate; the half-lives of these initiators are given in Table 4.2 [185].

L. Oxamide

Oxamide is produced by reacting hydrogen cyanide with oxygen and $Cu(NO_3)_2$ in an aqueous solution [115]:

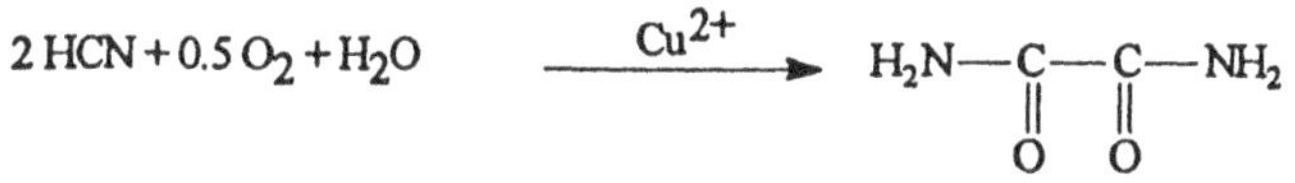

$$2\,HCN + 0.5\,O_2 + H_2O \ \xrightarrow{Cu^{2+}} \ H_2N-\underset{O}{\overset{}{\underset{\|}{C}}}-\underset{O}{\overset{}{\underset{\|}{C}}}-NH_2$$

This process was developed by Hoechst and is used at a 10,000-ton/yr plant in Italy. The oxamide is used as a slow-release fertilizer because of its low water solubility.

II. CARBON DISULFIDE

A. Production Processes

There are three methods to produce carbon disulfide. Two of these processes involve the direct reaction of carbon (in the form of coal) and sulfur:

$$C + 2S \longrightarrow CS_2 \tag{4.1}$$

These processes are the retort process and the electrothermal process [116]. The third process is the methane-sulfur process.

Retort Process

Carbon and sulfur are reacted (Reaction 4.1) in the absence of air in oval or cylindrical vessels called retorts. The vessels are approximately 3 ft in diameter and 10 ft in height [117] and are constructed from chrome alloy steel or cast iron. Usually, 1–4 retorts are installed in a furnace [118]. The furnace is heated by coal, gas or oil. Coal is intermittently added from the top of the retort while vaporized sulfur is continuously fed in from the bottom. Carbon disulfide is formed while the sulfur vapor works its way through the hot coal (800–1000°C) to the top of the retort. The reacted gases exit the top of the retort through a duct. Nonreactive ash and coal dust are periodically removed as they make their way to the bottom while fresh coal is added. Deposits are also removed from the inside walls of the retort, usually on a monthly or bimonthly basis. Because of the corrosive sulfur vapor, the retorts must be replaced every 1–2 years.

The reacted gases exiting the retort consist of carbon disulfide, sulfur, hydrogen sulfide, and carbonyl sulfide. Sulfur vapor is removed from the gas stream by condensation and is recycled. Carbon disulfide then is condensed and distilled to yield a pure product.

Electrothermal Process

In the electrothermal method, coal is fed into an electrical furnace. A coal bed is formed at the base of the furnace. Electrodes at the base of the furnace heat the furnace by passing electrical currents through the coal bed. Molten sulfur fed throughout the bed is vaporized and heated to 800–1000°C [119]. The sulfur reacts with the coal near the base of the furnace and forms carbon disulfide. Heat transfers from the vapors to the incoming coal as the product rises. The reacted gases are purified in the same manner as in the retort process.

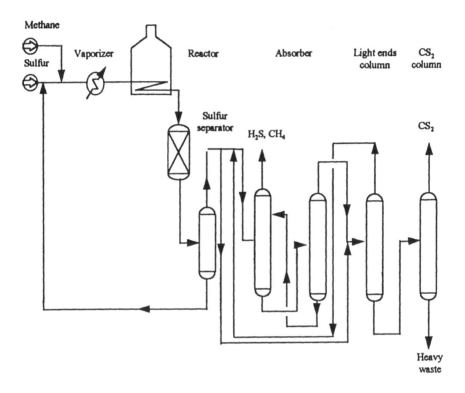

Figure 4.7 Carbon disulfide via the methane-sulfur process. *Source*: [121].

One advantage of the electrothermal process over the retort process is that large capacity reactors are possible because the heat is produced internally instead of having to be conducted through the walls [120].

Methane-Sulfur Process

The most advanced process used to produce carbon disulfide is to react methane with sulfur:

$$CH_4 + 4S \longrightarrow CS_2 + 2H_2S$$

Figure 4.7 is a flow diagram of this process [121]. Molten sulfur is vaporized and mixed with high purity (99+%) methane in a furnace [122]. Ethane and other olefins, such as propylene, can be used [123]. The reactants then are flowed into a reactor filled with activated alumina or clay catalyst. The reactor temperature is around 500–700°C. Usually excess methane (5–10%) is re-

Table 4.3 End-use of Carbon Disulfide in
the United States (Units: tons per year)

Rayon	46,000
Carbon tetrachloride	33,000
Rubber	12,000
Cellophane	5,000
Agricultural and miscellaneous uses	12,000

quired to give a conversion of 90–95% (sulfur). The product is passed to a scrubber where unreacted sulfur is removed by mixing with liquid sulfur and recycled to the furnace. The sulfur-free gases are fed into an absorption column where the carbon disulfide is extracted by mineral oil. The overhead, hydrogen sulfide and unreacted methane, is sent to a Claus sulfur recovery facility where hydrogen sulfide is burned and reacted to recover the sulfur. The oil phase is pumped to a stripper and carbon disulfide is removed. Then the carbon disulfide is sent to a two-phase distillation where the purified product yields are 85–90% methane and 91–92% sulfur. Some older plants operate at lower temperatures and conversions so that methane is recycled. The product is stored under water to minimize fire hazards [124].

Most countries are currently using the methane-sulfur process. Worldwide production of carbon disulfide is about 235,000 ton/yr in 1991 [125]. This is the only process used in the United States [126]. The United States used a total of 108,000 tons in 1990 [127]. A typical methane-sulfur plant tends to produce 50,000–200,000 ton/yr of carbon disulfide as compared to a coal-sulfur plant which produces about 5,000 ton/yr [128].

Carbon disulfide is used to produce intermediates in the manufacture of rubber vulcanization accelerators, rayon, carbon tetrachloride, cellophane, and agricultural and pharmaceutical chemicals. The end products for the U.S. production of carbon disulfide are given in Table 4.3 [129]. Production of carbon disulfide is heavily related to rayon and cellophane which accounts for about 55–65% of its usage [130].

CS_2 is highly flammable in air and also is a very reactive molecule. It is used to prepare carbon tetrachloride industrially:

$$CS_2 + 3Cl_2 \longrightarrow CCl_4 + S_2Cl_2$$

Important reactions of CS_2 involve nucleophilic attacks on carbon by the ions OR^- and SH^- and by primary or secondary amines, which lead, respectively, to thiocarbonates, xanthates, and dithiocarbamates:

$$SCS + :SH^- \dashrightarrow S_2CSH^- \xrightarrow{\quad OH^- \quad} CS_3^=$$

$$SCS + :OCH_3^- \dashrightarrow CH_3OCS_2^-$$

$$SCS + :NHR_2 \xrightarrow{\quad OH^- \quad} R_2NCS_2^-$$

Under high pressures, CS_2 polymerizes to a black solid having the structure

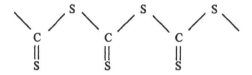

This is a scientifically very important chemical species with C-S bonds. Carbon disulfide has very unique solubility toward chemicals found in coal matrix and is very frequently used as an extracting solvent.

B.   Viscose

Viscose is a solution of cellulose xanthate in caustic soda. This solution can be processed into rayon or cellophane. The production of viscose consists of two steps. First, certain forms of cellulose, mainly from sulfite pulp and cotton linters, are reacted with pure caustic soda. This reaction converts the -OH groups on the cellulose unit to -ONa. The caustic soda must be pure and is obtained from mercury cells. Caustic soda from diaphragm cells contains sodium chloride which makes it unusable without purification. This solution must be aged for 2–3 days. Aging reduces the length of the cellulose chain which makes it easier to dissolve into caustic soda. The second step involves the reaction of the alkali cellulose and carbon disulfide in excess caustic soda. The net reaction of viscose is:

$$Cell–OH + CS_2 + NaOH \dashrightarrow Cell–O–\overset{\displaystyle S}{\overset{\displaystyle \|}{C}}–S–Na$$

On the average, 0.7 out of 3 -OH groups on each glucose unit reacts with carbon disulfide [131].

The viscose solution then is placed in a bath containing sulfuric acid and sodium sulfate to promote coagulation. The solution decomposes into cellulose fiber, sodium sulfate, and a number of sulfur compounds including carbon disulfide. Part of the carbon disulfide can be recovered by absorption.

Rayon is made by sending the viscose solution through spinnerettes. Cellophane only differs by extrusion through a slot [132]. Cellulose sponges can be made by adding crystals of $Na_2SO_4 \cdot 10H_2O$ into the viscose solution before adding the sulfuric acid [133].

Rayon fibers and filaments are used in the textile industry [134]. Conventional rayon has low wet-modulus strength, poor resistance to abrasion and to caustic materials, a high elongation, and it cannot be preshrunk [135]. Polynosic rayon, a blend with cotton, improves on the rayon properties and make it more competitive with other fibers such as polyester, cotton, acrylic and polyolefin fibers. Challis cloth, a free-flowing woven fabric made in the mid-1980s, promoted the production of rayon [136]. On the other hand, permanent-press resin treatments of polyolefin fibers [137], very popular in sportware, posed a threat to rayon-cotton blends. Rayon also is used in carpets and household goods.

Cellophane is used for high-speed packaging where synthetics, such as polyethylene, cannot compete due to a lack of adequate machinery [138]. However, it is expected that synthetics will assume an even greater part of the cellophane market.

World production of viscose rayon is estimated at 2.3 million tons [139]. The largest rayon manufacturers are American Viscose, American Enka, and Beaunit Mills.

C. Carbon Tetrachloride

Chlorination of carbon disulfide is the most important method in producing carbon tetrachloride:

$$CS_2 + 3Cl_2 \xrightarrow{\quad 30^\circ C \quad} CCl_4 + S_2Cl_2$$

$$2S_2Cl_2 + CS_2 \xrightarrow{\quad 60^\circ C \quad} CCl_4 + 6S$$

This process is used in large-scale production.

The process flowsheet for carbon tetrachloride via carbon disulfide is given in Figure 4.8 [140]. Carbon disulfide is mixed with chlorine and fed to a cooled reactor where the first reaction takes place. The product goes to a column where carbon tetrachloride is removed from the overhead and S_2Cl_2 is removed from the bottom. The S_2Cl_2 reacts catalytically with more CS_2 to produce more carbon tetrachloride and sulfur. The sulfur is removed by crystallization and can be recycled. The liquid product from the second reactor is recycled back to the CCl_4 fractionation column. Lime is used as a treatment to remove any unreacted S_2Cl_2 from the product. The product then is purified by fractionation. This process can be integrated with the production of carbon

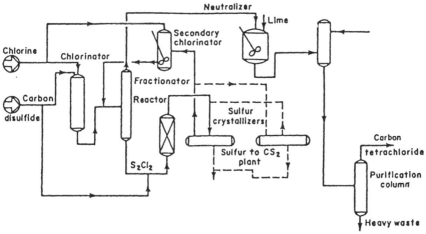

Figure 4.8   The carbon tetrachloride process. *Source*: [140].

disulfide. The manufacturer of carbon disulfide can produce CCl_4 and recycle the byproduct sulfur to the CS_2 plant [141].

Other methods for producing carbon tetrachloride includes reacting of 2 mols of phosgene [142] together:

$$2COCl_2 \longrightarrow CCl_4 + CO_2$$

and as byproducts from the production of perchloroethylene or other chlorinated methanes.

Carbon tetrachloride is a very common solvent. It decomposes photochemically and often readily transfers chlorine to various substrates. Therefore, it often is used to convert oxides to chlorides. The properties of several simple carbon compounds are given in Table 4.4 [183].

Carbon tetrachloride is used in spot removers, fire extinguishers, dry cleaning products, fumigants, solvents, and pesticides [143]. It is used to clean metal (primarily on aluminum parts) in machinery maintenance and as a raw material for certain oil additives.

Carbon tetrachloride reacts with ethylene at high pressure (15–20 MPa) to produce nylon 7 ($n = 3$) and nylon 9 ($n = 4$):

$$CCl_4 + nC_2H_4 \longrightarrow CCl_3(CH_2)_{2n}Cl$$

$$CCl_3(CH_2)_{2n}Cl + 2H_2O \longrightarrow HOOC(CH_2)_{2n}Cl + 3HCl$$

$$HOOC(CH_2)_{2n}Cl + 2NH_3 \longrightarrow HOOC(CH_2)_{2n}NH_2 + NH_4Cl$$

Table 4.4 Properties of Simple Carbon Compounds

Compound	Melting point ($^{\circ}$C)	Boiling point ($^{\circ}$C)	Remarks
CF_4	−185	−128	Very stable
CCl_4	−23	76	Moderately stable
CBr_4	93	190	Decomposes slightly on boiling
CI_4	171	—	Decomposes before boiling; can be sublimed under low pressure
COF_2	−114	−83	Easily decomposed by H_2O
$COCl_2$	−118	8	Phosgene; highly toxic
$COBr_2$	—	65	Fumes in air; $COBr_2 + H_2O \longrightarrow CO_2 + 2HBr$
$CO(NH_2)_2$	132	—	Isomerized by heat to NH_4NCO
CO	−205	−190	Odorless and toxic
CO_2	−57 (5.2 atm)	⸱ −79	
C_3O_2	—	6.8	Evil-smelling gas
COS	−138	−50	Flammable; slowly decomposed by H_2O
CS_2	−109	46	Flammable and toxic
$(CN)_2$	−28	−21	Very toxic; colorless; water soluble
HCN	−13.4	25.6	Very toxic; high dielectric constatnt (116 at 20°C) for the associated liquid

Source: [183]

The biggest use of carbon tetrachloride (90%) is the production of chloro-fluorocarbons (CFCs) [144]. Carbon tetrachloride reacts with hydrogen fluoride to produce freon:

$$CCl_4 + HF \longrightarrow HCl + CCl_3F \quad \text{(Freon 11)}$$

$$CCl_4 + 2HF \longrightarrow HCl + CCl_2F_2 \quad \text{(Freon 12)}$$

Chlorofluorocarbons are used as an aerosol propellant, as a refrigerant, and as a blowing agent for urethane foams.

Worldwide production of carbon tetrachloride in 1988 was 1.08 million tons [145]. The U.S. demands of carbon tetrachloride in 1994 and 1995 were 120 and 80 million pounds, respectively. The demand will virtually disappear by 1999. DuPont announced the dismantling of the production facility at Corps Christi, Texas, based on the conclusion that it is more advantageous to buy carbon tetrachloride, chloroform, and perchloroethylene, elsewhere

[146,147]. Carbon tetrachloride is no longer produced in the United States by disulfide because of a decrease in demand [146,148].

Because of environmental concerns, worldwide limitation on chlorofluoro-carbons (or even total abandonment) will decrease carbon tetrachloride usage considerably [149]. Less toxic products such as 1,1,1–trichloroethane (TCA) have replaced it in many applications. Perchloroethylene (PCE) replaced CCl_4 in the dry-cleaning field [150], trichloroethylene (TCE) replaced CCl_4 in industrial degreasing [151], and chlordane replaced CCl_4 in grain fumigant [152]. However, both 1,1,1–trichloroethane (TCA) and trichloroethylene (TCE) are currently under the government's regulatory control of phase down.

Producers of carbon tetrachloride in the United States include Allied Chemical Corp, Dow Chemical Company, DuPont, FCM Corp, Stauffer Chemical Co., and Vulcan Material Corp [153]. As of 1995, Dow and Vulcan are the only U.S. producers, whose combined capacity was 295 million pounds per year. The future outlook of carbon tetrachloride is literally nonexistant.

D. Perchloromethyl Mercaptan

Perchloromethyl mercaptan is produced from carbon disulfide and chlorine in the presence of iodine.

$$2CS_2 + 3Cl_2 \longrightarrow 2Cl_3C\text{–}S\text{–}Cl + S_2Cl_2$$
$$\text{perchloromethylmercaptan}$$

Unlike the carbon tetrachloride process, this reaction does not use the iron catalyst.

Perchloromethyl mercaptan is an intermediate for making two important fungicides, "Captan" and "Folpet."

"Captan"

Perchloromethyl mercaptan also can be reduced with stannous chloride (or tin) and hydrochloric acid to form thiophosgene (thiocarbonyl chloride), $CSCl_2$ [154]. Thiophosgene is used to produce many organic compounds.

E. Xanthates

Xanthates are produced by reacting alcohols, carbon disulfide, and a hydroxide:

$$\text{ROH} + \text{CS}_2 + \text{NaOH} \longrightarrow \overset{\displaystyle S}{\overset{\displaystyle \|}{\text{RO--C--S--Na}^+}}\ \text{H}_2\text{O}$$

These compounds are used to separate valuable metallic compounds found in ores. Zinc-xanthates are also used as accelerators for natural rubber.

F. Ammonium Thiocyanate

Reacting ammonia with carbon disulfide produces ammonium thiocyanate ($\$1.35$/lb) with ammonium sulfide [155] (Reaction 4.2) or ammonium trithio-carbonate [156] (Reaction 4.3) as a byproduct:

$$\text{CS}_2 + 4\text{NH}_3 \longrightarrow \text{NH}_4\text{SCN} + (\text{NH}_4)_2\text{S} \tag{4.2}$$

$$2\text{CS}_2 + 4\text{NH}_3 \longrightarrow \text{NH}_4\text{SCN} + (\text{NH}_4)_2\text{CS}_3 \tag{4.3}$$

Reaction 4.3 is promoted by alumina catalyst.

Ammonium thiocyanate is used as a corrosion inhibitor in ammonia-handling systems. At 160°C, the ammonium thiocyante is converted to thiourea.

III. HALOGENATED METHANES

A. Chloromethanes

Background

Chloromethanes are important industrially with their order of importance being methylene chloride (CH_2Cl_2), chloroform (CH_3Cl), methyl chloride (CHCl_3), and carbon tetrachloride (CCl_4), The U.S. consumption of 1993 was 591 million pound for CHCl_3, 237 million pounds for CH_2Cl_2, 448 million pounds for CHCl_3, and 340 million pounds for CCl_4. They are used because of their excellent solvent abilities and low flammability, except for CH_3Cl which is flammable [158,159].

Production

All four chlorinated methanes are produced using methane chlorination or catalytic oxychlorination of methane. The methane chlorination is initiated thermally or photochemically. The strongly exothermic free-radical reaction is conducted without external heat at 400–450°C at slightly raised pressure [158].

$$\text{CH}_4 + \text{Cl}_2 \xrightarrow{\hspace{2cm}} \text{CH}_3\text{Cl} + \text{CH}_2\text{Cl}_2 + \text{CHCl}_3 + \text{CCl}_4$$

Different ratios of Cl_2/CH_4 lead to different products. If methyl chloride is preferred a low ratio (1/10) is used; otherwise, a higher ratio is used. Higher chlorination also can be achieved by recycling the lower chlorinated products such as methyl chloride and methylene chloride back to the reactor. A selectivity of 97% chlorinated methanes can be achieved, with byproducts of hexachloroethane and trichloroethylene. Because the chlorination of methane reaction is highly exothermic, care must be taken during operation of the reaction. Usually a low Cl_2/CH_4 ratio is used with recycling to prevent explosions and to achieve higher chlorinated methane [158,159].

The chlorinated methanes are purified by first scrubbing the HCl produced in the reaction with water; second, condensing them to separate them from methane; and finally isolating them into pure form using pressurized distillation [158,159].

The second route to chlorinated methanes involving methane is oxychlorination. Direct oxychlorination has not been used yet commercially; however, an indirect chlorination method is used.

$$CH_4 + HCl + O_2 \xrightarrow[\;[cat.]\;]{-H_2O} CH_3Cl + CH_2Cl_2 + CHCl_3 + CCl_4$$

The process operates with a melt consisting of $CuCl_2$ and KCl, which act simultaneously as a chlorine source and catalyst. The melt first produces the chlorinated methanes and is fed into an oxidation reactor, where it is rechlorinated in an oxychlorination reaction with hydrogen chloride or hydrochloric acid and air. The chlorinated methanes then are recovered in a similar manner to the methane chlorination method [158,159].

Uses

The main use of methyl chlorides is for the production of higher chlorinated methanes. It also is used as a solvent and for many chemical reactions. It is used in the methylation or esterification of cellulose, which is used as industrial gum. Chloroform also is used in the esterification of alcohols. The scheme is as follows:

$$RCOOH + CH_3Cl \xrightarrow{H^+} R\overset{\overset{\displaystyle O}{\|}}{C}OCH_3$$

Mild conditions are used (30–50°C) with an acid such as H_2SO_4. The methylation of phenol to anisole (methyl phenyl ether, boiling point = 155°C) also can be accomplished this way. Anisole is almost as reactive as phenol but will not undergo facial oxidation reactions. This allows for different chemistry to be accomplished using phenol [158,159].

Chloroform also will react with Si-Cu to give dimethyldichlorosilane, which is polymerized to produce silicone. Noncrosslinked silicone is used as heat transfer agents, most notably as a replacement for polychlorinated biphenyls

(PCBs). Silly-Putty®, a children's toy, is a moderately crosslinked silicone. Highly crosslinked silicone is used as automobile tubing [159].

Chloroform also will react with amines to form quaternary ammonium salts, which are used as bactericidal surface-active agents. It also will react with hydrazine to form dimethyl hydrazine and tetramethyl hydrazine, both used as rocket fuels [158,159].

Methylene chloride (CH_2Cl_2) is used as an aerosol propellant, paint remover, and a blowing agent in polymers. It also is used as a solvent in the metals and in the electronics industry. Methyl chloride will react with benzene to form dimethyldibenzyl methane, an ingredient in the perfume industry because of its orange color [159].

Both $CHCl_3$ and CCl_4 are used as dry cleaning solvents and for the production of fluorocarbons. Approximately 90% of $CHCl_3$ is used to produce tetrafluoroethylene with most of the rest used to make fluorocarbon 22, CHF_2Cl. Carbon tetrachloride (CCl_4) is used to make fluorocarbon 11 and 12, $CFCl_3$ and CF_2Cl_2, respectively [158–160].

Chlorinated methanes are on the regulatory list by Occupational Safety and Health Administration (OSHA) as a suspected carcinogen. They are also considered to take part in the destruction of the earth's ozone layer [159].

Outlook

Chloromethanes are becoming less important because they are suspected carcinogens and some of their derivatives play a potential role in the depletion of the ozone layer. However, their use will not completely disappear because it has been difficult to find replacements for chlorinated methanes. The CH_2Cl_2 market has dropped 55% from 1980 to 1993 and will continue to drop as other solvents are used to replace CH_2Cl_2. The CCl_4 market also has dropped 46% and will continue to drop as CFCs are phased out. However, the use of CH_3Cl rose 68% during the same period and should continue to rise because of its use to make higher chloromethanes and silicones. The use of $CHCl_3$ also rose 33% over the same period and should continue to rise because of its use in the production of tetrafluoroethylene [161].

B.   Halochlorocarbons

Background

Multiple-halogen substituted methanes usually contain both chlorine and fluorine, better known as chlorofluorocarbons (CFCs). Chlorofluorocarbons are gases that are nonflammable, nontoxic, and extremely stable. Because of their properties, CFCs are used mainly as refrigerants and fire extinguishing agents, although some are used as solvents. The total production of CFCs for 1990 was 1.1 million tons [162].

Production

The chlorofluoromethanes are produced using two chloromethanes, $CHCl_3$ and CCl_4, as starting materials, depending on the desired product:

$$CHCl_3 + HF \xrightarrow{\ [cat.]\ } CHFCl_2 + CHF_2Cl + CHF_3$$

$$CCl_4 + HF \xrightarrow{\ [cat.]\ } CFCl_3 + CF_2Cl_2 + CF_3Cl$$

The chloromethanes are reacted with HF to form the chlorofluoromethanes and HCl. The reaction takes place in the gas phase at 150°C on fixed bed catalysts containing AlF_3, CrF_3, or CrOF. Halogen exchange also can take place in the liquid phase. Pressures of 2–5 bar and temperatures about 100°C are used [162].

The gaseous products then are separated from the bulk of the HCl and washed with alkali solution until acid free. The washed gases are dried, liquified by compression, and then separated by distillation. The major products are CF_2Cl_2 and $CFCl_3$. This process is considered a two-step process with methane as the starting material [162].

A one-step process for producing CFCs from CH_4 also has been developed. In this process, the chlorination and fluorination of methane occur simultaneously at 370–470°C and 4–6 bars over a fluidized-bed catalyst.

$$CH_4 \xrightarrow[\ [cat.]\ ]{\ Cl_2,\, HF\ } CF_2Cl_2 + CFCl_3$$

The major products are CF_2Cl_2 and $CFCl_3$ with HCl. The purification process is the same as the two-step process. Byproducts such as CCl_4 are recycled to the reactor [162].

Increasing the fluorine content of the chlorofluoromethanes increases their thermal and chemical stability. In addition, they are nonflammable and nontoxic. These properties have lead to their use as aerosol propellants, spray and foam agents, refrigerants, and dry cleaning solvents [162–164].

The excellent stability of chlorofluorocarbons also is a problem. Their stability allows them to rise high into the earth's atmosphere and they are suspected to act as a catalyst in the destruction of ozone. For this reason, self-restrictions and bans have been placed on their use and, as a result, their production is being phased out by the early 21st century.

Use

Difluorochloromethane is becoming increasingly important because it is used in the production of tetrafluoroethylene. Tetrafluoroethylene is produced from CHF_2Cl at high temperature (700–900°C) with a byproduct of HCl. Tetrafluoroethylene can be polymerized to form polytetrafluoroethylene or more commonly called Teflon®. Teflon® has great nonsticking properties, so it has a wide industrial applications. CHF_2Cl is used to produce hexafluoropropylene which also is an important monomer [162].

Other halogenated methane compounds contain bromine. An example of this is Halon 1301, CF_3Br. Halons are highly effective fire extinguishing agents, especially with high quality goods. However, they are thought to pose an even greater danger to the ozone layer and are also being phased out [162].

Fluoroplastics are a class of paraffinic polymers that have some or all of the hydrogen replaced by fluorine. These include polytetrafluoroethylene (PTFE), fluorinated ethylene propylene (FEP) copolymer, perfluoroalkoxy (PFA) resin, polychlorotrifluoroethylene (PCTFE), ethylene-chlorotrifluoroethylene (ECTFE) copolymer, ethylene-tetrafluoroethylene (ETFE) copolymer, polyvinylidene fluoride (PVDF), and polyvinylfluoride (PVF) [186].

Polytetrafluoroethylene is a completely fluorinated polymer manufactured by free-radical polymerization of tetrafluoroethylene. With a linear molecular structure of repeating $-CF_2-CF_2-$ units, PTFE is a crystalline polymer with a melting point of 326.7°C. Its specific gravity is 2.13–2.19. Polytetrafluoroethylene has exceptional resistance to chemicals. Its dielectric constant (2.1) and loss factor are low and stable across a wide range of temperature. It has useful mechanical properties from cryogenic temperatures to 260°C. In the United States, PTFE is sold as "Halon," "Algoflon," "Teflon," "Fluon," "Hostaflon," and "Polyflon."

Fluorinated ethylene propylene (FEP) is produced by copolymerization of tetrafluoroethylene and hexafluoropropylene, and consists of predominantly linear chains.

$$-CF_2-CF_2-CF_2CF-$$
$$|$$
$$CF_3$$

Fluorinated ethylene propylene has a crystalline melting point of 290°C and a specific gravity of 2.15. It is a relatively soft plastic with lower tensile strength, wear resistance, and creep resistance than many other engineering plastics [186].

Outlook

The future outlook for CFCs is bleak. With decreasing demand and production over the last ten years and an almost total phase-out planned, their future is limited, though not nonexistent. Because of their use for producing halogenated monomers, they will continue to play an important but smaller role in the chemical industry.

IV. DIRECT CONVERSION OF METHANE TO METHANOL

The direct conversion of methane into methanol, hydrocarbons, or oxygenated fuels has attracted a great deal of interest. At the beginning of the 20th century, one of the first patents was granted in 1905 to Lance and Elworthy [165]. These inventors claimed that oxidizing methane with hydrogen peroxide in the presence of ferrous sulfate could form methanol, formaldehyde, and formic acid.

In the 1920s, the partial oxidation of methane to methanol was studied. At low pressures, methanol was not formed either as a reaction intermediate or a final product. The reaction has been observed only under high pressures, indicating a radical chain mechanism that is apparently catalyzed by the reactor walls. In 1934, the work by Wiezevich and Frolich [166] and later by Boomer and Broughton [167], carried out partial oxidation at high pressures using various catalysts. The copper wall of the reactor served as one catalyst, while silver on asbestos as the other. The selective yield of meth- anol was found to be high when the concentration of oxygen was kept low. It also was observed that copper catalyst contained cuprous oxide, whereas the silver catalyst stayed reduced without containing any silver oxide. They concluded that each catalyst had a different role in reaction, with methyl formate produced over the silver catalyst and esters formed over the copper catalyst.

In recent years, molybdenum trioxide has been studied as the catalyst for the partial oxidation of methane. Dowden and Walker [168] developed a series of multicomponent oxides based on molybdenum. They also found it was advantageous to support the catalyst, for example, using alumina/silica. But the surface area was quite low, approximately in the order of 0.1 m^2/g. They reacted a substoichiometric mixture of methane and oxygen (97:3) at around 50 bar and in the range of 130–500°C. They obtained high selectivities toward methanol over several catalysts. The most active catalyst among the studied was $Fe_2O_3(MoO_3)_3$ which yielded 869 g/kg-cat/hr of methanol.

Unsupported two-component oxide systems were used by Stroud in 1975 [169]. In their composition, the first component was preferably molybdenum oxide and the second cupric oxide (i.e., $MoO_3 \cdot CuO$). The reaction conditions were 20 bar and 485°C, and the yield was 490 g/kg-cat/hr of oxygenated products, including methanol, formaldehyde, ethanol, and acetaldehyde. The work by Stroud used oxygen as the oxidant. Liu et al. [170] used nitrous oxide as the oxidant at 1 bar over the 1.7% Mo/SiO_2 catalyst. A combined selectivity of 84.6% towards methanol and formaldehyde was obtained with a conversion of 8.1%. They also used a different catalytic system of 1.7% MoO_3 supported on Cab-O-SilM-5 silica. Their kinetic study obtained a power law rate expression of the Arrhenius plot for CH_4 concentration was

$$\frac{d(CH_4)}{dt} = -k(N_2O)^m(CH_4)^n$$

which is linear over a temperature interval of 550–594°C and yields an activation energy of 176 ± 8 kJ/mol. Based on the ESR spectroscopy, they concluded that the more reactive O^- species is responsible for the initiation of the selective oxidation cycle.

Khan and Somorjai [171] successfully reproduced the Liu's results, and were able to specify the rate expressions distinctly for formation of methanol and formaldehyde. The rates were measured over 480–590°C. The activation energies for the methanol formation and the formaldehyde formation were 172 ± 17 kJ/mol (below 520°C) and 344 ± 17 kJ/mol (below 540°C), respectively. Zhen et al. [172] also showed that silica-supported vanadium pentoxide catalyst resulted in up to 100% selectivity to methanol and formaldehyde at 460°C with the conversion of 0.2%. They also showed that the V_2O_5/SiO_2 system was more active for methanol production than MoO_3/SiO_2 system; however, the product selectivity was generally poorer.

Metal-based catalysts also were used for methane oxidation. Especially over metals such as platinum and palladium, trace amounts of methanol, formaldehyde, and formic acid can be found. Organic halides increased the yield of partial oxidation products and inhibited the complete combustion of methane [173]. Inhibition effects of dichloromethane was observed. Mann and Dosi [174] used a Pd/Al_2O_3 catalyst and found that the addition of halogen compounds reduced the conversion of methane in the following order:

$$CH_2Cl_2 < CHCl_3 < CH_3Cl < CCl_4$$

They also were able to show from the X-ray powder diffraction of unused and used catalysts that the addition of halogen modifier actually increased the amount of crystalline PdO in the catalyst system.

It is clear that significant research efforts have been made in this subject and a few clear guidelines have been established including the roles of molybdenum and vanadium oxides as well as hydrogen modifiers to novel metal catalysts.

V. PYROLYSIS OF METHANE

A. General Background

Methane can be converted directly to C_2 hydrocarbons by thermal coupling or pyrolysis [175]. In most cases, pyrolysis means thermal decomposition that yields lighter hydrocarbons than the original. In this case, however, C_1 chemicals are pyrolytically converted to C_2 chemicals. The reaction is highly endothermic and therefore heat must be supplied at elevated temperatures.

As discussed in Chapter 1, the proven world gas and oil resources show an increase in gas reserves and a leveling off of the oil resources. Most of these reserves are located distantly from the major consumption markets. This is one of the reasons why there is a great incentive to convert natural gas into petrochemicals and clean liquid fuels.

The direct conversion of methane includes various routes including oxidative coupling, partial oxidation, and pyrolysis. Wheeler and Wood [176] and Fischer and Pichler [177] were the first to show that by controlling the residence time of reaction, products other than carbon and hydrogen could be obtained. Until this time the pyrolysis of methane had been the subject of decomposition yielding carbon, hydrogen, and trace amounts of aromatics. Many processes for ethylene production have been developed and even operated commercially with some limited success [175]. These processes include the use of an electric arc (Huels, DuPont), regeneration (Wulff process), flame techniques, partial oxidation (BASF, SBA, Tsutumi), and an admixture with hot combustion gases (Hoechst HTP) [178]. However, recent developments have focused mainly on the coproduction of ethene and ethyne.

All these processes interestingly but understandably have three things in common: high temperature, short residence time, and rapid quenching.

B. Thermodynamics

Figure 4.9 shows the standard free energy of formation of several hydrocarbons as functions of temperature [175]. Methane is particularly stable at lower temperatures compared to other hydrocarbons (C_2H_4, C_6H_6, and C_2H_2, but not C_2H_6). This can be concluded from the comparison of the free energy of formation at a given temperature.

Figure 4.10 shows the mole fraction of 14 species in gas phase equilibrium calculated from thermodynamic data [180]. The figure shows that the equilibrium yield of ethyne is low below 1373 K, but the yield increases strongly with increasing temperature. The figure also shows that the ethene yield also is low at all temperatures, less than 5% over the entire temperature range [179].

Figure 4.11 shows the heat of formation of several hydrocarbons from the elements [175]. The ΔH_f of saturated hydrocarbons is negative, whereas the ΔH_f of ethylene and acetylene is positive. As a consequence, a large amount of energy is required to convert to ethylene or acetylene, in a form of endothermic heat of reaction.

Figures 4.9, 4.10, and 4.11 show that the pyrolysis of methane to C_2 hydrocarbons has several thermodynamics restrictions.

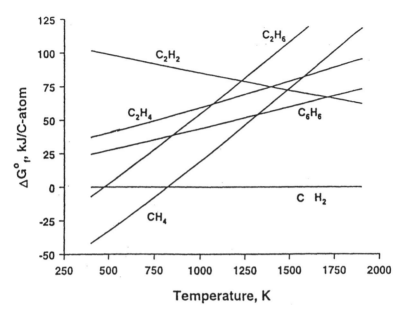

Figure 4.9   Gibbs free energy of formulation of several simple hydrocarbons. *Source*:
[175].

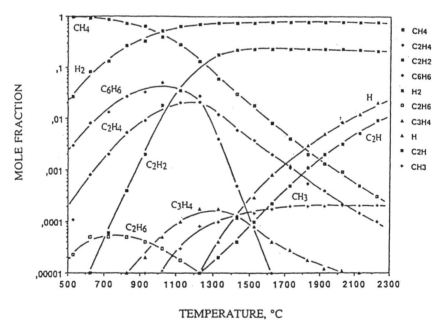

Figure 4.10   Mole fraction of species in gas-phase equilibrium. Calculated using
total pressure at 1 bar and H/C = 4. *Source*: [180].

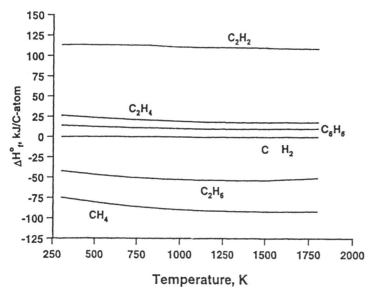

Figure 4.11   Heat of formation of several hydrocarbons. *Source*: [178].

C. Reaction Mechanisms

The reaction mechanisms for thermal coupling of methane has been studied by various researchers [175]. The overall reaction in thermal coupling of methane can be described as a stepwise dehydrogenation at high temperature.

$$2CH_4 \longrightarrow C_2H_6 + H_2$$
$$C_2H_6 \longrightarrow C_2H_4 + H_2$$
$$C_2H_4 \longrightarrow C_2H_2 + H_2$$
$$C_2H_2 \longrightarrow 2C + H_2$$

The formation of products is explained by a free-radical formation. The initiation step and the formation of ethane are described by:

$$CH_4 \longrightarrow CH_3 \cdot + H \cdot$$
$$CH_4 + H \cdot \longrightarrow CH_3 \cdot + H_2$$
$$2CH_3 \cdot \longrightarrow C_2H_6$$

The first reaction, the free-radical formation, is the rate determining step and serves as the only primary source of free radicals [175].

The secondary reactions of ethane via unimolecular decomposition can be written as:

$$C_2H_6 \longrightarrow 2CH_3\cdot$$

Also, the free radical mechanism is described by:

$$C_2H_6 + H\cdot \longrightarrow C_2H_5\cdot + H_2$$
$$C_2H_6 + CH_3\cdot \longrightarrow C_2H_5\cdot + CH_4$$
$$C_2H_5\cdot \longrightarrow C_2H_4 + H\cdot$$

The ethyne formation is by:

$$C_2H_4 + H\cdot \longrightarrow C_2H_3\cdot + H_2$$
$$C_2H_4 + CH_3\cdot \longrightarrow C_2H_3\cdot + CH_4$$
$$C_2H_3\cdot \longrightarrow C_2H_2 + H\cdot$$

At temperatures higher than 1350 K, the radical chain reactions are dominating for ethyne formation at both high and low conversions. The production of propylene and C_4 hydrocarbons can be explained analogously. However, the formation of benzene requires an isomerization step in addition to the free radical reaction. Westmoreland et al. [181] showed that benzene is formed by chemically activated addition and isomerization reactions.

$$C_2H_2 + C_4H_5\cdot \longrightarrow C_6H_6 + H\cdot$$

The last stage of decomposition, coke formation, is not yet fully understood mechanistically. Ethyne plays an important role in the coke formation and polynuclear aromatics (or polycyclic aromatic hydrocarbon, PAH) appear to be involved in one way or another [175]. A detailed discussion of the reaction mechanism can be found in Holman et al. [175].

To alleviate the thermodynamic restrictions, an oxidant such as O_2 or Cl_2 can be used. The free energies of such reactions are highly negative and the formation of the products is favored.

$$2CH_4 + O_2 \longrightarrow C_2H_4 + 2H_2O$$

$$2CH_4 + \frac{3}{2}O_2 \longrightarrow C_2H_2 + 3H_2O$$

However, it should be noted that the total combustion of methane to CO_2 also is favored. Therefore, the basic challenge is to find a system including a catalytic system that kinetically favors the formation of hydrocarbons compared to the total combustion reaction.

VI. PURE HYDROGEN BY STEAM REFORMING OF METHANE

The Polybed pressure swing adsorption (PSA) process developed by UOP, can produce hydrogen of any purity, typically 90% to 99.9999+ %. Impurities removed include: N_2, CO, CH_4, CO_2, H_2O, Ar, C_2–C_8+, CH_3OH, NH_3, and H_2S. Typical feed and product temperature is 40–120°F. This process also can be used for other separations such as methane from ethane, CO_2 from nitrogen, etc. [182].

The feed to this process is typically steam reformer or catalytic reformer net gas, but is not limited to them. Feed pressures up to 1,000 psig have been commercially demonstrated. Figure 4.12 shows a schematic of the Polybed PSA process [182]. The recovery of H_2 ranges between 60% and 98% depending upon the feed gas compositions and pressures. Typical feed and product temperatures are 40–120°F. The H_2 purity can be 99.9999+%.

Purification is based on advanced PSA technology. The process units contain 3–12 adsorber vessels. One or more adsorbers are on the adsorption step, while the others are undergoing various stages of regeneration. No feed treatment other than entrained liquid removal is required. Materials for piping and adsorbers are carbon steel. According to the licensor of the technology, UOP,

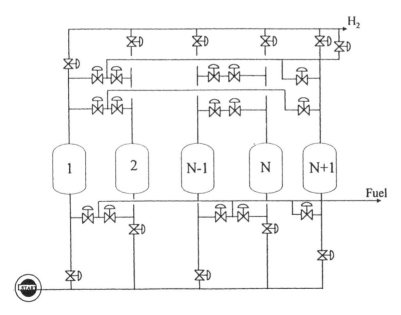

Figure 4.12 A schematic of the Polybed PSA process. *Source*: [182].

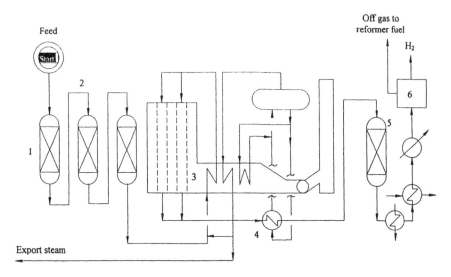

Figure 4.13 The steam reforming process by Howe-Baker Engineers, Inc. (1) hydrotreater, (2) desulfurizer, (3) reformer, (4) process steam generator, (5) water-gas shift reaction, (6) PSA purification system.

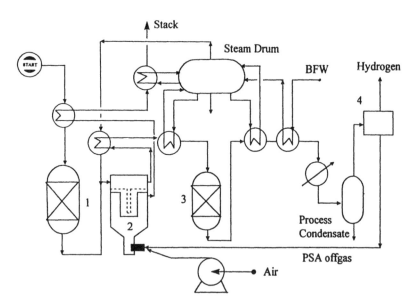

Figure 4.14 A flowsheet of the Halder Topsøe TCR hydrogen process. 1) desulfurizer, 2) convection reformer, 3) shift converter, 4) PSA. *Source*: [182]

hydrogen is purified at a cost in the range of $0.30–1.50 per 1000 ft^3 of feed, depending on the capacity, feed/tail gas fuel values, and operating conditions [182].

Hydrogen for refinery hydrotreating and hydrocracking, as well as for other petrochemical use, is very frequently obtained via steam reforming process. Feedstock for this process is preferably light hydrocarbons such as natural gas, refinery fuel gas, liquefied petroleum gas (LPG), butane, and light naphtha. Figure 4.13 shows a schematic of steam reforming process by Howe-Baker Engineers, Inc. The feed is heated in the feed preheater and passed through the hydrotreater (1). The hydrotreater converts sulfur compounds to hydrogen sulfide and saturates unsaturated hydrocarbons in the feed. The next stage (2) is the desulfurizer where hydrogen sulfide is absorbed from the gas. The desulfurizer feed gas is mixed with steam and superheated in the feed preheat coil. The mixture now passes through catalyst-filled tubes in the reformer (3). The catalyst is nickel and the reaction between feed gas and steam produces hydrogen and carbon oxides. The reforming reaction is endothermic and heat is provided by carefully controlled external firing in the reformer. The gas stream leaving the reformer is cooled by the process steam generator (4). The next stage (5) is for water gas shift reaction that converts CO and water vapor to hydrogen and carbon dioxide. The shift converter effluent gas is cooled in a feed preheater, a BFW preheater, and a DA feed water preheater. The condensate is separated and the gas then is sent to the PSA purification system (6). The pressure swing adsorption (PSA) purification system is automatic, requiring very little operator attention. Hydrogen from the PSA unit is sent off plant and a small hydrogen stream then is split off and recycled to the front of the plant for hydrotreating. Over 125 plants worldwide of various capacities ranging from less than 1 million scfd to over 70 million scfd, are being operated [182].

Haldor Topsøe A/S developed a process called Topsøe Convection Reformer (TCR). The process produces 99.5–99.999+% hydrogen from natural gas, hydrogen-rich off gas, LPG, naphtha, or kerosene using a compact heat-exchange reformer. Typical design capacity ranges from 0.2 to 20 million scfd. Figure 4.14 shows a flowsheet of TCR hydrogen process [182]. First, the feed gas is desulfurized and then flows to the conversion reformer (2) having a single burner. Both flue gas and process gas are cooled to about 600°C before leaving the reformer. The remaining waste heat is used for feed preheating and for steam production. All steam is used as process steam. Process gas goes through shift conversion and is purified in a PSA unit (4). Offgas from PSA is used as reformer fuel. The TCR process is claimed to offer low investment and operating costs, short erection time, high flexibility, safety, and reliability. The total energy consumption for this process is claimed to be 3.1–3.6 Gcal/1,000 Nm^3 H$_2$.

REFERENCES

1. Szmant H. H., *Organic Building Blocks of the Chemical Industry*, John Wiley & Sons, New York, 1989, p. 33.
2. Hahn, A. V., *The Petrochemical Industry: Market and Economics*, McGraw-Hill, New York, 1970, p. 151.
3. Szmant H. H., *Organic Building Blocks of the Chemical Industry*, John Wiley & Sons, New York, 1989, p. 65.
4. Weissermel, K. and Arpe, H., *Industrial Organic Chemisty*, VCH, Weinheim, 1993, p. 44.
5. Szmant H. H., *Organic Building Blocks of the Chemical Industry*, John Wiley & Sons, New York, 1989, p. 173.
6. McKetta, J. J., *Inorganic Chemical Handbook: Vol. 2*, Marcel Dekker, New York, 1993, p. 1032.
7. Hahn, A. V., *The Petrochemical Industry: Market and Economics*, McGraw-Hill, New York, 1970, p. 152.
8. Faith, W. L., Lowenheim, F. A., and Moran, M. K., *Faith, Keyes, and Clark Industrial Chemicals*, 4th. ed., John Wiley & Sons, New York, 1975, p. 483.
9. Hahn, A. V., *The Petrochemical Industry: Market and Economics*, McGraw-Hill, New York, 1970, p. 152.
10. Weissermel, K. and Arpe, H., *Industrial Organic Chemisty*, VCH, Weinheim, 1993, p. 45.
11. Weissermel, K. and Arpe, H., *Industrial Organic Chemisty*, VCH, Weinheim, 1993, p. 45.
12. Weissermel, K. and Arpe, H., *Industrial Organic Chemisty*, VCH, Weinheim, 1993, p. 45.
13. McKetta, J. J., *Inorganic Chemical Handbook: Vol. 2*, Marcel Dekker, New York, 1993, p. 1035.
14. McKetta, J. J., *Inorganic Chemical Handbook: Vol. 2*, Marcel Dekker, New York, 1993, p. 1035.
15. Hahn, A. V., *The Petrochemical Industry: Market and Economics*, McGraw-Hill, New York, 1970, p. 153.
16. Kent, J. A. (Ed.), *Riegel's Handbook of Industrial Chemistry*, 8th ed. VNR, New York, 1983, p. 202.
17. Weissermel, K. and Arpe, H., *Industrial Organic Chemisty*, VCH, Weinheim, 1993, p. 45.
18. Hahn, A. V., *The Petrochemical Industry: Market and Economics*, McGraw-Hill, New York, 1970, p. 153.
19. McKetta, J. J., *Inorganic Chemical Handbook: Vol. 2*, Marcel Dekker, New York, 1993, p. 1034.
20. Weissermel, K. and Arpe, H., *Industrial Organic Chemisty*, VCH, Weinheim, 1993, p. 45.
21. Hahn, A. V., *The Petrochemical Industry: Market and Economics*, McGraw-Hill, New York, 1970, p. 154.
22. Hahn, A. V., *The Petrochemical Industry: Market and Economics*, McGraw-Hill, New York, 1970, p. 153.

23. Weissermel, K. and Arpe, H., *Industrial Organic Chemisty*, VCH, Weinheim, 1993, p. 45.

24. Hahn, A. V., *The Petrochemical Industry: Market and Economics*, McGraw-Hill, New York, 1970, p. 154.

25. Szmant H. H., *Organic Building Blocks of the Chemical Industry*, John Wiley & Sons, New York, 1989, p. 173.

26. Weissermel, K. and Arpe, H., *Industrial Organic Chemisty*, VCH, Weinheim, 1993, p. 45.

27. Usachev, N. Y. and Minachev, K. H. M., Methane as feedstock for the chemical industry, *Petroleum Chemistry*, 33(5): 369–396 (1993).

28. Szmant H. H., *Organic Building Blocks of the Chemical Industry*, John Wiley & Sons, New York, 1989, p. 174.

29. Weissermel, K. and Arpe, H., *Industrial Organic Chemisty*, VCH, Weinheim, 1993, p. 45.

30. Weissermel, K. and Arpe, H., *Industrial Organic Chemisty*, VCH, Weinheim, 1993, p. 46.

31. Weissermel, K. and Arpe, H., *Industrial Organic Chemisty*, VCH, Weinheim, 1993, p. 44.

32. Faith, W. L., Lowenheim, F. A., and Moran, M. K., *Faith, Keyes, and Clark Industrial Chemicals*, 4th. ed., John Wiley & Sons, New York, 1975, p. 484.

33. Weissermel, K. and Arpe, H., *Industrial Organic Chemisty*, VCH, Weinheim, 1993, p. 44.

34. Szmant H. H., *Organic Building Blocks of the Chemical Industry*, John Wiley & Sons, New York, 1989, p. 174.

35. Szmant H. H., *Organic Building Blocks of the Chemical Industry*, John Wiley & Sons, New York, 1989, p. 174.

36. Faith, W. L., Lowenheim, F. A., and Moran, M. K., *Faith, Keyes, and Clark Industrial Chemicals*, 4th. ed., John Wiley & Sons, New York, 1975, p. 547.

37. Faith, W. L., Lowenheim, F. A., and Moran, M. K., *Faith, Keyes, and Clark Industrial Chemicals*, 4th. ed., John Wiley & Sons, New York, 1975, p. 547.

38. Hahn, A. V., *The Petrochemical Industry: Market and Economics*, McGraw-Hill, New York, 1970, p. 157.

39. Faith, W. L., Lowneheim, F. A., and Moran, M. K., *Faith, Keyes, and Clark Industrial Chemicals*, 4th. ed., John Wiley & Sons, New York, 1975, p. 548.

40. Hahn, A. V., *The Petrochemical Industry: Market and Economics*, McGraw-Hill, New York, 1970, p. 156.

41. Weissermel, K. and Arpe, H., *Industrial Organic Chemisty*, VCH, Weinheim, 1993, p. 279.

42. Weissermel, K. and Arpe, H., *Industrial Organic Chemisty*, VCH, Weinheim, 1993, p. 279.

43. Szmant H. H., *Organic Building Blocks of the Chemical Industry*, John Wiley & Sons, New York, 1989, p. 86.

44. Szmant H. H., *Organic Building Blocks of the Chemical Industry*, John Wiley & Sons, New York, 1989, p. 87.

45. Billmeyer, F. W., Jr., *Textbook of Polymer Science*, 3rd ed., John Wiley & Son, Inc., 1984, p. 387–388.

46. Faith, W. L., Lowenheim, F. A., and Moran, M. K., *Faith, Keyes, and Clark Industrial Chemicals*, 4th. ed., John Wiley & Sons, New York, 1975, p. 550.

47. Szmant H. H., *Organic Building Blocks of the Chemical Industry*, John Wiley & Sons, New York, 1989, p. 179.

48. Szmant H. H., *Organic Building Blocks of the Chemical Industry*, John Wiley & Sons, New York, 1989, p. 179.

49. Weissermel, K. and Arpe, H., *Industrial Organic Chemisty*, VCH, Weinheim, 1993, p. 46.

50. Hahn, A. V., *The Petrochemical Industry: Market and Economics*, McGraw-Hill, New York, 1970, p. 162.

51. Weissermel, K. and Arpe, H., *Industrial Organic Chemisty*, VCH, Weinheim, 1993, p. 47.

52. Weissermel, K. and Arpe, H., *Industrial Organic Chemisty*, VCH, Weinheim, 1993, p. 47.

53. Weissermel, K. and Arpe, H., *Industrial Organic Chemisty*, VCH, Weinheim, 1993, p. 47.

54. Szmant H. H., *Organic Building Blocks of the Chemical Industry*, John Wiley & Sons, New York, 1989, p. 321.

55. Szmant H. H., *Organic Building Blocks of the Chemical Industry*, John Wiley & Sons, New York, 1989, p. 210.

56. Szmant H. H., *Organic Building Blocks of the Chemical Industry*, John Wiley & Sons, New York, 1989, p. 210.

57. Hahn, A. V., *The Petrochemical Industry: Market and Economics*, McGraw-Hill, New York, 1970, p. 162.

58. Szmant H. H., *Organic Building Blocks of the Chemical Industry*, John Wiley & Sons, New York, 1989, p. 210.

59. Hahn, A. V., *The Petrochemical Industry: Market and Economics*, McGraw-Hill, New York, 1970, p. 162.

60. Hahn, A. V., *The Petrochemical Industry: Market and Economics*, McGraw-Hill, New York, 1970, p. 162.

61. Hahn, A. V., *The Petrochemical Industry: Market and Economics*, McGraw-Hill, New York, 1970, p. 162.

62. Hahn, A. V., *The Petrochemical Industry: Market and Economics*, McGraw-Hill, New York, 1970, p. 162.

63. Szmant H. H., *Organic Building Blocks of the Chemical Industry*, John Wiley & Sons, New York, 1989, p. 210.

64. Hahn, A. V., *The Petrochemical Industry: Market and Economics*, McGraw-Hill, New York, 1970, p. 163.

65. Szmant H. H., *Organic Building Blocks of the Chemical Industry*, John Wiley & Sons, New York, 1989, p. 184.

66. Hahn, A. V., *The Petrochemical Industry: Market and Economics*, McGraw-Hill, New York, 1970, p. 164.

67. Hahn, A. V., *The Petrochemical Industry: Market and Economics*, McGraw-Hill, New York, 1970, p. 164.
68. Weissermel, K. and Arpe, H., *Industrial Organic Chemisty*, VCH, Weinheim, 1993, p. 301.
69. Hahn, A. V., *The Petrochemical Industry: Market and Economics*, McGraw-Hill, New York, 1970, p. 163.
70. Lewis, G. R., *1001 Chemicals in Everyday Products*, Van Nostrand Reinhold, New York, 1994, p. 154.
71. Hahn, A. V., *The Petrochemical Industry: Market and Economics*, McGraw-Hill: New York, 1970, p. 164.
72. Hahn, A. V., *The Petrochemical Industry: Market and Economics*, McGraw-Hill, New York, 1970, p. 164.
73. Lewis, G. R., *1001 Chemicals in Everyday Products*, Van Nostrand Reinhold, New York, 1994, p. 154.
74. Lewis, G. R., *1001 Chemicals in Everyday Products*, Van Nostrand Reinhold, New York, 1994, p. 154.
75. *Kirk-Othmer Encyclopedia of Chemical Technology* (Kroschwitz, J. I. and Howe-Grant, M., Eds.), John Wiley & Sons, New York; 4th ed. Vol. 7, 1993, p. 769.
76. *Kirk-Othmer Encyclopedia of Chemical Technology* (Kroschwitz, J. I. and Howe-Grant, M., Eds.), John Wiley & Sons, New York; 4th ed. Vol. 7, 1993, p. 770.
77. *Kirk-Othmer Encyclopedia of Chemical Technology* (Kroschwitz, J. I. and Howe-Grant, M., Eds.), John Wiley & Sons, New York; 4th ed. Vol. 7, 1993, p. 769.
78. *Kirk-Othmer Encyclopedia of Chemical Technology* (Kroschwitz, J. I. and Howe-Grant, M., Eds.), John Wiley & Sons, New York; 4th ed. Vol. 7, 1993, p. 766.
79. *Kirk-Othmer Encyclopedia of Chemical Technology* (Kroschwitz, J. I. and Howe-Grant, M., Eds.), John Wiley & Sons, New York; 4th ed. Vol. 7, 1993, p. 773.
80. *Kirk-Othmer Encyclopedia of Chemical Technology* (Kroschwitz, J. I. and Howe-Grant, M., Eds.), John Wiley & Sons, New York; 4th ed. Vol. 7, 1993, p. 773.
81. *Kirk-Othmer Encyclopedia of Chemical Technology* (Kroschwitz, J. I. and Howe-Grant, M., Eds.), John Wiley & Sons, New York; 4th ed. Vol. 7, 1993, p. 769.
82. Szmant H. H., *Organic Building Blocks of the Chemical Industry*, John Wiley & Sons, New York, 1989, p. 210.
83. Szmant H. H., *Organic Building Blocks of the Chemical Industry*, John Wiley & Sons, New York, 1989, p. 436.
84. Szmant H. H., *Organic Building Blocks of the Chemical Industry*, John Wiley & Sons, New York, 1989, p. 597.
85. *Concise Science Dictionary*, Oxford University Press, Oxford, 1991, p. 262.
86. Büchner, W., Schliebs, R., Winter, G., and Büchel, K. H., *Industrial Inorganic Chemistry*, VCH, Weinheim, 1989, p. 547.
87. Büchner, W., Schliebs, R., Winter, G., and Büchel, K. H., *Industrial Inorganic Chemistry*, VCH, Weinheim, 1989, p. 547.
88. Büchner, W., Schliebs, R., Winter, G., and Büchel, K. H., *Industrial Inorganic Chemistry*, VCH, Weinheim, 1989, p. 547.
89. Büchner, W., Schliebs, R., Winter, G., and Büchel, K. H., *Industrial Inorganic Chemistry*, VCH, Weinheim, 1989, p. 548.

90. *Kirk-Othmer Encyclopedia of Chemical Technology* (Kroschwitz, J. I. and Howe-Grant, M., Eds.), John Wiley & Sons, New York; 3rd ed. Vol. 13, 1981, p. 766.

91. Hahn, A. V., *The Petrochemical Industry: Market and Economics*, McGraw-Hill, New York, 1970, p. 165.

92. Hahn, A. V., *The Petrochemical Industry: Market and Economics*, McGraw-Hill, New York, 1970, p. 166.

93. Szmant H. H., *Organic Building Blocks of the Chemical Industry*, John Wiley & Sons, New York, 1989, p. 624.

94. Hahn, A. V., *The Petrochemical Industry: Market and Economics*, McGraw-Hill, New York, 1970, p. 166.

95. Hahn, A. V., *The Petrochemical Industry: Market and Economics*, McGraw-Hill, New York, 1970, p. 166.

96. Szmant H. H., *Organic Building Blocks of the Chemical Industry*, John Wiley & Sons, New York, 1989, p. 185.

97. *Kirk-Othmer Encyclopedia of Chemical Technology* (Kroschwitz, J. I. and Howe-Grant, M., Eds.), John Wiley & Sons, New York; 4th ed. Vol. 2, 1992, p. 376.

98. Szmant H. H., *Organic Building Blocks of the Chemical Industry*, John Wiley & Sons, New York, 1989, p. 177.

99. Szmant H. H., *Organic Building Blocks of the Chemical Industry*, John Wiley & Sons, New York, 1989, p. 177.

100. *Encyclopedia of Chemical Processing and Design* (McKetta, J. J., and Cunningham, W. A., Eds.), Marcel Dekker, New York, 1977; Vol 3, p. 167.

101. *Encyclopedia of Chemical Processing and Design* (McKetta, J. J., and Cunningham, W. A., Eds.), Marcel Dekker, New York, 1977; Vol 3, p. 167.

102. *Encyclopedia of Chemical Processing and Design* (McKetta, J. J., and Cunningham, W. A., Eds.), Marcel Dekker, New York, 1977; Vol 3, p. 167.

103. Hahn, A. V., *The Petrochemical Industry: Market and Economics*, McGraw-Hill, New York, 1970, p. 167.

104. Szmant H. H., *Organic Building Blocks of the Chemical Industry*, John Wiley & Sons, New York, 1989, p. 532.

105. Szmant H. H., *Organic Building Blocks of the Chemical Industry*, John Wiley & Sons, New York, 1989, p. 340.

106. *Kirk-Othmer Encyclopedia of Chemical Technology* (Kroschwitz, J. I. and Howe-Grant, M., Eds.), John Wiley & Sons, New York; 4th ed. Vol. 2, 1992, p. 377.

107. *Kirk-Othmer Encyclopedia of Chemical Technology* (Kroschwitz, J. I. and Howe-Grant, M., Eds.), John Wiley & Sons, New York; 4th ed. Vol. 2, 1992, p. 377.

108. *Kirk-Othmer Encyclopedia of Chemical Technology* (Kroschwitz, J. I. and Howe-Grant, M., Eds.), John Wiley & Sons, New York; 4th ed. Vol. 2, 1992, p. 384.

109. *Kirk-Othmer Encyclopedia of Chemical Technology* (Kroschwitz, J. I. and Howe-Grant, M., Eds.), John Wiley & Sons, New York; 4th ed. Vol. 2, 1992, p. 384.

110. Hahn, A. V., *The Petrochemical Industry: Market and Economics*, McGraw-Hill, New York, 1970, p. 167.

111. Szmant H. H., *Organic Building Blocks of the Chemical Industry*, John Wiley & Sons, New York, 1989, p. 175.

112. Szmant H. H., *Organic Building Blocks of the Chemical Industry*, John Wiley & Sons, New York, 1989, p. 318.

113. Szmant H. H., *Organic Building Blocks of the Chemical Industry*, John Wiley & Sons, New York, 1989, p. 176.

114. Hahn, A. V., *The Petrochemical Industry: Market and Economics*, McGraw-Hill, New York, 1970, p. 167.

115. Weissermel, K. and Arpe, H., *Industrial Organic Chemisty*, VCH, Weinheim, 1993, p. 47.

116. Faith, W. L., Lowenheim, F. A., and Moran, M. K., *Faith, Keyes, and Clark Industrial Chemicals*, 4th. ed., John Wiley & Sons, New York, 1975, p. 227.

117. *Kirk-Othmer Encyclopedia of Chemical Technology* (Kroschwitz, J. I. and Howe-Grant, M., Eds.), John Wiley & Sons, New York; 4th ed. Vol. 5, 1993, p. 60.

118. *Kirk-Othmer Encyclopedia of Chemical Technology* (Kroschwitz, J. I. and Howe-Grant, M., Eds.), John Wiley & Sons, New York; 4th ed. Vol. 5, 1993, p. 60.

119. *Kirk-Othmer Encyclopedia of Chemical Technology* (Kroschwitz, J. I. and Howe-Grant, M., Eds.), John Wiley & Sons, New York; 4th ed. Vol. 5, 1993, p. 61.

120. *Kirk-Othmer Encyclopedia of Chemical Technology* (Kroschwitz, J. I. and Howe-Grant, M., Eds.), John Wiley & Sons, New York; 4th ed. Vol. 5, 1993, p. 61.

121. Hahn, A. V., *The Petrochemical Industry: Market and Economics*, McGraw-Hill, New York, 1970, p. 168.

122. *Kirk-Othmer Encyclopedia of Chemical Technology* (Kroschwitz, J. I. and Howe-Grant, M., Eds.), John Wiley & Sons, New York; 4th ed. Vol. 5, 1993, p. 61.

123. Faith, W. L., Lowenheim, F. A., and Moran, M. K., *Faith, Keyes, and Clark Industrial Chemicals*, 4th. ed., John Wiley & Sons, New York, 1975, p. 227.

124. *Kirk-Othmer Encyclopedia of Chemical Technology* (Kroschwitz, J. I. and Howe-Grant, M., Eds.), John Wiley & Sons, New York; 4th ed. Vol. 5, 1993, p. 67.

125. McKetta, J. J., *Inorganic Chemical Handbook: Vol. 1*, Marcel Dekker, New York, 1993, p. 531.

126. *Kirk-Othmer Encyclopedia of Chemical Technology* (Kroschwitz, J. I. and Howe-Grant, M., Eds.), John Wiley & Sons, New York; 4th ed. Vol. 5, 1993, p. 71.

127. *Kirk-Othmer Encyclopedia of Chemical Technology* (Kroschwitz, J. I. and Howe-Grant, M., Eds.), John Wiley & Sons, New York; 4th ed. Vol. 5, 1993, p. 66.

128. *Kirk-Othmer Encyclopedia of Chemical Technology* (Kroschwitz, J. I. and Howe-Grant, M., Eds.), John Wiley & Sons, New York; 4th ed. Vol. 5, 1993, p. 71.

129. Faith, W. L., Lowenheim, F. A., and Moran, M. K., *Faith, Keyes, and Clark Industrial Chemicals*, 4th. ed., John Wiley & Sons, New York, 1975, p. 229.

130. Hahn, A. V., *The Petrochemical Industry: Market and Economics*, McGraw-Hill, New York, 1970, p. 171.

131. Hahn, A. V., *The Petrochemical Industry: Market and Economics*, McGraw-Hill, New York, 1970, p. 171.

132. Szmant H. H., *Organic Building Blocks of the Chemical Industry*, John Wiley & Sons, New York, 1989, p. 141.

133. Hahn, A. V., *The Petrochemical Industry: Market and Economics*, McGraw-Hill, New York, 1970, p. 171.

134. Hahn, A. V., *The Petrochemical Industry: Market and Economics*, McGraw-Hill, New York, 1970, p. 171.

135. Szmant H. H., *Organic Building Blocks of the Chemical Industry*, John Wiley & Sons, New York, 1989, p. 142.

136. Hahn, A. V., *The Petrochemical Industry: Market and Economics*, McGraw-Hill, New York, 1970, p. 171.

137. Szmant H. H., *Organic Building Blocks of the Chemical Industry*, John Wiley & Sons, New York, 1989, p. 142.

138. Hahn, A. V., *The Petrochemical Industry: Market and Economics*, McGraw-Hill, New York, 1970, p. 172.

139. Szmant H. H., *Organic Building Blocks of the Chemical Industry*, John Wiley & Sons, New York, 1989, p. 141.

140. Hahn, A. V., *The Petrochemical Industry: Market and Economics*, McGraw-Hill, New York, 1970, p. 173.

141. Hahn, A. V., *The Petrochemical Industry: Market and Economics*, McGraw-Hill, New York, 1970, p. 173.

142. Hahn, A. V., *The Petrochemical Industry: Market and Economics*, McGraw-Hill, New York, 1970, p. 173.

143. Lewis, G. R., *1001 Chemicals in Everyday Products*, Van Nostrand Reinhold, New York, 1994, p. 53.

144. McKetta, J. J., *Inorganic Chemical Handbook: Vol. 1*, Marcel Dekker, New York, 1993, p. 534.

145. Weissermel, K. and Arpe, H., *Industrial Organic Chemisty*, VCH, Weinheim, 1993, p. 50.

146. Szmant H. H., *Organic Building Blocks of the Chemical Industry*, John Wiley & Sons, New York, 1989, p. 68.

147. Szmant H. H., *Organic Building Blocks of the Chemical Industry*, John Wiley & Sons, New York, 1989, p. 54.

148. *Kirk-Othmer Encyclopedia of Chemical Technology* (Kroschwitz, J. I. and Howe-Grant, M., Eds.), John Wiley & Sons, New York; 4th ed. Vol. 5, 1993, p. 57.

149. Weissermel, K. and Arpe, H., *Industrial Organic Chemisty*, VCH, Weinheim, 1993, p. 54.

150. Faith, W. L., Lowenheim, F. A., and Moran, M. K., *Faith, Keyes, and Clark Industrial Chemicals*, 4th. ed., John Wiley & Sons, New York, 1975, p. 234.

151. Faith, W. L., Lowenheim, F. A., and Moran, M. K., *Faith, Keyes, and Clark Industrial Chemicals*, 4th. ed., John Wiley & Sons, New York, 1975, p. 234.

152. Faith, W. L., Lowenheim, F. A., and Moran, M. K., *Faith, Keyes, and Clark Industrial Chemicals*, 4th. ed., John Wiley & Sons, New York, 1975, p. 234.

153. Faith, W. L., Lowenheim, F. A., and Moran, M. K., *Faith, Keyes, and Clark Industrial Chemicals*, 4th. ed., John Wiley & Sons, New York, 1975, p. 234.

154. *Kirk-Othmer Encyclopedia of Chemical Technology* (Kroschwitz, J. I. and Howe-Grant, M., Eds.), John Wiley & Sons, New York; 4th ed. Vol. 5, 1993, p. 57.

155. Hahn, A. V., *The Petrochemical Industry: Market and Economics*, McGraw-Hill, New York, 1970, p. 178.

156. *Kirk-Othmer Encyclopedia of Chemical Technology* (Kroschwitz, J. I. and Howe-Grant, M., Eds.), John Wiley & Sons, New York; 4th ed. Vol. 5, 1993, p. 57.

157. *Kirk-Othmer Encyclopedia of Chemical Technology* (Kroschwitz, J. I. and Howe-Grant, M., Eds.), John Wiley & Sons, New York; 4th ed. Vol. 5, 1993, p. 57.

158. Szmant, H., *Organic Building Blocks of the Chemical Industry*, John Wiley & Sons, Inc., New York, 1989.

159. Weissermel, K. and Arpe, H, *Industrial Organic Chemisty*, 2nd Ed. (Linley, C., trans.), New York, 1993.

160. Kent, J. A., (Ed.), *Riegel's Handbook of Industrial Chemistry*, 8th ed., Van Nostrand Reinhold Co., New York, 1983.

161. Hileman, B., Long, J., and Kirschner, E., Chlorine industry running flat out despite persistent health fears, *Chemical and Engineering News*, 72(47): 12–26 (1994).

162. Weissermel, K. and Arpe, H, *Industrial Organic Chemisty* 2nd Ed., (Linley, C., trans.), New York, 1993.

163. Szmant, H., *Organic Building Blocks of the Chemical Industry*, John Wiley & Sons, Inc., New York, 1989.

164. Kent, J. A., (Ed) *Riegel's Handbook of Industrial Chemistry*, 8th ed. Van Nostrand Reinhold Co., New York, 1983.

165. Lance, D. and Elworthy, E. G., Brit. Patent 7,297,1906.

166. Wiezevich, P. J. and Frolich, P. K., Direct oxidation of saturated hydrocarbones at high pressures, *Ind. Eng. Chem.*, 26: 267–276 (1934).

167. Boomer, E. H. and Broughton, J. W., The oxidation of methane at high pressures, *Can. J. Res.*, 15B: 375–382 (1937).

168. Dowden, D. A. and Walker, G. T., Oxygenated hydrocarbon production, Brit. Patent 1,244,001 (1971).

169. Stroud, H. J. F., Improvement in or relating to the oxidation of gases which consist principally of hydrocarbons, Brit. Patent, 1,398,385 (1975).

170. Liu, H. F. et al., Partial oxidation of methane by nitrous oxide over molybdenum on silica, *J. Am. Chem. Soc.*, 106: 4117–4121, (1984).

171. Khan, M. M. and Somorjai, G. A., A kinetic study of partial oxidation of methane with nitrous oxide on a molybdena-silica catalyst, *J. Catal.*, 91: 263–271 (1985).

172. Zhen, K. J., Khan, M. M., Lewis, K. B., and Somorjai, G. A., Partial oxidation of methane with nitrous oxide over V_2O_5–SiO_2 catalysts, *J. Catal.*, 94: 501–507 (1985).

173. Cullis, R. S., Keene, D. E., and Trimm, D. L., Studies of the partial oxidation of methane over heterogeneous catalysts, *J. Catal.*, 19: 378–385, (1970).

174. -2Mann, R. S. and Dosi, M. K., Partial oxidation of methane to formaldehyde over halogen modified catalysts, *J. Chem. Tech. Biotechnol.*, 29: 467–479, (1979).

175. Holman, A., Olsvik, O., and Rokstad, O. A., Pyrolysis of natural gas: Chemistry and process concepts, *Fuel Processing Tech.*, 42: 249–268, (1995).

176. Wheeler, R. V. and Wood, W. L., The pyrolysis of methane, *Fuel*, 7: 535–539 (1928).

177. Fischer, F. and Pichler, H., Uber die thermische zersetzung von methane, *Brennst-Chem.*, 13(20): 381–383, (1932).

178. Miller, S. A., *Acetylene. Its Properties, Manufacture and Uses, Vol. 1*, Ernest Benn Ltd., London, 1965.
179. Rokstad, O. A., Olsvik, O., Jenssen, B. and Holmer, A., Ethylene, acetylene, and benzene from methane pyrolysis, in: *Novel Production Methods for Ethylene, Light Hydrocarbons and Aromatics* (L. F. Albright, B. L. Cryness, and S. Nowak, Eds.), Marcel Dekker, New York, 1992, pp. 256–272.
180. Gueret, C. Valorisation du methane on hydrocarbures superieurs: Etude parametrique et modile cinetique. Ph.D. Thesis, Institut National Polytecnique de Lorraine, 1993.
181. Westmoreland, P. R., Dean, A. M., Howard, J. B. and Longwell, J. P., Forming benzene in flames by chemically activated isomerzation, *J. Phys. Chem.*, 93: 8171–8180, 1989.
182. Gas processes, *Hydrocarbon Processing*, April: 69–116 (1994).
183. Cotton, F. A. and Wilkinson, G., *Advanced Inorganic Chemistry*, Wiley Interscience, New York, 1988.
184. Lee, S. and Rengarajan, R., U.S. Patent 5,079,302 (1992).
185. Odian, G., *Principles of Polymerization*, Wiley-Interscience, New York, 1991.
186. Fifoot, R. E., Fluoroplastics, *Modern Plastics Encyclopedia 89*, McGraw-Hill Co., New York, October 1988.

5

Natural Gas Engineering

I. TRANSPORTATION OF NATURAL GAS

Efficient transportation of natural gas either "upstream" or "downstream" is an important step in the natural gas industry that affects all the consumers. Here, upstream is defined as the portion in which the crude natural gas as obtained from the gas reserve is transported to the refineries for treatment. Downstream is the portion wherein the processed natural gas is transported to the ultimate destinations (e.g., households, industries, etc.).

Natural gas (gas phase), liquefied natural gas (LNG), substitute natural gas (SNG), and liquefied petroleum gas (LPG) are considered to be integral parts of the gas transport industry. The criterion for selection of the mode of gas transportation depends on several factors among which the following are very important [1]:

- the distance over which the gas has to be transported
- the geographical and geological characteristics of the terrain over which the gas has to be moved
- the complexity of the distribution system (few or many source points, few or many distribution points)
- environmental factors directly associated with the gas transport mode
- the physical characteristics of the gas to be transported (i.e., the phase, gaseous or liquid)
- the construction and projected operating costs of the transportation system

A system can be designed to carry either the gaseous or liquid phase allowing greater flexibility in the transport of the gas although this might not be an economically ideal design. Pipeline design, selection and construction for the transport of natural gas is a large industry. The design considerations for these projects are [2,5]:

the composition and properties of the material transported (since this varies from one field to the other)
the weather in the areas to be covered
the geographical constraints in the area, such as railroad crossings, rivers, highways, minor streams, etc.
future demands and changes

The first major natural gas pipelines in the United States were constructed between 1928 and 1932 from the Panhandle gas field in Texas to Chicago and Detroit and were about 1570 and 1375 km long, respectively. Later on Tennessee gas trunk lines of about 2000 km long were constructed to transport gas from the Texas gas fields to West Virginia. The single longest pipeline in the United States spans over 2944 km, carrying gas from Texas to New York and Philadelphia. The overall length of pipelines in the United States is now about 250,000 km and there are more gas pipelines than railway lines [4]. Generally, gas pipelines are constructed underground to minimize the effect of variation in seasonal temperatures. Usually, 1 m below the ground level is an ideal situation because at lower depths water may be found.

The trenches for gas pipelines are dug by large excavators, the pipelines are brought section by section and welded together at site. The finished pipelines are coated with bitumen or Brisol or other special insulation material to prevent and protect them against corrosion and then laid into the trenches and covered up. These operations are usually done by special pipe surface cleaning and insulating machines. The quality of the insulation material is checked by flaw detectors which create an electric spark between the detector electrode and pipe if the insulation is faulty; precautions are taken if the necessity arises. The pipelines have to be made corrosion proof on both the outer and inner surfaces. The major contributor to the inner surface corrosion is the hydrogen sulfide in the natural gas; hence sulfur removal from the natural gas becomes mandatory for its safe and efficient transport. There are difficulties involved in laying the pipelines across rivers and canals; special equipment needed for this purpose generally increases the cost of construction for these projects.

The diameter of the pipeline to be used is also a major concern for the economic and efficient transportation of the natural gas. As the diameter of the pipeline increases the inlet pressure of the gas can be increased and in turn the throughput capacity of the pipeline is increased. However, with the larger pipelines the associated pressure drop is also higher and hence more compressor stations have to be installed which might increase the cost of the project. Since inlet gas pressures of about 100–150 atm are more common and since there will be a pressure drop as the gas is transported in the pipeline, so more compressor stations are needed for maintaining steady flow rates of gas. However, careful prior planning can result in the best trade-off for economically most efficient transporting of natural gas.

The first gas pipeline in the erstwhile Soviet Union [3] was from Saratov to Moscow which was about 325 mm (about 13 in.) in diameter. Based on this throughput capacity, new flow capacity value for 1000 mm, 1200 mm and 1400 mm diameter pipe was calculated to be 10, 15 and 20 times this original value, respectively. The consumption of metal per unit volume of gas transported, the capital investments, and operation and maintenance expenses can be reduced by using pipelines of larger diameter.

The usage of plastics, glass-fibers, asbestos-cement, and other materials for the construction of these pipelines is being considered in place of the conventional steel [6]. The advantages of the steel are the high strength and leak-proof nature, but as mentioned earlier they are prone to corrosion by some common components of unclean natural gas which can be overcome by substituting the nonmetal pipelines. However, the nonmetals have their own disadvantages and the usage of these is still under investigation. Possible leaks in the gas pipeline are found through aerial inspection by checking the color of the soil and vegetation surrounding the pipeline. The refractive index (RI) of the air also is routinely checked as the gas escaping into air changes the RI.

The gas-pipeline construction costs in 1983 showed a 1.05% increase for the composite gas pipeline cost index. Even though some of the materials (pipes, valves, and fittings, etc.) were discounted, some other costs such as labor, construction machinery and equipment increased and hence the total pipeline construction cost index showed an increase. Bigger pipelines (10–36 in.) contributed to most of the gas transportation in the United States. A comparison of costs of construction for land and offshore pipelines indicated that the offshore construction is three times as expensive as onshore projects. An annual cost index (AI) of 200 was published for the year 1983 for pipeline installation and an AI of 203 for compressor equipment and drive units in the compressor stations. The high pressure gas pipeline cost index in 1983 was the same as the previous year.

Natural gas is of crucial importance to Southeast Asia for both fuel and feedstock uses. The interest has become particularly intense since 1980. The countries that have gas supplies are developing them for both purposes as well as for export in the form of liquefied natural gas (LNG) [73].

A natural gas pipeline that would traverse the countries of Southeast Asia was proposed. The countries are specifically those of the Association of South East Asian Nations (ASEAN). This new pipeline would mean an entrance into the chemical and petrochemical business for these nations [73].

II. MULTIPHASE BEHAVIOR IN GAS AND OIL SYSTEMS

Methane, ethane, nitrogen, carbon dioxide, and hydrogen sulfide are some of the major components of any natural gas stream [7], even though the composition of these constituents may be different for various gas fields. As discussed in the following sections, the presence of carbon dioxide in the natural gas stream is not desirable because it can lead to solid formation in the process piping, valves, and heat exchangers during cryogenic processing of the gas. Carbon dioxide not only causes process complications as mentioned previously, but also lowers the heating value of the gas stream. For any chemical mixture (especially for a multiphase mixture) to be stored or handled, it is very important to know the thermodynamic behavior of that particular system. Various studies are available describing the multicomponent mixture behavior for the natural gas systems and a review of some of the principles and data is presented in this section.

To understand the thermodynamic behavior of a multicomponent system, one needs to specify the temperature, pressure, and composition of the various components of the mixture. It is very important to know the critical point of a mixture to determine the phase behavior of any system. In a phase diagram, the location of the critical point is very valuable because it fixes the shape of the quality lines which govern the vapor-liquid ratio at a given temperature and pressure within the phase envelope. The maximum pressure at which liquid may exist is called a "Cricodenbar" and most of the experimental data show that the critical point occurs very close to this for naturally occurring hydrocarbon mixtures [8]. The composition of the mixture being examined also determines the shape and location of the phase envelope. Natural gas is generally represented as a pseudobinary mixture and the line that passes through the critical point of the lightest and heaviest components of this mixture is called the critical locus of that mixture.

Four types of reservoirs have been identified depending upon the temperature and pressure in the wellbore:

1. A *black oil reservoir* is one whose temperature is less than the critical temperature and is undersaturated. Gas will form at the wellbore in this reservoir only when the pressure reaches the bubble point.
2. A *volatile oil reservoir* also occurs before the critical temperature; however, the gas-oil ratio is higher than in black oil reservoir. As mentioned earlier, even though gas forms at the wellhead normally, the bubble point pressure has to be reached for gas to form in the reservoir.
3. A *gas condensate reservoir* occurs between the critical temperature and cricodentherm (the maximum temperature at which the vapor and liquid

can coexist in equilibrium). Liquid forms in this reservoir when the pressure falls below the dew point and this liquid does not flow to the wellbore until it reaches a critical saturation in the pore space.

4. A *gas reservoir* occurs at temperatures above the cricodentherm and no liquid forms at any pressure in this well. However, if the wellhead conditions prevail in the phase envelope, some liquid may form in the wellbore and appear at the surface.

Hence, it is not easy to judge the type of reservoir available by looking at the surface, and the fluid sample from the well has to be analyzed to obtain the necessary experimental phase data. In designing the pumping equipment and pipelines to service a mainly gas phase, it is thus critical to construct the phase diagrams because when operating closer to the phase boundary small changes in temperature and pressure can cause large variations in vapor-liquid ratios, which could be detrimental to the pumps as well as transportation of natural gas.

The general approach to constructing a phase diagram is to determine (calculate) the critical point, cricodenbar and cricodentherm points. The following equations are proposed to calculate the critical temperature and critical pressure, respectively.

$$\frac{T_c}{T_c'} = 1.0 + (0.03)(\text{molecular weight of gas} - 16)$$

where T_c' is pseudocritical temperature from Kay's rule ($T_c' = \Sigma y_i T_{ci}$, where y_i is mole fraction of the component and T_{ci} is the critical temperature available from standard tables).

$$\log P_c = 0.0752 - 1.078 X_{2-4} + 1.7 \log(\text{MW}_7) + \log(1 - g)$$

where

X_{2-4}	=	total mol fraction of ethanes, propanes, and butanes in the mixture
(MW_7)	=	molecular weight of C_{7+} fraction
g	=	aromaticity factor
	=	$0.025\ K_7^2 + K_7 - 49.875 + 0.104125\ \text{MW}_7 - 0.00875\ K_7 \text{MW}_7$
K	=	Watson characterization factor
	=	11.9 for all paraffinic liquids

Other literature is available for predicting the cricodenbar and cricodentherm values which should be confirmed with experimental data. Also, literature is available for the prediction of physical properties of gas mixtures that are typically encountered in natural gas reservoirs [8]. Several approaches have been suggested to calculate the compressibility, density, viscosity, molecular refraction, etc., of these gas mixtures. In fact, composition specific methods are available for the prediction of these properties:

The Katz method and Stewart-Burkhardt-Voo methods are suitable for lean, sweet (S-free) gas

The Stewart-Burkhardt-Voo (SBV) method also may be used for reservoir gases

Prediction of pressure-volume-temperature (P-V-T) relations for sour gases (containing large quantities of sulfur) by any of the methods is quite unreliable and it is suggested that several approaches be taken to compare the results and decide the basis for judgment. Several publications are available with the data on binary and multicomponent mixtures of most of the components of generally occurring natural gas. In the following discussion, vapor-liquid behavior of multiphase fluids (vapor-liquid in particular) is presented.

The equilibrium vaporization (K) of a vapor-liquid mixture is calculated based on fugacity calculations and computer solutions are being sought for this purpose. A concept of "convergence pressure" is used to indirectly measure the composition of a multicomponent, multiphase mixture which utilizes the K value. This approach is based on the fact that for any composition of a particular mixture, the loci of log K versus pressure curves converge at K = 1.0. If this graph (log K versus pressure) is drawn at the critical temperature, the convergence pressure is the critical pressure. Several other methods are used for the prediction of convergence pressure. Once the K value is obtained, the bubble point (the condition at which a multicomponent system is all liquid with only one drop of vapor), dew point (the system is all vapor except for one small droplet of liquid), and flash calculation (vapor-liquid behavior inside the phase envelope) are calculated. The following procedure and equations are generally used for flash calculations [8].

1. Find K at the pressure and temperature of the two-phase system (separation condition).
2. Assume V, L, V/L or L/V (where V is the mol fraction of the vapor and L is the mol fraction of the liquid)
3. Solve the flash equation to obtain the composition and quantity of the streams, which are

$$V = V \sum y_i$$

$$= \sum \left[\frac{z_i}{1 + (1/K)(L/V)} \right] \text{ if the value of } L/V \text{ is assumed}$$

or

$$L = L \sum x_i$$

$$= \sum \left[\frac{z_i}{1 + K(V/L)} \right] \text{ if the value of V/L is assumed}$$

where

z_i = mole fraction of a constituent in the feed

4. If the assumed values of V and L are satisfied, the calculation is complete. Otherwise, the steps are repeated until convergence is reached. Thus, the state of the system present is determined with reasonable accuracy.

III. ADVANCES IN GAS TREATMENT

Natural gas as extracted has many impurities and undesirable compounds which not only reduce the heating value, but also create problems during its transport, including but not limited to clogging and corroding the valves, fittings, and pipelines. Hence natural gas has to be refined, conditioned, or treated before it can be used efficiently and economically. All the processes that contribute to upgrading the natural gas can be classified under this category.

The most common unwanted compounds that are part of natural gas are water, nitrogen, carbon dioxide, hydrogen sulfide, carbonyl sulfide, helium, and other heavier hydrocarbons. Any or all of these compounds may be found in natural gas from almost any gas field at various concentrations.

The presence of water, as mentioned earlier, can have several detrimental results among which is the formation of gas hydrates—snowlike, crystalline compounds composed of small amounts of methane, ethane, propane, or iso-butane and water. The formation of these hydrates is aided by the presence of liquid water and areas of turbulence. The formation of these hydrates increases the pressure drop along the pipeline, thereby decreasing its capacity; the presence of liquid water also can contribute to some corrosion. The formation and inhibition of these hydrates will be discussed in Section XII. In this section about gas treatment, the removal of hydrogen sulfide and other sulfide forms from the natural gas is discussed along with removal of carbon dioxide. A number of processes have been commercialized in this area and a few of them will be described here.

Propylene carbonate is an efficient medium for the removal of carbon dioxide from high pressure natural gas streams with reasonably high carbon dioxide contents. The application of propylene carbonate for this application is based on several of the following aspects:

1. low degree of solubility for carbon dioxide

2. low heat of reaction with carbon dioxide
3. low vapor pressure at operating temperature
4. low solubility for hydrogen and other low molecular weight hydrocarbons of the gas stream
5. nonreactivity to all other components of the gas
6. low viscosity
7. low hygroscopicity (lower absorption of water
8. high stability under operating conditions
9. inertness toward common metals

The process is widely accepted because of its simplicity, low thermal energy and pumping requirements and overall low cost of operation. Propylene carbonate is an efficient solvent to remove carbonyl sulfide [10].

A typical carbon dioxide removal system consists of an absorber where the feed gas is introduced at the bottom and the lean propylene carbonate solution is contacted with the rising gas in a countercurrent manner. The carbon dioxide content of the treated gas depends upon the initial content of CO_2 in the lean gas. The rich gas (containing the removed CO_2 and other compounds) is passed through an intermediate flash tank from where some of the low molecular hydrocarbons are recycled to the absorber. The stripped solvent is then passed through a low pressure flash tank where the carbon dioxide is flashed to the atmosphere and the lean gas is pumped back to the absorber. This process can be modified further to achieve lower CO_2 exit concentrations in the treated natural gas by adding strippers operating at atmospheric pressure followed by vacuum strippers. Power requirements for any of these units are very low, thus keeping the process very efficient and economical.

Among various processes used for carbon dioxide removal, the Giammarco-Vetrocoke, a hot potassium carbonate and water wash process, is the most efficient when the inlet CO_2 concentration was about 50% and the exit gas contained 2%. It was also found that when the exit stream contains a higher CO_2 level, the propylene carbonate process becomes the most economical. An amine-cleanup process was recommended for units where very low exit carbon dioxide concentrations are required.

Natural gas can be classified [9] into five distinct categories based on its composition:

1. wet gas containing condensable hydrocarbons like propane, butane, and pentane
2. lean gas that is devoid of all the condensable hydrocarbons
3. dry gas that has been dehydrated by various processes
4. sour gas that has hydrogen sulfide and other sulfur compounds
5. sweet gas that has been stripped of all sulfur related compounds.

Natural gas sold to the household consumer can be classified as lean, dry, and sweet gas.

Gas treatment also includes the recovery of various hydrocarbons associated with natural gas. There are several processes to achieve this goal: lean gas absorption, various types of refrigeration, turboexpansion, and others. However, since there were certain disadvantages in these processes, there was need for a refined process which can eliminate some of these limitations to economically and efficiently recover the valuable hydrocarbons from natural gas. One of the major limitations of the conventional processes was the inflexibility to meet the market demand. A process called the Mehra Process has been patented to solve some of these problems [9], in which a physical solvent is used for the removal and recovery of only desirable hydrocarbons from the gas stream. Basically, this is an extraction process since the volatility of hydrocarbons is increased on contact with a physical solvent. The equipment used for this process are arranged in two different configurations: extractive-flashing and extractive-stripping.

The first operation in the extractive-flashing (EF) mode is to pass the gas stream through an extractor and bring the gas in contact with a lean solvent where the methane gas is taken out from the overhead stream and the rich solvent (carrying all the natural gas liquids) is taken from the bottom of the extractor column; then the pressure is reduced in multiple flashing stages to separate the natural gas liquids from the solvent. The undesirable components still in the rich solvent can be selectively separated by the flashing pressures and recycling if necessary. The stream coming out of the final flashing stage is expanded, cooled, and condensed before being sent into a stripper to produce the specification-grade product. Thus by varying the flashing pressures the market demand for a particular component at a particular time can be met. The flexibility in this process is achieved by selecting the flow rate of the solvent with respect to the flow rate of the inlet gas stream, the flashing pressure of one or more successive flashing stages, selectively recycling the desirable gases to the extraction column, and rejecting selected components of liquid product by operating the stripper column in either the demethanizing, de-ethanizing, depropanizing, or debutanizing mode by varying the pressure and bottoms temperature of the stripper column.

In the extractive-stripping (ES) mode the extractor and stripper are combined together. The extractor section is contacted with the lean gas and the rich gas flows down into the stripper column coming in contact with countercurrent stripped hydrocarbons from the reboiler at the bottom of the ES column. The recovery of the desired compounds is controlled by the lean gas flow rate, stripper bottom temperatures and operating pressures. The rich solvent leaving from the ES column is expanded to the pressure of operation in the product column, which is essentially a distillation column. Here the

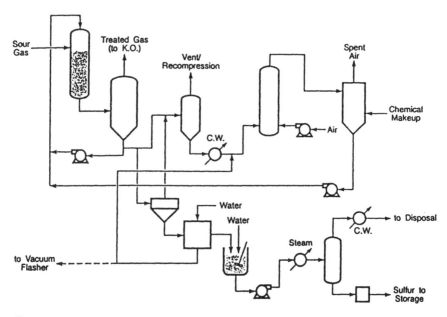

Figure 5.1   A schematic of the SulFerox process. *Source*: [76].

desired compounds are withdrawn as the overhead compounds, and, to minimize the loss of solvent, the liquid product also is refluxed with the condensed hydrocarbons. This mode of operation of Mehra process simplifies the
process and is more economical in investment and operating costs than the
EF configuration.

   The conventional lean-gas absorption, refrigeration, and turboexpansion
methods are described in [9–12].

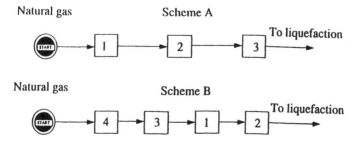

Figure 5.2   A schematic of the Calgon mercury removal process. 1) Amine treating,
2) gas dehydration, 3) removal with GAC, 4) LNG/liquids recovery. *Source*: [77].

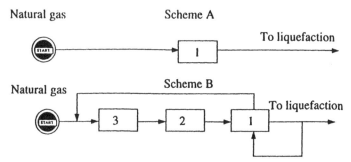

Figure 5.3 A schematic of the UOP mercury removal process. 1) HgSIV absorbent, 2) acid gas removal, 3) mercury removal bed. *Source*: [77].

Traditionally, acid gas treatment has required two separate processes. One is to remove hydrogen sulfide and the other is to convert the concentrated hydrogen sulfide stream to sulfur. A new process called the SulFerox process, developed by Shell Oil Company and The Dow Chemical Company, now offers a single process that handles both steps. The process has high degree of flexibility and requires a smaller capital investment than the Amine/Claus process. Chelated iron compounds are the heart of the process and the chemistry allows an aqueous solution of iron in high concentrations. As a result, circulation rates are low and equipment size is very small for the capacity. Figure 5.1 shows a schematic of the SulFerox process [76].

Natural gas streams contain small amounts (typically, in ppb levels) of mercury. Mercury has to be removed to protect downstream cryogenic heat exchangers and catalysts. The mercury exists in both elemental and organic forms. There are several commercial processes being operated, such as The Calgon Carbon Corporation mercury removal process and the UOP mercury removal process. In the Calgon process (Figure 5.2) mercury reacts chemically and adsorbs on impregnated granular activated carbon (GAC). The process can achieve effluent mercury levels of less than 1 ng/NM^3. In the UOP process (Figure 5.3) the mercury is removed in a conventional drier by the inclusion of some HgSIV adsorbent [78]. Mercury is chemisorbed on the HgSIV adsorbent during the drying stage. As a result, a totally dry and mercury-free effluent is provided. Since the sorption sites can be rejuvenated each cycle, the HgSIV adsorbent maintains its very high mercury removal efficiency [78].

IV. CRYOGENIC GAS PROCESSING

Natural gas reserves are generally contaminated with undesirable (nonburning) components which contribute to the reduction of the heating value of the gas.

If the calorific value (CV) of these components falls below specifications, minimum standards, or contract heating-value requirements, they are termed as low-Btu gases. These compounds not only reduce the calorific value of the gas, but also make it unsuitable for the transporting and distributing. Hence, they must be removed before the gas can be efficiently used. Some of the methods for the gas treatment were mentioned under the previous topic; here, upgrading of the natural gas by the removal of these low-Btu gases is discussed. The basic principle of cryogenic processing is the application of very low subambient temperatures to create a two-phase mixture, thereby allowing the phases, which contain widely varying heating values, to be withdrawn. The relative volatilities between the two components of the mixture result in selective mass-transfer between the phases in which one phase becomes concentrated with hydrocarbons and the other phase is stripped of the hydrocarbons. As can be predicted, the component with higher hydrocarbon content has higher calorific value than the incoming gas as well as the second phase. It has also been found by experimentation that several stages are required to achieve the separation necessary to produce gas with higher calorific value and high hydrocarbon recovery. Even though the cryogenic processing can be applied to separate CO_2 and H_2S, this process has been applied commercially only to remove nitrogen and helium. This also is one of the major processes for commercial production of helium gas.

The low-Btu gas has to be preconditioned (i.e., to remove any component of the gas that could adversely affect the operation of the cold sections of the plant) before it is charged to the cryogenic unit. Some of the compounds that cause concerns are water vapor, carbon dioxide, and heavier hydrocarbons having high solidification temperatures and low solubilities. These components might form solid components in the cryogenic temperature range and might deposit in the form of scales on the heat exchanger units resulting in reduced performance, possible blockage, and total shutdown of the plant. Special attention has to be paid to detect the components with high freezing points in the low-Btu gases. A bank of special absorption and adsorption processes are involved in pretreatment of these gases. Typical allowable inlet impurity levels for the cryogenic unit are 5 ppmv of carbon dioxide, 1 ppmv of water vapor, 30 ppmv of pentanes, and 1 ppmv of C_6+ hydrocarbons.

In a typical cryogenic processing unit the inlet gas has high nitrogen and varying helium contents. It might have small quantities of carbon dioxide, water and heavier hydrocarbons. This gas is passed through monoethanolamine (MEA) absorption columns to remove CO_2, molecular sieve adsorption columns to remove moisture and activated carbon beds for the removal of heavier hydrocarbons. This gas then will be expanded to form a vapor-liquid mixture which is passed through a high-pressure fractionator. This unit separates the gas into methane and nitrogen, and also produces a liquid nitrogen reflux for the low-pressure fractionator. These are the feed and reflux streams for the low pressure fractionator and the methane rich fraction is drawn from the base of the high pressure fractionator. The upgrading of the gas is completed in the low pressure fractionator and the exiting methane-rich gas from this unit is evaporated and superheated against incoming low-Btu gases. Thus, in this process by the use of two distillation columns alone, the separation of methane and nitrogen can be achieved using only the pressure energy available in the low-Btu gas. The methane-rich gas has a typical calorific value of about 940–980 Btu/ft^3. Most of the operating units also recover helium from the feed gas. The refrigeration required for the plant is provided in part by the efficient expansion of nitrogen contained in the feed and as the nitrogen in feed increases, the net power to compress product methane decreases.

The materials of construction are generally aluminum, stainless steel, and 9% nickel steel. The refrigeration units (cold boxes) are insulated by filling the columns with expanded perlite and purged with nitrogen. Refrigeration normally is achieved by mechanical compression-type cycle. The most common refrigerants in current use are ammonia and propane. In larger plants these columns are free standing and insulated with polyurethane and clad with aluminum sheeting.

The advantages of this cryogenic process are:

1. efficient utilization of the gas
2. improved gas burning characteristics and improved transport properties (low-Btu gases have higher density and viscosity resulting in higher pumping costs)
3. increased thermal efficiency in pipelines
4. high methane recovery with minimal power consumption at high inlet nitrogen content
5. allowance for extraction of heavier hydrocarbons and natural gas liquids
6. higher hydrocarbon partial pressures when the gas is used for chemical reactions

There are, however, some disadvantages in the cryogenic processing of natural gas:

1. Since the two-column cycle performs for most of the operation at low pressures and therefore lower temperatures, the system can only accommodate gas with < 20 ppmv of carbon dioxide without solid formation.
2. Also, the double column requires extra power and is more complex to obtain high methane recoveries at feed compositions of less than 20% nitrogen. Since the feed to the low pressure fractionator is entirely different from the high pressure fractionator, at low inlet concentrations of nitrogen, reflux generated in the process is insufficient to reduce the methane concentration in the nitrogen product to levels consistent with high recoveries.

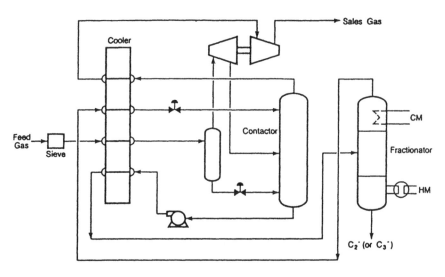

Figure 5.4   A schematic of the Cryoplus process. *Source*: [76].

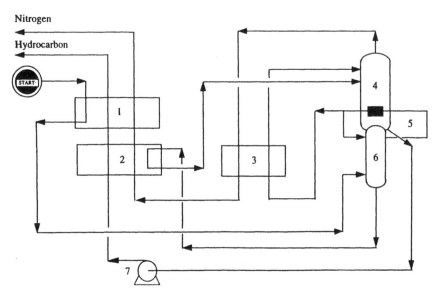

Figure 5.5   A schematic of the Costain nitrogen removal process. 1) heat exchanger, 2) heat exchanger (subcooling), 3) subcooler by low pressure nitrogen, 4) low pressure column, 5) condenser/reboiler, 6) high pressure distillation column, 7) hydrocarbon pump. *Source*: [77].

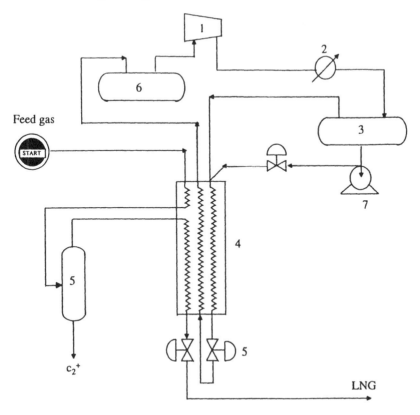

Figure 5.6 A schematic of the PRICO (LNG) process. 1) compressor, 2) partial condenser, 3) accumulator, 4) refrigerated heat exchanger, 5) Joule-Thomson valve, 6) refreigerant suction drum, 7) centrifugal pump, 8) partial condenser. *Source*: [77].

Cryogenic plants can be designed to accept a wide variation of inlet gas compositions and flow rates, to yield high hydrocarbon recovery, to recover natural gas liquids as an integral part of the process, to produce liquefied natural gas (LNG) for peak shaving, and to produce liquid cryogens such as liquid nitrogen and helium. Various cryogenic units, process parameters, and the end product specifications can be found in [13–15].

Figure 5.4 shows a schematic of the Cryoplus process [76] for the removal of nonhydrocarbon contaminants, and Figure 5.5 shows a flow diagram of the Costain nitrogen removal process [77]. The Costain double-column process is sufficiently flexible to handle natural gas with nitrogen concentrations of 5–80 mol%. Feed natural gas above 27 bar can be directly processed without any compression.

Figure 5.6 shows a schematic of the PRICO (LNG) process [77] which converts natural gas to liquid form for transportation and/or storage at atmospheric pressure by using the PRICO mixed-refrigerant loop. Applications range from large LNG units to small peak-sharing units. The refrigeration loop includes a compressor, a partial condenser, an accumulator, a refrigerant heat exchanger, a Joule-Thomson expansion valve, and a refrigerant suction drum. Accumulator liquids are directed to the refrigerant heat exchanger with a low head centrifugal pump. The refrigerant heat exchanger is a plate-and-fin brazed aluminum core.

V. UNCONVENTIONAL GAS RECOVERY

The dwindling energy sources and increased dependence on foreign imports has prompted research to develop alternate sources to increase the availability of natural gas. The ratio of gas reserves to production has been continually fluctuating since World War II and in 1995 this was estimated to be 31. This means that the present gas reserves without the addition of any new sources would be exhausted in 31 years. Thus it is important to find ways to add to the natural gas resources we already have.

One of the alternate technologies developed and commercialized is the conversion of crude oil, especially heavy crude, to a substitute natural gas (SNG). In this section some of the available oil gasification technologies will be discussed, including the process and cost analysis. The processes to be discussed in this section include: pyrolysis, partial oxidation, refinery processes (e.g., hydrocracking, hydrogasification, etc.), and gasification. In the subsequent discussion, processes developed by different companies for each of these techniques are presented.

Gasification of petroleum fractions has been carried out to produce two types of fuel gases: one with heating values in the range of 450–550 Btu/scf, termed town gas, and the other with calorific values of about 900–1100 Btu/scf. The lower heating value gases contain high concentrations of hydrogen and carbon monoxide and have a high burning velocity. The higher heating-value gases contain mostly methane, the major constituent of natural gas.

Pyrolysis (thermal cracking) is employed for the reduction of molecular weight of petroleum feeds. This process generally requires enough heat to be supplied to break the C-H bonds of the longer chain hydrocarbons in the feedstock. When proper control of the process variables (temperature, pressure and residence time) could not be achieved, lower selectivity to the desired compounds occurs and in extreme cases carbon black and hydrogen are produced. More often than not, the desired products in the pyrolysis are the lower molecular weight hydrocarbons of the feedstock. The operating conditions for this unconventional recovery of natural gas are more severe than the refinery petrochemical and pyrolysis processes. Temperatures in the range of 1300–1700°F, residence time of 1–5 sec and atmospheric pressure have been found to produce inert-free product gas (natural gas) with a heating value range of 900–1600 Btu/scf.

In the cyclic pyrolysis process, the conversion of distillates and residual oils is conducted in a refractory-lined or refractory-filled vessel with periodic air blasting to remove the deposited coke. The heat required for the pyrolysis is supplied by the combustion heat of oil and a part of the heat also is supplied by the periodic air blasting. This is followed by a steam purge in which process oil is sent to the pyrolysis zone, a second steam purge and a blast purge. This process is about 3–8 min long with the oil introduction period of 1–4 min. The thermal efficiency of this process was calculated to be about 85%. The high-Btu oil gases produced from the cyclic pyrolysis process contain considerable quantities of nitrogen, carbon dioxide, and aromatics [17]. Hence, this gas has to be conditioned before its use and one of the processes used to reduce the quantity of low molecular weight aromatics is scrubbing with absorption oil.

Various types of hydrogenolysis processes also have been employed to convert the oil feedstocks to high-Btu gases. The basic difference in the principle of operation of this process compared to the pyrolysis process is the methanation of the gases to achieve high production of methane rich natural gas [16]. A gas recycle hydrogenator (GRH) has been designed to convert the 650°F or lower fraction of crude into synthetic natural gas which operates at 1100 psig and 1400°F. The mixture of distillate vapor and hydrogen are introduced into the GRH and the momentum of the jet causes gaseous reactants to circulate in the tube and annulus of the GRH. The reaction products contain aromatics, which are sent to an aromatic removal unit which operates at 500 psig and 100°F and the essentially aromatic-free gas is then desulfurized to a level of < 0.2 ppm of sulfur. The aromatics recovered in this unit are used as the feedstock to produce hydrogen (by steam reforming) for the GRH. The rich gas (aromatics and sulfur clean) is then sent to the CRG (catalytic rich gas) reactor for low temperature steam reforming of the ethane component to methane, carbon dioxide and hydrogen. This gas is cooled before it is sent to the methanation unit where excess hydrogen available from the GRH reactor is used to convert most of the carbon dioxide to methane. The methane-rich gas is cooled to remove excess water and further processed to clean the gas of traces of carbon dioxide. The final methane-rich CO_2–free gas is compressed to about 1000 psig and dried to meet the pipeline water specifications.

There is a second GRH process which employs almost all the same principles for the production of natural gas except that the hydrogen is generated by partial oxidation rather than steam reforming [16]. The specifications of the gas produced from these processes are essentially the same:

Heating value	983 Btu/scf
Methane	96.79 vol%
CO	0.03 vol%
Carbon dioxide	2.14 vol%
Hydrogen	1.04 vol%

Another process developed to overcome the limitation of coke deposition during the GRH process is called the fluidized-bed hydrogenation (FBH). The essential difference between these two processes is that the GRH unit must be shut down intermittently to remove the built-up solids. In the FBH process, the fluidized bed of coke is circulated in the reactor with the removal of excess coke without interrupting the process. The advantages of this process over the GRH technology are:

1. better control of the temperature throughout the bed to achieve uniform hydrogenation rates, thus preventing hot spots and subsequent carbon deposition
2. reduction of the bed agglomeration by heavier oils by providing adsorption surface of coke

Partial oxidation is another important method to convert the petroleum feedstocks to methane rich synthetic gas [20]. There are three well-developed commercial processes available in the market: the Shell Gasification Process (SGP), the Texaco Gasification Process (TGP), and the Ube Process. In the following text a description of the SGP is presented and this process is compared with the TGP. A description of the Ube process is not included.

The general process sequence of any partial oxidation process can be divided into the following:

1. partial oxidation of heavier feedstocks into mostly CO, H_2, CO_2, etc.
2. cooling the hot gases from the gasification reactor
3. carbon removal
4. CO shift reaction
5. acid gas purification
6. sulfur recovery
7. methanation and drying

The steps involved in the SGP are:

1. gasification of feedstock with oxygen
2. heat recovery from partial oxidation hot gases
3. carbon and ash removal
4. sulfur and hydrogen sulfide removal
5. water-gas shift reaction to produce a 3:1 hydrogen to carbon monoxide ratio

6. removal of carbon dioxide
7. methanation of carbon oxides with hydrogen to produce essentially all methane
8. product drying to meet pipeline gas specifications

The reactions occurring in the partial oxidation (or gasification) reactor can be represented as:

$$C_nH_m + \left(n + \frac{m}{4}\right)O_2 \dashrightarrow nCO_2 + \frac{m}{2}H_2O$$

$$C_nH_m + nCO_2 \dashrightarrow 2nCO + \frac{m}{2}H_2$$

$$C_nH_m + nH_2O \dashrightarrow nCO + \left(n + \frac{m}{2}\right)H_2$$

The water-gas shift reaction converts most of the CO produced to carbon dioxide shown as:

$$CO + H_2O \dashrightarrow CO_2 + H_2$$

The shift converter reaction can be represented as follows:

$$CO_2 + 4H_2 \dashrightarrow CH_4 + 2H_2O$$

After most of the carbon dioxide is converted from the above reaction, the remaining carbon oxides (after purification) are reacted to form methane over a fixed bed of catalyst in an adiabatic reactor.

Even though both the SGP and the TGP essentially produce the same exit composition of synthetic natural gas, the former is designed primarily for synthesis gas whereas the latter can be easily altered to produce either hydrogen or synthesis gas. The following major differences can be pointed out between the SGP and TGP:

1. The SGP recovers most of the heat of the gasification effluent gases by inclusion of a waste heat boiler, whereas the gasification product gases in the TGP are quenched with water which results in a reduced thermal efficiency.
2. The SGP could remove the soot formed in the gasification system either by a pelletizing system or by a closed carbon-recovery system in which the soot-laden water is contacted with naphtha to form a slurry which is subsequently distilled off to recover naphtha and fuel oil. In the TGP only the second method is used.
3. The SGP advocates the removal of sulfur before converting CO to hydrogen by the shift reaction over a sulfur sensitive catalyst, whereas in the TGP

the CO-to-hydrogen shift takes place on a sulfur-resistant cobalt-molybdenum catalyst.

4. The TGP operates at a higher pressure (80–100 atm) than the SGP (50–60
   atm), which translates into lower operating costs in the former process.

5. In SGP the hydrocarbon feed, steam and oxygen are all separately preheated and then mixed at the burner exit. In the TGP feedstock is mixed
   with oxygen and steam and the mixture is preheated before being fed to
   the reactor.

Apart from the two processes described above, there are several other (at
least 25) processes developed for the production of synthetic natural gas and
many plants employing these processes are in operation. Even though the
basic principles of most of these processes are same, there are slight variations
in the plant design which result in varying operating costs and small variation
in product spectrum. Most of the processes produce a synthetic gas consisting
of 96–98+% of methane, with a calorific value of about 900–1200 Btu/scf.
References [16–21] discuss some of the aspects mentioned here in detail.

VI. STORAGE AND PEAK-SHAVING

The storage of extracted gas is an important step in the better use of natural
gas. As the usage of gas has variations according to the weather conditions,
the availability of the gas should be made to meet the requirements. Essentially, the demand for gas for industrial and domestic purposes fluctuates from
hours of the day and months of the year. The domestic usage is maximum
during the winter and is a minimum in the summer. For example, the variations in daily consumption in cities and towns can amount to millions of cubic
feet of natural gas. Since the gas fields cannot produce more than their
capacity during the peak demand seasons, the gas produced in less demanding
periods is stored to meet the variations. This process of storage of natural gas
for supplementing peak period demands is called "peak-shaving" of natural
gas. This process not only provides a continuous supply of the gas but also
helps maintain the economic effectiveness of the pipelines by letting them
operate at a higher load factor. For example, the transport costs of 1000 ft^3
of gas over 100 miles at a load factor of 100% is $0.011; a 25% load factor
raises the cost to $0.043 cents. As the volume of gas transported increases,
the cost of transportation decreases.

The density of methane is approximately 1/1000 of that of petroleum.
Hence, to store enough gas to have a calorific value that is equal to that of
1,000 tons of petroleum a storage capacity of 1 million cubic feet is required.
Since it is very difficult to find suitable materials of construction to store the
highly compressed natural gas (to reduce the volume), alternative sources of

storage have to be found. Depleted oil and gas fields are generally used for this storage which can be 1,000–8,500 ft. deep. In Illinois and Iowa, the gas is stored in nonpotable water-bearing zones which are called aquifer reservoirs. Several improvements in the storage and handling technology have helped to handle the high gas flow rates and repetitive storage cycles enabling an efficient and interruption-free supply of gas during the cold season. Generally, the excess gas supplied during the low demand season (summer) is stored 30–150 miles from the city markets. In these places, medium capacity thick-walled storage facilities (cylinders) are used to ensure an uninterrupted supply. Normally, gas pipeline companies and gas distributors maintain and oversee the storage facilities. Underground storage facilities around the United States are shown in the map of reference [25].

Porous media, generally limestone or dolomite, have a porosity of about 12–25% which is the space for storage. Permeability is the ease with which the gas or water can flow through the rock. The water that sticks to the sandstone even in the presence of natural gas is termed as interstitial or connate water. The temperature and pressure in the earth increases with depth; typical temperatures in underground storage reservoirs in California and Louisiana at a depth of about 8,500–9,000 ft are 185–190°F.

Natural gas deviates from ideal behavior at increased pressures, but the deviation decreases at higher temperatures. Compressibility factors applicable to the deviation can be predicted at given temperatures and pressures if the composition is known. Gas gravity [(molecular weight of the sample/29) × density of air] is substituted for gas composition for gases that are almost free of nonhydrocarbon components. Gas viscosity can be predicted if the temperature, pressure, and gas composition are known. Viscosity of the natural gas decreases as the temperature decreases and increases as the gas gravity decreases.

The following major items have to be dealt with before establishing a storage reservoir:

1. data acquisition of the production capacity, properties of the gas, and the geology and physical conditions of the reservoir land
2. economic evaluation of the project
3. an entire engineering design for cost estimate and construction bids

A schematic representation of different reservoir types is given by Speight [76] and reproduced here in Figure 5.7.

The Telearc process, a simple liquefaction process, was introduced for peak shaving of liquefied natural gas (LNG), which can provide high thermodynamic efficiency at a low investment and operating costs. A review of ten different LNG-peak shaving plants was published which describes all the parameters used to determine the suitability of a liquefaction process for

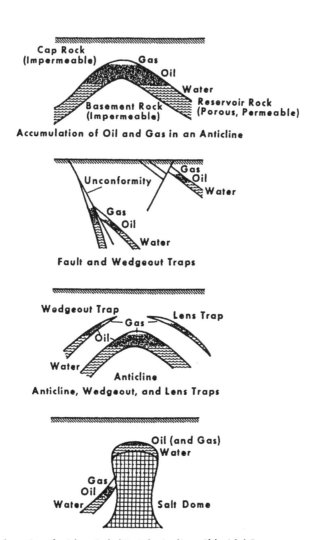

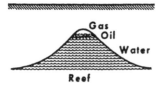

Figure 5.7   Schematic representation of different reservoir types. *Source*: [76].

peak-shaving plants. The variables considered were investment costs, operating costs, plant simplicity, flexibility, degree of automation, feed gas data, tail gas utilization and limitations, compressor driver systems, LNG requirements, and liquefaction capacity. Three different liquefaction processes—the mixed-refrigerant cycle process, the cycle expansion process, and the natural gas expansion process—were evaluated in the context of these parameters. Various peak-shaving processes have been invented and put to use in different facilities around the United States. Some of these processes are described in detail in references [22–27].

In this section, emphasis is placed on the underground storage that can be classified as market storage and field storage. However, there is always an option for surface storage (i.e., above ground storage). In surface storage methods, typically three types are considered, namely, wet gas holders, dry gas holders, and liquified storage.

VII. NATURAL GAS FUEL CELL

A fuel cell is a galvanic energy conversion device which circumvents the limitations of the Carnot cycle resulting in high conversion efficiency. The space program largely aided the research and development of fuel cell technology. The advantages of fuel cells are their low temperature operation, low emissions, and few moving parts; they are a maintenance-free, quiet, reliable source of electric power. The compact size and the amount of power generated by a fuel cell are some of the most accepted factors for the continuing research on them.

Electric current is generated by reaction on the electrode surfaces which are in contact with an electrolyte. Fuel and oxidant are not considered to be an integral part of the fuel cell; however, they are supplied as required by the current load and the reaction products are continuously removed. Hydrogen combustion with air or oxygen has been one of the most widely-used reactions in fuel cells for power generation. However, the use of natural gas in the fuel cell has significance, because unlike many other fuel sources, natural gas need not be converted into hydrogen. It can be combusted as it is and the heat energy supplied. A picture of a hydrogen-air fuel cell is shown in reference [28]. The system consists of a pair of porous catalyzed electrodes separated by an acid electrolyte. The hydrogen is oxidized to hydrated protons with the release of electrons on the anode and oxygen reacts with the protons consuming the electrons produced at the anode to form water vapor at the cathode. Electrons flow from the anode through the external load to the cathode and this circuit is closed by ionic current transport through the electrolyte. Conversely, in an acid cell the current is carried by the protons. The cell is operated at a fairly high temperature to vaporize the water formed at the cathode. The catalyzed layer of the electrode acts as the surface where the electrolyte solution and the gases react; because of the porous nature of the electrode the reactants and products can be removed rapidly from the surface. Other than this the electrode also acts as a path for the flow of current to the terminals. The electrolyte not only provides ionic conduction, but also separates the reactants. Typical electrolyte systems include, but are not limited to, potassium hydroxide (KOH) and concentrated phosphoric acid (H_3PO_4). Researchers around the world have been studying the use of various electrodes and catalytic materials that can be coated on the electrode and some of those results are presented here.

Scientists at ONSI Corporation (South Windsor, CT) have designed, installed, and commissioned a phosphoric acid PC25 fuel cell plant using natural gas as fuel. It is claimed to be an on-site cogeneration that can be conducted at a commercial building level. It is a 200 kW installation and does not emit any pollutants because it is an electrochemical process with no combustion; it is very quiet and highly efficient. The rated electrical efficiency of this system was found to be 40% and with 760,000 Btu/hr heat generation at 200 kW facility the efficiency could be as high as 80%. This system can be transported in a truck and weighs about 60,000 lbs and fits in a room of 24 × 10 × 11.5 ft. It can be seen from all the description given above that this might be a very convenient source of power generation (probably as an uninterruptible power source) in office buildings, computer data facilities, etc. Research also is under progress in this company for a higher level power generator (3–4 megawatt) [28].

Another interesting development in natural gas fueled fuel cell area is the Direct Fuel Cell developed by Energy Research Corporation (ERC; Danbury, CT). This is a carbonate fuel cell which could use directly either natural gas or coal gas without the necessity of external supply (usually by steam reforming of hydrocarbon fuel) of hydrogen. Since the galvanic combustion of methane is essentially a zero entropy process, the maximum theoretical efficiency could be 100%. A carbonate fuel cell was chosen because:

1. The operating temperature of about 650°C is high enough to be able to operate the process on nonnoble metal electrodes.
2. A significant amount of reforming could take place at this temperature.
3. Since steam is generated at the negative electrode and fuel also is consumed there, the electricity generation process ensure the complete conversion of fuel into reactive species within fuel cells (based on favorable mass transfer and nonequilibrium reaction conditions arising from current production).

Nickel catalysts are used for the catalytic coating on the electrodes. However, the overall conversion efficiency (based on lower heating value of the fuel) was about 60% in a 2-megawatt plant without employing the bottoming cycle. The NO_x levels from this power plant are < 1 ppm and direct fuel-cell stack operating times are approximately 10,000 hr and cell operating times are about 20,000 hr.

Another Ni-based solid oxide fuel cell (SOFC) electrode was developed on which a YSZ (yttria-stabilized zirconia) cermet and Lanthanum chromite were deposited by a slurry coating method. It was also suggested that a plasma spraying process can be used for the cermet deposition on the electrodes. The following reactions are expected to take place in a fuel cell employing a natural gas source, where internal reforming takes place on the Ni-YSZ electrode:

$$CH_4 + H_2O \longleftrightarrow 3H_2 + CO \qquad \Delta H = 206 \text{ kJ/mol} \qquad (5.1)$$

$$CO + H_2O \longleftrightarrow H_2 + CO_2 \qquad \Delta H = -41 \text{ kJ/mol}$$

$$2CO \longleftrightarrow C + CO_2 \qquad \Delta H = -172 \text{ kJ/mol}$$

This internal reforming reaction of natural gas (Reaction 5.1) is a very efficient process because it uses the heat released by ohmic and nonohmic losses and TΔS term in the fuel cells.

Natural-gas fueled fuel cells are more efficient than the conventional hydrogen-oxygen (air) fuel cells since they reduce capital costs and simplify the process by eliminating the external source of hydrogen supply. Acceptable efficiencies also have been found for these processes. With the developing technology, new electrode materials and catalytic materials could be used to further improve this technique of alternate electricity generation from natural gas [28].

VIII. MEMBRANE GAS SEPARATION

The acid gas removal and upgrading the heating value of the natural gases can be divided into three broad categories:

1. bulk removal of carbon dioxide, hydrogen sulfide, and water
2. cleaning the gas to meet pipeline requirements
3. upgrading the calorific value of the low-Btu gases

Separation or purification of gases can be achieved by employing membranes which separate a mixture of gases by the difference in their permeability. The selective passage of gases through a membrane (permeation) is a physical phenomenon, in which the gas mixture is introduced on one side of the membrane and is separated into two compartments on the other side. Usually, the permeation of a gas in a particular membrane material depends on its mobility (or solubility) and diffusivity. However, there is an exception in the case of microporous membranes wherein only Knudsen (free molecular) diffusion takes place, assuming the pores are very small. Absorption of the gases also has a role in the permeation process which, in turn, is dependent on the pressure, temperature, and surface flow. The overall permeation process can thus be classified as a process occurring due to the molecular interactions of the components of the gas mixture and the membrane material. Basically, the components of the mixture with higher permeability in a particular membrane would pass through it at a higher rate than the components with lower permeability.

A membrane should have a reasonable separation factor (efficiency) and a high flow rate of more permeable gases to be a useful separation technique in any process industry. As could be expected, different gases react differently with the membrane and hence the permeation flux is different for different components. The membrane technology is relatively new compared to some of the separation techniques used in the natural gas industry today. Realizing the considerable advantages (lower capital and operating costs, less severe operating conditions, etc.) the process industry as well as academia have shown interest in developing new membranes for various applications and optimizing the separation process. The membrane separation technology has been developed and is being used in the separation of carbon dioxide and hydrogen sulfide from natural gas, dehydration of natural gas streams, recovery of methane from biogas sources (unconventional recovery), synthesis gas ratio adjustment, recovery of helium from natural gas well heads, etc. Only natural gas-related applications are mentioned here, but there are several other processes in use for various other gas separations. In the following text the various factors affecting this separation process are discussed.

The most important step in the membrane separation process is the selection of the membrane material, its properties, and its selectivity to the different components of the gas stream. The membrane area and operational power consumption are the major factors to be considered in the design of a membrane gas-separation system once the suitable membrane is selected. Capital investment is affected by the membrane area, while the operating cost could be altered by the power requirements. However, the individual (specific to the separation problem) design parameters (e.g., pressure ratio between feedstock and permeate stream, reflux fraction for a recycle permeator, and relative areas for a cascade system) must be considered during the design process.

Suitability of a membrane for a particular process is decided based on the following factors:

1. adaptability of the membrane to changes in the flow rates of feedstock
2. capability to accommodate variation in concentration of permeable gases
3. feed pressure
4. feed temperature
5. product specifications (recovery of impurities and purity of desired component)

In the applications where high pressure of residual gases (nonpermeate stream) is required, membranes are well suited. The transport steps involved in a membrane system are very similar to the transport steps involved in a heterogeneous reaction system where first the reactant should be adsorbed on the surface, then diffused into the pellet, and finally the products are desorbed. The permeating component of the gas stream is sorbed in the membrane, diffused through the membrane, and then desorbed on the other side of the membrane. Fundamental research in the area of development of suitable membranes for different gas processing and petrochemical-related industry applications is being carried out in various academic institutions.

Several separators may be connected in series or in parallel depending on specific requirements. A schematic of a typical two-stage process is given in Figure 5.8.

Monsanto Co. developed a membrane system under the tradename PRISM. This system is very useful in the separation of acid gases from the natural gas and landfill gas. Carbon dioxide, hydrogen sulfide, and water vapor are the "fast" species compared to methane, nitrogen, and other hydrocarbon gases. Fast species are the components of the gas mixture that have high rate of permeation across the hollow fibers enclosed in pressure vessels. The PRISM separation system has the geometry of a shell-and-tube heat exchanger, in which the feed stream enters at one end of the shell side and the products exit at the other end of the shell. The acid gases permeating to the bores of hollow fibers pass through the internal tube sheet and are discharged through the closure flange on the pressure vessel. A series of these separators can be used if the separation achieved in one pass is not sufficient.

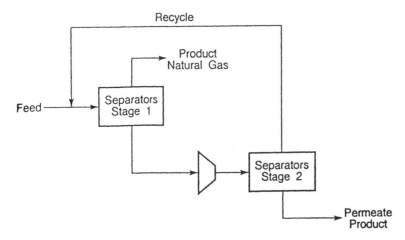

Figure 5.8 A schematic of a typical two-stage process. *Source*: [76].

In the natural gas application using the Monsanto PRISM separator, a high partial pressure differential is applied to the faster permeating components, which is essentially what is required to clean the natural gases of carbon dioxide and hydrogen sulfide. The pressure gradient across can be achieved in this process even at low feed pressures of 200–300 psig because the permeate gases are collected at atmospheric pressure. The higher this pressure gradient, the higher the removal of acid gases, translating into smaller separation areas. The PRISM separator process also can be used for a retrofit application. A combination of this membrane separation technique with the conventional absorption or adsorption techniques can result in lowered capital and operating costs for new facilities. This synergism of combined processes also could be used to gain significant flexibility in processing streams containing significant flow rate and composition fluctuations. For older absorption or adsorption plants that have to accept a wider range of acid gas compositions, addition of a membrane process is the easiest way of removing bottlenecks. Single or multiple-stage separations are designed to meet product specification.

Cellulose ester membranes (e.g., cellulose acetate, cellulose diacetate, cellulose triacetate, cellulose propionate, cellulose butyrate, cellulose cyanoethalate, cellulose methacrylate, and mixtures of these) can be employed for the acid components of a natural gas stream. The membranes can be either flat films or hollow fibers.

The membrane separation process offers significant savings in energy costs; for example the heat energy savings in an ammonia plant using this type of separator system was 450 MMBtu/day and this resulted in an increase of about 3% of ammonia production without additional energy use. Several other publications are available describing the simplicity in the installation, commissioning, and operation as well as the advantages of the membrane gas separation technique. Thus, it can be seen that membrane gas separation is an important growing technology with significant economic and operational advantages. Extensive literature is available on the membrane separation technology and a few of those studies are given in references [29–47].

IX. NO$_X$ AND SO$_X$ REMOVAL

The nitrous oxides (NO$_x$) and sulfurous oxides (SO$_x$) result from the combustion of any petroleum, coal, and natural gas fuels. Various industrial processes and almost a majority of the households employ natural gas for energy and heating purposes. Natural gas-fueled vehicles also are NO$_x$ and SO$_x$ producers. These compounds are the main contributors to the greenhouse effect and acid rain. Hence, it is imperative that methods have to be developed to save the environment and future generations from this problem. There are many processes that are already in operation and many more are being explored. The front-end techniques strive to produce lesser quantities of these pollutants by changing the design of the burners, fuel to air ratios, and control of temperatures. The downstream processes concentrate on the elimination of these pollutants in the effluent streams before they are released to the environment. It can be seen from various published literature that catalytic decomposition/destruction of these gases is a very viable, important, and economic method.

It is important for the engineer to know the chemistry of the formation of NO_x gases (NO and NO_2) before designing front-end NO_x reduction systems.

$O_2 <\text{-->} 2O$ Initiation and termination

$N_2 + O <\text{-->} NO + N$ Propagation, $T > 2800°F$ $\hspace{1cm}$ (5.2)

$O_2 + N <\text{-->} NO + O$ Propagation, $T > 1500°F$

It can be seen from these reactions that oxygen concentration and furnace temperature are the two most important factors in the formation of NO_x. This implies that the philosophy of NO_x reduction levels revolves around proper burner and furnace design along with the control of stoichiometric oxygen. When there is not enough excess air, the formation of oxygen atoms decreases and the reactant for (Reaction 5.2) decreases; hence the formation of NO comes down. Controlling the temperature in the furnace below $2800°F$ also decreases the possibility of formation of NO which is accomplished by proper design of furnace and burner.

The furnace-cooling surface should be increased in order to decrease the NO_x production, which is done by increasing horizontal and vertical burner spacing and/or providing water-cooled division walls in the burner zone. If the furnace is too deep, the stoichiometric air must be increased for complete combustion and if the furnace is shallow, the flames may impinge on the walls causing incomplete combustion, high tube-metal temperatures, tube corrosion, or slagging. The furnace should be designed such that a maximum furnace-cooling surface is utilized, which is accomplished by selecting a furnace in which the flame is directed along a wall. If the walls of the furnace are kept clean (avoid formation of ash) by providing wall blowers, maximum heat energy can be recovered, thus avoiding hot spots in the furnace and the high temperatures that cause NO_x to form. Double-register burners are used for natural gas fuel, which have the flexibility of operating with high swirl for stable off-stoichiometric firing or with low swirl for diffusion burning. If the fuel is injected into the burner perpendicularly to a secondary air stream and the flue gases are recirculated or if a two stage combustion is employed, NO_x levels can be reduced by 20–30%.

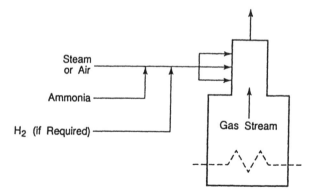

Figure 5.9   A schematic of the Thermal DeNOx process. *Source*: [76].

Several processes have been developed to catalytically reduce nitric oxide to nitrogen since nitric oxide is the main contributor to smog. One of the processes developed for this reduction utilized a Catalytic Combustion Corporation catalyst and reducing agents like hydrogen, carbon monoxide, methane, and other hydrocarbon vapors. This process operates at atmospheric pressure and about 1000°F. However, this process also can be conducted at elevated pressures in Catridge-type reactors. In these processes, the gas containing nitric oxide enters the reactor through a side nozzle and flows upward through an annular space surrounding the main shell. At the top of the vessel, the gas flow turns downward into a second annulus where it contacts the hollow cylindrical catalyst bed. The treated gas is discharged from the bottom of the reactor at the exit concentrations of less than 200 ppm. The catalyst's life period is about 20,000 operating hours. Typical catalysts are noble metals, copper-zinc-chrome, barium-promoted copper chromite, and chromium-promoted iron oxide, etc.

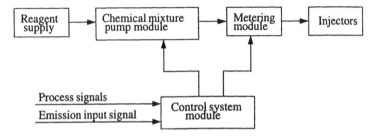

Figure 5.10   Noncatalytic NO_x out process. *Source*: [78].

There are a number of processes developed for noncatalytic NO_x removal. The thermal $DeNO_x$ process (Figure 5.9) and the NO_x out process (Figure 5.10) are typical examples. In the first, the gas phase reaction between NO_x and ammonia occurs at 870–1200°C using air or steam as a carrier gas:

$$NO_x + NH_3 = N_2 + H_2O$$

Instead of ammonia, hydrogen may be used and the temperature can be lowered down to 750°C. In the NO_xout process, removal of NO_x is achieved by using chemical sprays to convert NO_x to nitrogen, carbon dioxide, and water [78].

Removal of SO_x from the flue gases is easier than NO_x removal because of the lower solubility of NO_x in the water. Wet scrubbing processes have been developed for SO_x removal in which the sulfur compound is reacted with the chemical system to form metal sulfite or metal bisulfite. Sodium-, magnesium-, and calcium-based systems are available for dissolving SO_x and to clean the effluent stream to environmentally acceptable levels. Regenerable or disposable systems are available which essentially either recover sulfur removed from the waste gas stream or throwaway the recovered sulfur compound, the choice depends on the value of the recovered product. Methods also have been developed for simultaneous removal of NO_x and SO_x. The Neville-Krebs process is one such system in which NO_x is reduced into nitrogen and small amounts of NO_2. Scrubbing solutions containing sodium compounds with small amounts of catalysts that have the dual function of fixing and reducing the NO_x gases as soon as they are in contact with the scrubbing solution have been specially formulated to react chemically with SO_x and NO_x simultaneously. The advantages of the simultaneous NO_x and SO_x removal processes are:

1. no production of scale-producing insoluble compounds in the scrubber
2. simple and inexpensive process for NO_x and SO_x removal
3. use of readily and easily available chemical for scrubbing
4. noncorrosive to the scrubber components because the solution operates at a pH of 6–8.

Multiple stages are also suggested in the simultaneous removal process to achieve acceptable levels of these pollutants. Further reading materials can be found in references [48–55].

X. SOUR GAS PROCESSING

Sour gas is natural gas that is contaminated with hydrogen sulfide (H_2S), carbonyl sulfide (COS), mercaptans (RSH), and other forms of sulfur. The sulfur in the gas has to be removed to acceptable levels in order to save the transmission pipelines from corrosion. When the inlet gas contains high amounts of sulfur-containing compounds, it must be treated in a conventional Claus-type reactor. These plants cannot vent the extracted sulfur in the form of sulfur dioxide because of the environmental regulations. Hence, whenever possible, these compounds have to be converted into elemental sulfur which has an end use. The natural gas can be desulfurized either by wet or dry methods. In the wet method usually an organic solvent is used to either absorb or react the sulfur compound, whereas in the dry method a solid medium is employed to either adsorb or react the acid gases. Common schemes employed in the sulfur removal/recovery from natural gas are:

1. amine systems
2. physical solution absorption
3. fixed-bed adsorption (molecular sieve)
4. membranes
5. nonregenerative chemical reaction

The amine systems are the most widely-used systems among all of these procedures. In the first two methods the principle of operation is similar to that in the lean-oil absorption system used for carbon dioxide removal. Membranes are used to treat gases with high sulfur content, in which the permeation of gases through a membrane with large pressure drop as the driving force. The nonregenerative chemical reaction process is used when the sulfur content is very low in the inlet gas and the chemical costs are acceptable. In summary, the first two processes can be classified as chemical and physical solvent processes, whereas the third one is a dry-bed process. The desulfurization of natural gas also is called sweetening.

Monoethanolamine (MEA) is the most widely-used chemical solvent along with diethanolamine (DEA). Diethylene glycol and triethylene glycol combined with one of these amines are also used for the acid gas treatment (glycol-amine process) which simultaneously dehydrate and remove carbon dioxide and hydrogen sulfide. Basic operation in these processes is contacting the solvent in a countercurrent manner with the incoming sour gas in a gas-liquid absorber. The rich solvent is passed through a steam-stripping still which operates at 115°C and 143 kPa and the solvent is recovered. Since the vaporization losses with MEA are higher and since it cannot effectively remove any sulfur containing compound other than hydrogen sulfide, DEA is employed when the inlet gas contains compounds like carbonyl sulfide because of its resistance to attack. However, diglycolamine (DGA) solvent has replaced DEA and MEA as the most widely-used agent for the sour gas processing because of its lower plant costs. The circulation rate of DGA is about half that of MEA, it can decompose carbonyl sulfide, and it can withstand freezing up to $-40°C$. The following equations represent the desulfurization by amine solutions:

$$RNH_2 + H_2S \longrightarrow RNH_3^+ + HS^-$$

The physical solvents most commonly used for sweetening of sour gases are Fluor solvent, Sulfinol, and Selexol. These processes are generally very useful while treating gases at very high pressures of the order of 300–1000 psig. These solvents can remove carbon dioxide and hydrogen sulfide, and are resistant to the degradation caused by carbonyl sulfide. Since these processes can treat natural gas with acid gases at low circulation, the plant investment and operating costs are considerably reduced. However, these solvents absorb pentane and higher hydrocarbons, which is not desirable. The Fluor solvent process employs propylene carbonate for the treatment of natural gas with high carbon dioxide (> 20%) and low hydrogen sulfide contents. This solvent can be regenerated by simply flashing the rich solvent to a vacuum through a hydraulic turbine. Poly(ethylene glycol) dimethyl ether is the solvent in the Selexol process; it also is suitable for large carbon dioxide and low hydrogen sulfide contents. However, the solubility of hydrogen sulfide in this solvent is 7–10 times higher than that of carbon dioxide and hence preferential removal of hydrogen sulfide takes place in this process. The Sulfinol process utilizes a solvent which is a mixture of a physical (sulfanone, tetrahydrothiophene-1,1-dioxide) and chemical (diisopropanolamine) solvents. This process can desulfurize the gas stream to meet the pipeline specifications while leaving some carbon dioxide in the stream.

One of the dry-bed processes is the iron-sponge process wherein the hydrogen sulfide is selectively removed. This is one of the oldest and simple processes for sour gas treatment. In this process the iron sponge consists of wooden chips coated with hydrated iron hydroxide which reacts with hydrogen sulfide to form iron sulfide. Although this bed can be regenerated periodically, it loses its effectiveness and must be replaced in short intervals making it an expensive process. That is why this process is now limited to small volumes of gases containing low hydrogen sulfide content. The following equations represent the iron oxide process:

$$Fe_2O_3 + 3H_2S \longrightarrow Fe_2S_3 + 3H_2O$$

$$Fe_2S_3 + \frac{3}{2}O_2 \longrightarrow Fe_2O_3 + 3S$$

Another chemical solvent process that was frequently used is the Giammarco-Vetrocoke process in which a mixture of aqueous solutions of an alkali metal carbonate, an alkali metal arsenite, and an arsenate is used as the absorption medium. When correct proportions of arsenite to arsenate and pH of the solution are chosen, both CO_2 and H_2S can be absorbed. The following reaction represent the desulfurization using the above principles:

$$M_2CO_3 + H_2S \dashrightarrow MHS + MHCO_3$$

$$Na_4As_2S_5O_2 + H_2S \dashrightarrow Na_4As_2S_6O + H_2O$$

$$Na_4As_2S_6O + \frac{1}{2}O_2 \dashrightarrow Na_4As_2S_5O_2 + S$$

These reactions are some of the processes currently employed in the sour gas processing industry. The selection of the process depends on the inlet composition of the gas, suitability of a particular process (based on the product specification), and the economics of the process. However, work still needs to be done wherein thiols (R-SH) in the feed gas are efficiently removed since they cause odor problems. Currently, considerable amount of thiols are removed in the oil-absorption plants and molecular sieves also have shown some promise for this purpose. A vast amount of literature is available on this topic and a few of them are listed in references [48–55].

XI. NATURAL GAS VEHICLES

Petroleum reservoirs are depleting, as the production surveys reveal, and the OPEC countries are increasing the cost of oil they supply to the fuel companies which, in turn, increases the cost of fuel to consumer. Especially in the United States, automobile use and gasoline and diesel fuel consumption are enormous. Another important factor in cutting down on conventionally-fueled automobiles is the global realization about the conservation of our environment. The solution for these problems is being worked out with natural gas vehicles. The potentially harmful emissions from the natural gas vehicles are as low as 10% of the conventional gasoline and diesel vehicles. Apart from this, the cost of natural gas is significantly (30–50%) cheaper than the conventional fuels depending on the geographical location and it is comparatively more abundant than petroleum.

It is not only just the environmental regulations that are driving the initiatives to find alternative fuel for automobiles, but also the utilities companies which plan to diversify into the natural gas fuel industry. Considerable efforts are being expended for the development of natural gas as fuel which can be seen from the growing number of government and industry fleet vehicles. It has been a practice to use liquefied natural gas (LNG) as fuel. However, efforts are now underway to use the compressed natural gas (CNG) as fuel.

There are two ways in which the implementation of natural gas vehicles (NGVs) is taking place: manufacture of more natural gas-fueled vehicles and the conversion of older vehicles to adapt to the new fuel. There are concerns in the NGV industry about how to make these more acceptable by the public and to provide more supply points for the fuel. In a recent study, it has been shown that the use of NGVs by the utility companies in some states is relatively high compared to the others. Studies also have highlighted the savings in fuel costs and the time to offset the conversion costs, which estimated a break-even point in about 3 yr.

Conventional spark ignition gasoline engines can be converted to natural gas fueled engines fairly easily, while the modification process for diesel engines is more complicated. The NGVs have proved to be more efficient in terms of energy savings because of the higher compression ratio and their exhaust contains lesser amounts of CO_2 than petroleum-fueled vehicles. A typical gasoline automotive conversion for use of natural gas fuel involves addition of storage tanks, pressure regulators, gas/air mixer, fuel switching mechanisms, and a natural gas refueling connector. The internal combustion engine operating on natural gas essentially operates on the same principles as the gasoline fueled engine, compression followed by combustion, expansion and exhaust.

In any internal combustion engine there are losses from pumping, timing, heat, exhaust, and friction. Since the methane flame speed is lower than gasoline flame, the timing loss for a natural-gas engine is larger than for a gasoline engine. By 1981, natural-gas engines producing 16% work output were developed, which under steady state conditions increase to 28%. The air/fuel ratio is 14.7 for gasoline engines while that on the natural gas engine is 17.2. The ratio of actual air/fuel ratio divided by the stoichiometric air/fuel ratio is termed λ. The air-fuel mixture is called a lean mixture when $\lambda < 1$ and a rich mixture when $\lambda > 1$. Natural gas engine power output is closer to the maximum when a rich mixture is available. The power loss at leaner mixtures is about 10–15% at 2500 rpm, which is due to the following:

1. More air is replaced by natural gas in the air intake than by liquid fuel.
2. The flame speed of natural gas is lower than gasoline (leading to more timing loss).
3. The air mass breathing capacity increases due to the increase in air density in liquid-fuel engines.

The power output and fuel economy of a natural gas plotted against λ shows that maximum fuel economy can be obtained at leaner mixtures since the thermal efficiency of the engine is greater. As mentioned earlier, since the air to fuel ratio is higher for natural-gas engines than the gasoline engines, it can be readily concluded that natural-gas engines would be more fuel efficient.

The amount of pollutants produced depends on air/fuel ratio and spark timing. When rich mixture is supplied there is not enough oxygen available to convert CO to CO_2. Similarly, NO_x production depends on the time and temperature history of the fuel supplied. When rich gas mixtures are supplied the formation of NO_x gases is lower since the maximum temperature reached is lower and there is insufficient oxygen for the reaction. Spark timing also affects the NO_x production because when the fuel mixture is ignited early and then compressed by the rising piston, the maximum temperature attained is higher which aids higher NO_x production. To obtain lower NO_x levels without sacrificing fuel economy the combustion should propagate quickly after ignition, reducing the required spark advance.

One of the limitations of NGVs is the limited range of these vehicles, which depends on the fuel properties, amount stored, and the fuel consumption rate. The fuel consumption rate depends on the power requirement of the vehicle and the engine efficiency. Research is under progress in many academic institutions and research institutions to improve both the performance and efficiency of the natural gas vehicles.

Various methods of natural gas refueling have been suggested to meet the increasing implementation of NGVs. A home-rechargeable natural gas commuter car has been demonstrated and the idea has been promoted. Since almost every household in North America has a natural gas connection, it would be easier to have the refueling done at home when the automobile is not in use. It usually takes about 2–3 h. Gas compressors (multiple stages) are being manufactured that could be installed around the house that can compress the natural gas supplied to a household to about 2500–3000 psi. Thus, slowly but steadily, we can switch to the cleaner and cheaper natural gas. The various designs for converting gasoline engines to natural gas and for natural gas refueling systems are shown in references [56–66].

XII. HYDRATES

A. Hydrates as Resources

Natural gas hydrates are becoming more and more important as an invaluable fossil fuel resource. As much as 10^{18} cubic meters of natural gas may exist in the hydrate state, which has not been exploited on a large scale. The amount of natural gas hydrates is enormous and enough to replace all other forms of fossil fuels. Much of Alaska, Northern Canada, and Siberia have climates suitable for natural gas hydrate formation. Virtually the entire ocean bottom also has conditions suitable for natural gas hydrate formation. Up to 3 m thickness of hydrates can be found [81].

B. Formation of Hydrates

Natural gas contains water before processing. High pressure, low temperature, or both aid the combination of water with light gases (methane to butane, carbon dioxide, and hydrogen sulfide) to form hydrates. Free water must be present for the hydrates to be formed. The hydrates are often found under water at depths deeper than 30 m, and under the permafrost and represent a potentially huge resource [74,75]. There has also been speculation that there may even be free natural gas trapped under the hydrate resource [76]. As mentioned earlier, the hydrates (solid crystals similar to wet snow) create problems by obstructing the flow of gas especially at the valves, fittings, joints, and other equipment. Hence, it is imperative to know what the conditions are under which these hydrates are formed; then designs can be found that have minimum disruption in the flow of gas and keep the operating costs at minimum. Turboexpander systems using methanol have been used to prevent the hydrate formation in the natural gas. However, more recent designs have adopted solid-bed dehydration even though the former process offers considerable cost advantages as well as the ability to handle gases with low water content. In this section, a few of the common dehydration processes and the principles are reviewed.

Thermodynamically, gas hydrates may be considered solutions of gas in crystalline solids. A loose association by molecular forces forms a solid crystalline structure of water and gas constituents, a clathrate [81]. Unlike solid crystals, liquid water has a mobile lattice structure with two different vacant lattice positions. Gas molecules occupy these vacant lattice positions and cause the water-gas system to solidify (Figure 5.11). Each unit cell of hydrate Structure I has two small and six large voids. Structure II has sixteen small and eight large voids. In general, CH_4, C_2H_6, CO_2, and H_2S are the guests of the host to form hydrate water structure I; C_3H_8, i-C_4H_{10}, CH_2Cl_2, and $CHCl_3$ form hydrate Structure II [80,81]. The C_5 and higher hydrocarbons are too large to fit into the larger cavities to form hydrates [81]. Therefore,

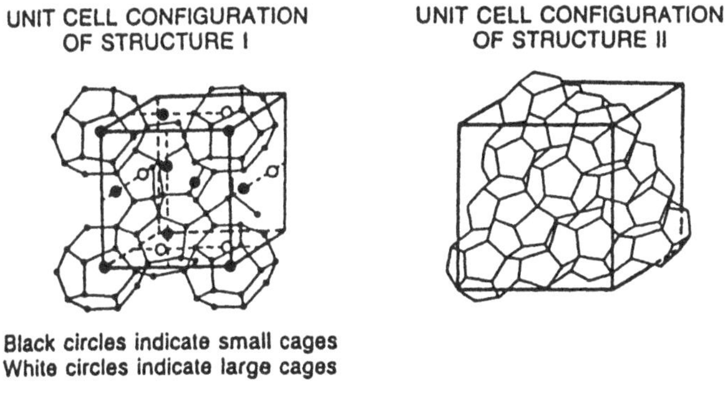

UNIT CELL CONFIGURATION   UNIT CELL CONFIGURATION
OF STRUCTURE I   OF STRUCTURE II

Black circles indicate small cages
White circles indicate large cages

Figure 5.11   Structures of a hydrate lattice. *Source*: [81].

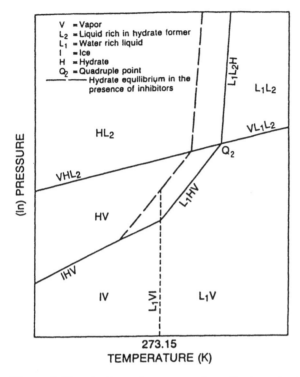

Figure 5.12 A schematic representation of hydrate-forming conditions for a subcritical forming gas. *Source*: [81].

it is important to realize that most hydrocarbons present in natural gas form hydrates when sufficient free liquid water is present. The general chemical formula assigned to hydrates is $C_nH_{2n+2}\cdot mH_2O$ where m and n differ depending on the hydrocarbons. For example [76],

$$C_3H_8 + 17H_2O = C_3H_8 \cdot 17H_2O \text{ (s)}$$

$$CH_4 + nH_2O = CH_4 \cdot nH_2O \text{ (s)}$$

Conditions for hydrate formation are given in Figures 5.12 through 5.16. Pure hydrate-forming substances form hydrate in the region above the three-phase, liquid water/liquid/gas line (Figure 5.12). Figure 5.12 serves as a key to reading Figures 5.13, 5.14, and 5.15. Figure 5.15 shows hydrate-forming conditions for various substances including CO_2, ethylene, acetylene, freon 12, hydrogen sulfide, sulfur dioxide, chlorine, and methyl chloride.

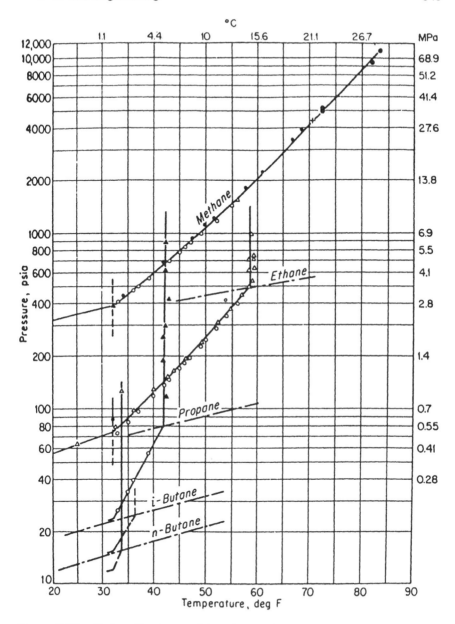

Figure 5.13 Hydrate-forming conditions for paraffin hydrocarbons. *Source*: [80].

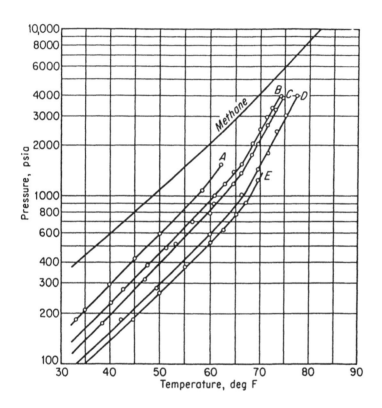

Figure 5.14 Hydrate-forming conditions for natural gas. *Source*: [80].

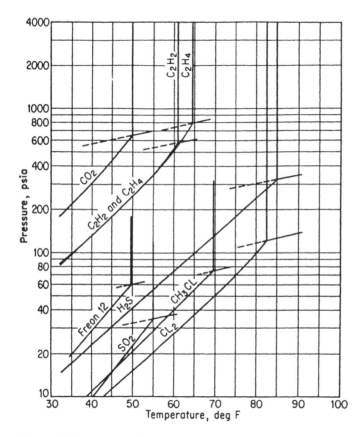

Figure 5.15 Hydrate-forming conditions for miscellaneous substances. *Source*: [80].

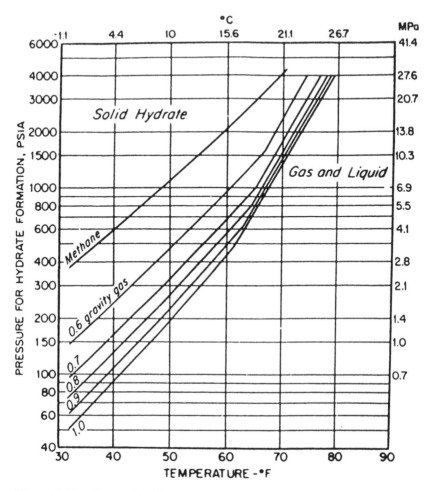

Figure 5.16   Hydrate-forming conditions for natural gases with various specific gravities. *Source*: [80].

   Primary or polyfunctional alcohols such as methanol and ethylene glycol have been used as freezing point suppressants. This could be seen routinely in the antifreezing windshield wash solution used for automobiles. Typically, a 25% methanol solution decreases the freezing point of water by about 36°F to about –4°F, which is called the depressed ice point. Similarly, a typical natural gas forms hydrates in the presence of free water at about 63°F, whereas it is depressed to 39°F (24°F hydrate depression) in the presence of a 25% methanol solution. Normal and depressed hydrate points are functions of

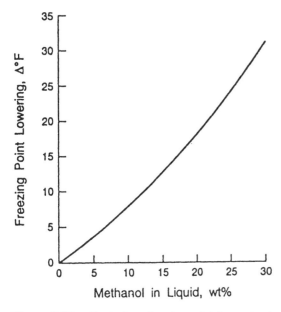

Figure 5.17 Gas hydrate freezing point depression in relation to methanol percentage (w/w) in liquid. *Source:* [76].

pressure, whereas the hydrate depression is not a function of pressure. It has also been found that hydrate depression is always smaller than the ice point depression. Figure 5.17 shows the gas hydrate freezing point depression in relation to the amount of methanol weight percentage in the liquid [76]. Methanol is relatively inexpensive and effective in freezing point depression, so the use of methanol for this purpose is economically justified.

The thermodynamics of this process are described in detail in references (67–72, 80,81). Let us examine a typical methanol injection system. In a typical methanol injection and recovery system for a cold-oil absorption or turboexpander plant, feed gas passes through a free-water knockout drum and into a gas-gas exchanger with methanol being sprayed on exchanger tube-sheets. Methanol inhibits hydrate formation and aqueous methanol condenses in the exchanger (and the chiller following it) and is pumped to a primary separator. The methanol-water solution is then flashed in a flash drum and filtered into a methanol still to recover methanol. Normally, methanol dissolves in the hydrocarbon liquids and is distilled as a mixture of propane and methanol. Some of the methanol is recovered as the overhead product; to recover the methanol dissolved in the heavier solution, the bottoms of the methanol still (propane product or hydrocarbon liquids from the demethanizer) are washed with water. Some of the methanol decomposes into methane and fresh solution has to be added to replenish the inlet. It is more economical to be able to recover and fractionate the methanol solution at the same location. In the following paragraph procedures are described in the design of a methanol hydrate control system.

Before a methanol hydrate control system is designed, the minimum temperatures in the plant process, liquid condensation rates, and gas and liquid compositions should be calculated. Then the amount of methanol and its concentration can be estimated. The hydrate point of the gas leaving the separator is calculated based on the pressure and gas composition leaving the primary separator using the Parrish-Prausnitz, Katz, or Peng-Robinson methods. The locus of the depressed hydrate point for various inlet methanol concentrations is plotted using modified Hammerschmidt equation. Then the methanol concentration leaving the primary separator is set and the hydrate depression point is found out. Generally, a 90% methanol concentration is set which gives a safety margin in the primary separator.

The rate of methanol injection into the gas-gas exchanger depends on the amount of free water condensed in the chiller, methanol solubility in the vapor phase, and the amount methanol dissolving in the hydrocarbon phase. From these data and the equations provided, the methanol lost can be calculated and from this the injection rate can be estimated. Once the rate of injection is estimated, the mode of injection should be determined. Since methanol has a much higher vapor pressure than glycols, it should be injected upstream of front-end exchangers so that it can be distributed and vaporized by the incoming gas. It is sprayed onto the tubesheets by spray nozzles which can be used to add excess methanol to flood the exchanger (as a backup system) during normal operation to prevent hydrate formation. Methanol injection points should also be installed at the inlet of turboexpander and any refrigerated condenser in downstream fractionation to avoid hydrate formation.

An aqueous phase may form in the liquid being fed to the fractionator which could retard good fractionation. This has been observed in deethanizers, which could be due to formation of a separate methanol liquid phase in the fluid feeding the bottom reboiler. Hence provisions should be made to remove this phase before it enters the fractionating column. As a general practice, withdrawal points should be arranged whenever there is a possibility of aqueous methanol phase formation. The methanol dissolved in the hydrocarbon phase can be recovered by washing the raw liquid or propane product with water. Generally, water washing recovers about 90% of dissolved methanol.

Comparing dehydration processes by activated alumina, molecular sieves, and methanol injection system, a methanol injection system is cheaper to build and operate than an activated alumina dehydration system when low water content natural gas is the feed. When natural gas with higher water content is to be treated the methanol injection system has lower capital cost than a molecular sieve system whereas the operating costs are about the same. Further, the plant overall performance would be better when hydrate control is achieved by methanol injection system than when it is conducted using solid beds along with improved economics.

A few other dehydration systems are also widely used which employ glycol injection and contact with the gas. Mono-, or di-, or triethylene glycol solution is used for this purpose. The principles are very similar to the methanol dehydration system. The disadvantages associated with the glycol methods are inadequate distribution in the gas-gas exchangers, difficulty to remove from the dehydrated gas, foaming, and mutual solubility between the hydrocarbon liquids and the glycol solution.

Fixed-bed dehydration is the most widely-used method even though it is not the most economical process since it can be employed to achieve very low concentration of water (< 1 ppm) in the exit gases. Silica gel also can be used as a dehydration agent in a solid-bed operation. Further reading material relevant to this topic are listed in references [67–72].

XIII. SAFETY AND FLAMMABILITY

Natural gas and air in proper proportions will ignite and liberate heat. The liberated heat is absorbed primarily by the products of combustion and air. The sudden rise of temperature of the mixture gases also causes a sharp increase in pressure. Under a confined situation, this instantaneous pressure increase can result in an explosion. This is why it is important that everyone involved with the natural gas industry understand the flammability limits and the nature of combustion.

Methane and air in the proper proportions do not react until some other source ignites the rapid process. Possible mechanisms of chemical reactions involved in the ignition process are given by Lewis and Von Elbe [79]. A number of oxygenated intermediates are formed along with free radicals like OH in the process of forming CO_2 and H_2O.

The combustion process is initiated by an ignition source converting some number of methane molecules into free radicals. Free radicals are in turn converted to OH free radical. Possible oxygenated compounds include aldehyde, alcohol, carboxylic acid, and oxide. The hydroxyl free radical then reacts with methane and is regenerated. The successive (chain type) combustion reaction is impeded by destruction of the OH radicals. Solid surfaces often destroy the OH radicals before they can react with hydrocarbons. The same effect is exploited in a porous-bed flame arrestor. In general, the combustion rates are very fast and nearly measurable with a few exceptional situations where time scales can be expanded to microseconds (10^{-6} s). The reaction rate increases very rapidly with increase in temperature [80]. As a rule of thumb, the reaction rate doubles for each 15°C of temperature rise.

There are two composition limits of flammability for air and a gaseous fuel under specified conditions [76]. The lower limit is the minimum concentration of combustible gas that will support combustion, while the higher limit is the maximum concentration. Table 5.1 shows the lower and higher limits for pure hydrocarbons in air at room temperature and atmospheric pressure (RTP) [76]. For methane in air, the flammability limit is 5–15 mol%. For ethane in air, the limits are 2.9–13.0 mol%. The limits become lower with increasing molecular weight. It also is interesting to note that the limits are the same for n-pentane and isopentane, and also for n-butane and isobutane.

Since natural gas is a mixture of various hydrocarbons, it is important to know how to estimate the flammability limits of the mixture gas. Furthermore, natural gas often contains hydrogen sulfide and its flammability limits are 4.3–45.5 mol%.

The lower flammability limits of gaseous mixtures can be predicted based on the lower limits of the pure constituents :

$$\frac{n_1}{N_1} + \frac{n_2}{N_2} + \frac{n_3}{N_3} = N$$

where subscripts 1, 2, 3 refer to constituents 1, 2, and 3; n is the mole fraction of each constituent in the total mixture at lower limit; and N is the lower flammability limit in mole percent for each constituent as a pure hydrocarbon.

Pressure influences the flammability limits. As the pressure increases, the upper limit rises very rapidly. At low pressure, the natural gas-air mixture becomes more difficult to combust. Dilution of a fuel-air mixture with inert constituents such as carbon dioxide or nitrogen also can change the flammability limit. By adding a diluent, the lower and higher limits are raised, due in part to the lower oxygen content. Most people have experienced the start-up problem with carburated engines. Natural gas is very combustible when mixed with pure oxygen or oxygen-enriched air. Therefore, in handling natural gas, safety issues are very important. Usually an odorant is added to natural gas to alert operators with leakage of gas, when the natural gas concentration in air reaches 20% of the lower flammability limit. Even with a very small concentration of such odorant, normal people can detect the odor.

REFERENCES

1. *Kirk-Othmer Encyclopedia of Chemical Technology*, (Kroschwitz, J. I. and Howe-Grant, M., Eds.) Vol. 10, The Interscience Encyclopedia Inc., New York, 1991, pp. 443–463.

2. *Kirk-Othmer Encyclopedia of Chemical Technology*, (Kroschwitz, J. I. and Howe-Grant, M., Eds.) Vol. 10, The Interscience Encyclopedia Inc., New York, 1991, pp. 164–192.

3. Sokolov, V., *Petroleum*, Mir Publishers, Moscow, 1972.

4. *Introduction to Oil and Gas Technology*, (Giuliano, F. A., Ed.), Printice-Hall, Engelwood Cliffs, NJ, 1989.

5. Szilas, A. P., *Production and Transport of Oil and Gas*, Part A, Elsevier, Amsterdam, The Netherlands, 1985.

6. Crossman, A. B., in *Energy Technology Handbook*, (Considine, D. M., Ed.), McGraw-Hill, New York, 1977.

7. Melvin, A., *Natural Gas—Basic Science and Technology*, British Gas plc, Adam Hilger, 1988.

8. Campbell, J. M., *Gas Conditioning and Processing*, Vol. 1, Campbell Petroleum Series, Norman, OK, 1978.

9. *Kirk-Othmer Encyclopedia of Chemical Technology*, (Kroschwitz, J. I. and Howe-Grant, M., Eds.) Vol. 10, The Interscience Encyclopedia Inc., New York, 1991, pp. 1–60.

10. *Gas Conditioning Fact Book*, The Dow Chemical Company, Midland, MI, 1962.

11. Fisch, E. J., in *Energy Technology Handbook*, (Considine, D. M., Ed.), McGraw-Hill, New York, 1977.

12. Riesenfeld, F. C. and Kohl, A. L., *Gas Purification*, Gulf Publishing Co., Houston, 1974.

13. Kevin Jones, J., in *Energy Technology Handbook*, (Considine, D. M., Ed.), McGraw-Hill, New York, 1977.

14. Campbell, J. M., *Gas Conditioning and Processing*, Vol. 1, Campbell Petroleum Series, Norman, OK, 1978.

15. Bogart, M. J., *Absorption Refrigeration in Cryogenic Services*, AIChE Symposium Series, 79(224): 57–64, (1982).

16. Pelofsky, A. H., (Ed.), *Heavy Oil Gasification*, Marcel Dekker, Inc., New York, 1977.

17. Kuhre, C. J., in *Energy Technology Handbook*, (Considine, D. M., Ed.), McGraw-Hill, New York, 1977.

18. McMahon, J. F., in *Energy Technology Handbook*, (Considine, D. M., Ed.), McGraw-Hill, New York, 1977.

19. Nogima, S., in *Energy Technology Handbook*, (Considine, D. M., Ed.), McGraw-Hill, New York, 1977.

20. Sykes, J. A., in *Energy Technology Handbook*, (Considine, D. M., Ed.), McGraw-Hill, New York, 1977.

21. Taylor, C. H., in *Energy Technology Handbook*, (Considine, D. M., Ed.), McGraw-Hill, New York, 1977.

22. Huntington, R. L., *Natural Gas and Natural Gasoline*, McGraw-Hill Book Co., New York, 1950.

23. Sokolov, V., *Petroleum*, Mir Publishers, Moscow, 1972.

24. Melvin, A., *Natural Gas—Basic Science and Technology*, British Gas plc, Adam Hilger, 1988.

25. Considine, D. M. (Ed.), *Energy Technology Handbook*, McGraw-Hill, New York, 1977.

26. Stebbing, R. F. and O'Brien, J. V., Liquefication of natural gas by PRICO process, *Natural Gas Processing and Utilization Conference*, Prichard-Rhodes, London, 1975, pp. 3.37–3.43.

27. *Kirk-Othmar Encyclopedia of Chemical Technology*, (Kroschwitz, J. I. and Howe-Grant, M., Eds.) Vol. 10, The Interscience Encyclopedia Inc., New York, 1991, pp. 200–214.

28. Adlhart, O. J., in *Energy Technology Handbook*, (Considine, D. M., Ed.), McGraw-Hill, New York, 1977.

29. Bollinger, W. A., MacLean, D. L. and Narayan, R. S., *Chem. Engr. Prog.*, 10(Oct.): 27–32, (1982).

30. Stookey, D. J., Patton, C. J. and Malcolm, G. L., *Chem. Engr. Prog.*, 11(Nov.): 36–40, (1986).

31. Schell, W. J., *Hydrocarbon Processing*, Aug.: 43–46 (1983).

32. Torrey, S. and Scott, J., *Membrane and Ultrafiltration Technology*, Noyes Data Corp., Park Ridge, NJ, 1984, pp. 90–92.

33. Stookey, D. J., Graham, T. E. and Pope, W. M., *Env. Prog.*, 3(3): 212–214 (August 1984).

34. Rautenbach, R. and Ehresmann, H. E., Membrane separation in chemical industries, *AIChE Symposium Series*, 85(272): 48–54 (1989)

35. Russel, F. G., *Hydrocarbon Processing*, Aug.: 55–57 (1983).

36. Schendel, R. L., Mariz, C. L. and Mak, J. Y., *Hydrocarbon Processing*, Aug.: 58–62 (1983).

37. Knieriem, H., Jr., *Hydrocarbon Processing*, July: 65–67 (1980).

38. Chen, S. and Kao, Y-K., Membrane separation in chemical industries, *AIChE Symposium Series*, 85(272): 55–61 (1989).

39. Hwang, S-T. and Kammermeyer, K., *Membrane in Separations*, Wiley-Interscience, John Wiley & Sons, New York, 1975, pp. 461–464.

40. Riesenfeld, F. C. and Kohl, A. L., *Gas Purification*, Gulf Publishing Co., Houston, 1974.

41. Groenendaal, W., in *Energy Technology Handbook*, (Considine, D. M., Ed.), McGraw-Hill, New York, 1977.

42. Williamson, J. A., in *Energy Technology Handbook*, (Considine, D. M., Ed.), McGraw-Hill, New York, 1977.

43. Johnson, S. A. et al., in *Energy Technology Handbook*, (Considine, D. M., Ed.), McGraw-Hill, New York, 1977.

44. Sterner, S. W., *Proc. 85th Annual Meeting Air and Waste Management Assoc.*, Kansas City, Missouri, June 21–26, 1992, pp. 1–20.

45. Ma, W. T., Haslbeck, J. L., Neal, L. G. and Yeh, J. T., *Separations Technology*, 1: 195–204 (1991).

46. Saitou, T., Memoirs of The School of Science and Engineering, Waseda Univ., No. 57, Tokyo, 1993, pp. 17–35.

47. Nakamura, T., Smart, J. P., and van de Kamp, W. L., *Institute of Energy—Combustion and Emissions Control*, Proceedings of Institute of Energy's International

Conference on Combustion and Emission Control, Institute of Energy, London, 1995, pp. 213–229.

48. Campbell, J. M., *Gas Conditioning and Processing*, Vol. 1, Campbell Petroleum Series, Norman, OK, 1978.

49. *Gas Conditioning Fact Book*, The Dow Chemical Company, Midland, MI, 1962.

50. Huntington, R. L., *Natural Gas and Natural Gasoline*, McGraw-Hill Book Co., New York, 1950.

51. Riesenfeld, F. C. and Kohl, A. L., *Gas Purification*, Gulf Publishing Co., Houston, 1974.

52. Van Scoy, R. W., in *Energy Technology Handbook*, (Considine, D. M., Ed.), McGraw-Hill, New York, 1977.

53. Vickery, D. J., Adams, J. T. and Wright, R. D., *IChemE Symposium Series*, 128: B241–B252 (1990).

54. Hise, R. E. et al., *AIChE Symposium Series*, 79(224): 51–56, (1982).

55. *Kirk-Othmer Encyclopedia of Chemical Technology* (Kroshwitz, J. I. and House-Grant, M., Eds.), Vol. 10, The Interscience Encyclopedia Inc., New York, 1991, pp. 630–652.

56. *Natural Gas Fuels*, 3(9): 17–35 (Sept. 1994).

57. Siuru, Jr., W. D., *Natural Gas Fuels*, 3(9): 30 (Oct. 1994).

58. Bechtold, R. L., Timbario, T. J., Tison, R. R. and Sprafka, R. J., Compressed natural gas as a motor vehicle fuel, *Proc. Soc. Auto Eng. Congr.*, P-129: 47–67, June 22–23, 1983, Soc. Auto Eng., Detroit.

59. Karim, G. A, Compressed natural gas as a motor vehicle fuel, *Proc. Soc. Auto Eng. Congr.*, P-129, pp. 71–79, June 22–23, 1983, Soc. Auto Eng., Detroit.

60. Seal, M. R., Compressed natural gas as a motor vehicle fuel, *Proc. Soc. Auto Eng. Congr.*, P-129, pp. 81–84, June 22–23, 1983, Soc. Auto Eng., Detroit.

61. Wright, J. E., in *Methane—Fuel for the Future*, (McGeer, P. and Durbin, E., Eds.), Plenum Press, New York, 1982, pp. 193–206.

62. Bellini, V., in *Methane—Fuel for the Future*, (McGeer, P. and Durbin, E., Eds.), Plenum Press, New York, 1982, pp. 183–192.

63. Buchanan, J. J., in *Methane—Fuel for the Future*, (McGeer, P. and Durbin, E., Eds.), Plenum Press, New York, 1982, pp. 173–181.

64. Axworthy, R. T, in *Methane—Fuel for the Future*, (McGeer, P. and Durbin, E., Eds.), Plenum Press, New York, 1982, pp. 131–138.

65. Karim, G. A., in *Methane—Fuel for the Future*, (McGeer, P. and Durbin, E., Eds.), Plenum Press, New York, 1982, pp. 113–129.

66. Born, G. J. and Durbin, E. J., in *Methane—Fuel for the Future*, (McGeer, P. and Durbin, E., Eds.), Plenum Press, New York, 1982, pp. 101–112.

67. Huntington, R. L., *Natural Gas and Natural Gasoline*, McGraw-Hill Book Co., New York, 1950.

68. Riesenfeld, F. C. and Kohl, A. L., *Gas Purification*, Gulf Publishing Co., Houston, 1974.

69. Sokolov, V., *Petroleum*, Mir Publishers, Moscow, 1972.

70. Campbell, J. M., *Gas Conditioning and Processing*, Vol 1, Campbell Petroleum Series, Norman, OK, 1978.

71. *Gas Conditioning Fact Book*, The Dow Chemical Company, Midland, MI, 1962.
72. *Kirk-Othmer Encyclopedia of Chemical Technology*, (Kroshwitz, J. I. and Howe-Grant, M., Eds.), Vol. 10, The Interscience Encyclopedia Inc., New York, 1991, pp. 144–163.
73. Haggin, J., Natural gas assumes growing role in Asia as fuel chemical feedstock, *Chemical & Engineering News*, Aug. 3: 7–13 (1992).
74. Holder, G. D., Angert, P. F., John, V. T., Yen, S. L., *J. Petrol. Technol.*, 34: 1127, (1982).
75. Holder, G. D., Katz, D. L., and Hand, J. H., Hydrate formation in subsurface environments, *AAPG Bull.* 60(6): 981–988 (1976).
76. Speight, J. G., *Gas Processing, Environmental Aspects and Methods*, Butterworth-Heinmann, Oxford, 1993.
77. Gas processes, 1994, *Hydrocarbon Processing*, April: 68–114 (1994).
78. Gas processes, 1990, *Hydrocarbon Processing*, April: 69–111 (1990).
79. Lewis, B. and Von Elbe, G., *Combustion, Flames, and Explosions of Gases*, Academic Press, New York, 1951.
80. Katz, D. L. and Lee, R. L, *Natural Gas Engineering*, pp. 107–108, McGraw Hill Publishing Co., New York, 1990.
81. Holder, G. D., Zetts, S. P., and Pradham, N., Phase behavior in systems containing clathrate hydrates, *Reviews in Chemical Engineering*, (1988); personal communication (1995).

6

Environmental Issues of Methane Conversion Technology

I. METHANE CONVERSION TECHNOLOGIES

Methane is a one-carbon paraffinic hydrocarbon that is not very reactive under normal conditions. Only a few chemicals can be produced directly from methane under relatively severe conditions. Chlorination of methane is only possible by thermal or photochemical initiation. Methane can be partially oxidized with a limited amount of oxygen or in presence of steam to a synthesis gas mixture. Many chemicals can be produced from methane via the more reactive synthesis gas mixture. Synthesis gas is the precursor for two major chemicals, ammonia and methanol. Both compounds are the hosts for many important petrochemical products. Figure 6.1 shows the important chemicals based on methane, synthesis gas, methanol, and ammonia [20].

A few chemicals are based on the direct reaction of methane with other reagents, including: carbon disulfide, hydrogen cyanide, chloromethanes, and synthesis gas mixture. A more detailed discussion for chemistry and synthesis can be found in Chapter 4. In this chapter, emphasis will be placed on the environmental aspects of the chemicals and their manufacturing technologies.

A. Carbon Disulfide

Methane reacts with sulfur, which is an active nonmetal of group 6A, at high temperatures to produce carbon disulfide (CS_2). The reaction is endothermic and an activation energy of approximately 160 kJ is required [47]. Activated alumina or clay is used as the catalyst at approximately 67°C and 2 atm. The

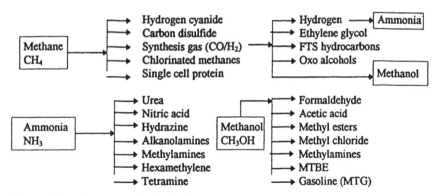

Figure 6.1   Important chemicals based on methane, synthesis gas, ammonia, and methanol.

process starts by vaporizing pure sulfur, mixing it with methane, and passing the mixture over the alumina catalyst. The reaction is:

$$CH_4 \text{ (g)} + 2S_2 \text{ (g)} \longrightarrow CS_2 \text{ (g)} + 2H_2S \text{ (g)} \qquad \Delta H = 160 \text{ kJ/mol}$$

Hydrogen sulfide, a coproduct, is used to recover sulfur by the Claus reaction. A CS_2 yield of 85–90% based on methane is estimated. An alternative route for CS_2 is by the reaction of liquid sulfur with charcoal. However, this method is not used very much. A detailed discussion of this synthesis is given in Chapter 4.

Carbon disulfide is used primarily to produce rayon and cellophane (regenerated cellulose); it also is used to produce carbon tetrachloride using iron powder as a catalyst at 30°C:

$$CS_2 + 3Cl_2 \longrightarrow CCl_4 + S_2Cl_2$$

Sulfur monochloride is an intermediate that is then reacted with carbon disulfide to produce more carbon tetrachloride and sulfur:

$$CS_2 + 2S_2Cl_2 \longrightarrow CCl_4 + 6S$$

The net reaction is:

$$CS_2 + 2Cl_2 \longrightarrow CCl_4 + 2S$$

This reaction involves environmentally hazardous and toxic chemicals, including reactants, intermediates, and products. Therefore, waste discharge as well as operational safety require special attention. Carbon disulfide also is used to produce xanthates ROC(S)SNa as an ore flotation agent and ammonium thiocyanate as a corrosion inhibitor in ammonia handling systems.

B.   Hydrogen Cyanide

Hydrogen cyanide (HCN) is a colorless liquid with a boiling point of 25.6°C. It is miscible with water, producing a weakly acidic solution. It is a highly toxic compound, but a very useful chemical intermediate with high reactivity. It is used in the synthesis of acrylonitrile and adiponitrile, which are important monomers for plastic and synthetic fiber production.

Hydrogen cyanide is produced via the Andrussaw process using ammonia and methane in the presence of air. The reaction is exothermic and the released heat is used to supplement the required catalyst-bed energy:

$$2CH_4 + 2NH_3 + 3O_2 \longrightarrow 2HCN + 6H_2O$$

A platinum-rhodium alloy is used as a catalyst at 1100°C. Approximately equal amounts of ammonia and methane with 75 vol% air are introduced to the preheated reactor. The catalyst has several layers of wire gauze with a special mesh size (approximately 100 mesh). The Degussa process, on the other hand, reacts ammonia with methane in the absence of air using a platinum aluminum-ruthenium alloy as a catalyst at approximately 1200°C. The reaction produces hydrogen cyanide and hydrogen, and the yield is over 90%. The reaction is endothermic and requires 251 kJ/mol.

$$CH_4 + NH_3 \longrightarrow HCN + 3H_2 \qquad \Delta H^o_{298} = 251 \text{ kJ/mol}$$

Hydrogen cyanide may also be produced by the reaction of ammonia and methanol in the presence of oxygen:

$$NH_3 + CH_3OH + O_2 \longrightarrow HCN + 3H_2O$$

A detailed discussion of its synthesis technology is given in Chapter 4.

Hydrogen cyanide is a reactant in the production of acrylonitrile, methyl methacrylates (from acetone), adiponitrile, and sodium cyanide. It also is used to make oxamide, a long-lived fertilizer that releases nitrogen steadily over the vegetation period. Oxamide is produced by the reaction of hydrogen cyanide with water and oxygen using a copper nitrate catalyst at about 70°C and atmospheric pressure:

$$4HCN + O_2 + 2H_2O \longrightarrow 2H_2N\text{-}\overset{\displaystyle O}{\overset{\displaystyle \|}{C}}\text{-}\overset{\displaystyle O}{\overset{\displaystyle \|}{C}}\text{-}NH_2$$

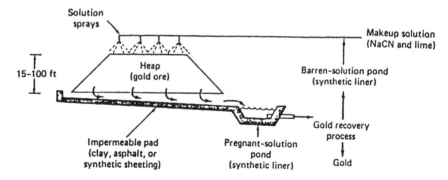

Figure 6.2 A schematic of the heap-leaching process. *Source*: [69].

In the process involving hydrogen cyanide, treatment of process waste water becomes an issue of environmental concern in addition to the air pollution control.

Heap Leaching of Gold

Gold has been an attractive investment for thousands of years of human history. Due to the low capital investment and also to fast payout, the production of gold by heap leaching process is very widely used. This process is capable of extracting gold from relatively low-grade ores, waste rocks, and discarded tailings from previous mining processes. In the state of Nevada there are approximately 60–70 active heap-leach operation.

Figure 6.2 shows a descriptive diagram of the heap-leaching process [69]. The process sprays and alkaline cyanide solution over ore that has been stacked on an inclined, impermeable pad. Gold is dissolved in the solution by the following reaction and flows off the pad to a lined impoundment.

$$2Au + 4CN^- + 2O_2 + 2H_2 \longrightarrow 2Au(CN)_2^- + H_2O_2 + 2OH^- \tag{6.1}$$

The gold is recovered from the pregnant solution, pumped from the impoundment. The barren solution with makeup sodium cyanide (NaCN) and lime (CaO) is returned to the ore to complete a closed loop. After the completion of leaching, the residue is rinsed with fresh water. Cyanide is the only commercially proven lixiviant used in heap leaching. Reaction 6.1 is typically called the primary leaching reaction of gold by cyanide, whereas Reaction 6.2 is known as secondary leaching reaction of gold by which additional gold is dissolved:

$$4Au + 8CN^- + O_2 + 2H_2O \longrightarrow 4Au(CN)_2^- + 4OH^- \tag{6.2}$$

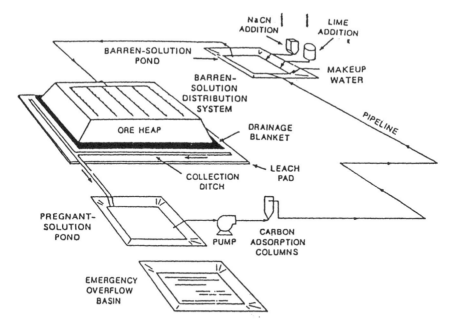

Figure 6.3   A typical heap-leach operation. *Source*: [69].

The rate of cyanidation reactions depends on the cyanide concentration as well as the alkalinity of the leach solution. The optimal pH of the solution is 10.3, typically controlled by caustic (NaOH). Gold is recovered either by adsorption on activated carbon or by zinc-dust precipitation.

As sodium cyanide (NaCN) is quite reactive with other metals, the process also generates various forms of cyanides depending upon the mineral constituents of ores. Such cyanides can be formed by reactions with realgar (As_2S_2), orpiment (As_2S_3), and stibnite (Sb_2S_3) [69].

Due to the environmentally hazardous and highly toxic nature of sodium cyanide solution, as well as potential sources of free cyanide at leaching operations, the process requires ultimate care in addressing to the issues of environmental safety and health. Cyanides in leach residues may occur in combinations of various metallic-cyanide complexes and free cyanide ions. Such complexes include: $Fe(CN)_6^{4-}$, $Co(CN)_6^{4-}$, $Au(CN)_2^{-}$, $Cu(CN)_2^{-}$, $Cu(CN)_3^{2-}$, $Ni(CN)_4^{2-}$, $Ag(CN)_3^{2-}$, $Zn(CN)_4^{2-}$, $Cd(CN)_3^{-}$, and $Cd(CN)_4^{2-}$.

Heap-leach operations are typically zero-discharge facilities [69]. A typical heap leach operation is shown in Figure 6.3. A weak cyanide solution is sprayed over the stacked ore and pregnant solution flows to a lined collection ditch, then to a pregnant solution pond, then to a recovery plant with carbon adsorption columns, and back to a barren solution pond. The barren solution is recycled back to the heap. Pads are constructed with asphalt or polyethylene liners over clays. The pipe for the solution is made of polyethylene.

The heap-leach residue, the barren ore remaining, is either left on the pad or hauled away to a dump site. At closure, such residue can be 1–50 acres in surface area and can be 16–200 ft in height.

The heap is rinsed with fresh water for several days. Nevada requires rinsing the residue heap until the pH reaches 8 or lower and cyanide of 0.2 ppm or less has been achieved in the rinse water. No discharge is permitted, only drying by evaporation is allowed.

Capping the residue pile and closure of operation must be done in an environmentally safe manner. This also is an expensive proposition. Free cyanides are left in the residue may volatilize to the atmosphere or be destroyed by ultraviolet radiation. However, surface water contamination, even after the closure, becomes an important issue. Monitoring of groundwater wells installed at the time of closure is continued through the postclosure period, typically over a period of 30 years. It seems to be an important research task to develop processes that can be used for decisive treatment of residues and remediation of sites.

Advanced Wastewater Treatment Involving Nitrogen Removal

Any process that follows conventional primary and biological treatment may be considered to be advanced treatment. Nitrogen exists in diverse forms in wastewater, including domestic and industrial wastes. As bacteria decompose waste, nitrogen in complex organic molecules is released as ammonia, which can exist in water either as ammonium ion (NH_4^+) or as ammonia gas (NH_3). The ammonia or ammonium concentration in water can be seriously affected by industrial discharge as well as by agricultural fertilizer runoff [1,21].

As the pH increases, the equilibrium is driven toward the less soluble ammonia gas:

$$NH_4^+ + OH^- \longrightarrow NH_3 \uparrow + H_2O \tag{6.3}$$

Ammonia stripping process is based on Reaction 6.3, in which the pH of treated waste water is increased to at least 10, thus forming more dissolved ammonia gas. In a stripping tower, the ammonia can be separated. Lime (CaO) is used to increase the pH and has to be removed periodically from the stripping surface. The lime scaling is often a headache to this process. However, the most basic concern is that the process simply transfers the pollution problem from one medium to another (i.e., water to air in this case) [24,26,27].

A second approach for nitrogen or ammonia control is based on the utilization of aerobic bacteria that convert ammonium ion to nitrate ion (i.e., nitrification). As a subsequent stage, different bacteria convert nitrates to nitrogen gas (i.e., denitrification). Therefore, the overall process in this scheme may be called the nitrification/denitrification process.

The nitrification step involves two stages: 1) conversion of ammonia to nitrites (NO_2^-) by *nitrosomonas*, and 2) oxidation of nitrites to nitrates (NO_3^-) by nitrobacter.

$$NH_4^+ + \frac{3}{2}O_2 \dashrightarrow NO_2^- + H_2O + 2H^+$$

$$NO_2^- + \frac{1}{2}O_2 \dashrightarrow NO_3^-$$

The denitrification step is anaerobically carried out by

$$2NO_3^- + \text{organic matter} \dashrightarrow N_2\uparrow + CO_2\uparrow + H_2O \qquad (6.4)$$

Reaction 6.4 is exothermic; however, there may not be enough organic matter left in wastewater stream to provide energy. Instead, an additional source like methanol is provided.

In most of modern petrochemical wastewater treatment, ammonia, phenolics, phosphorus, polycyclic aromatic hydrocarbons (PAH), and chlorinated solvent are very common contaminants/pollutants.

C.   Chloromethanes

The successive substitution of methane hydrogens with chlorine produces a mixture of four chloromethanes, namely, monochloromethane (methyl chloride, CH_3Cl), dichloromethane (methylene chloride, CH_2Cl_2), trichloromethane (chloroform, $CHCl_3$), and tetrachloromethane (carbon tetrachloride, CCl_4). Each of these four compounds has many industrial applications.

Production of Chloromethanes

Methane is the most difficult alkane to chlorinate. The reaction is initiated by chlorine free radicals obtained via the application of heat (thermal) or light (*h*n). Thermal chlorination (more widely used industrially) occurs at approximately 350–370°C and atmospheric pressure. A typical product distribution for CH_4/Cl_2 feed ratio of 1.7 is mono- (58.7%), di- (29.3%), tri- (9.7%), and tetra- (2.3%) chloromethanes.

The highly exothermic chlorination reaction produces approximately 95 kJ/mol of HCl. The first step is the breaking of the Cl-Cl bond (bond energy = +584.3 kJ), which forms two chlorine free radicals:

$$Cl_2 \dashrightarrow 2Cl\cdot$$

The Cl radical attacks methane and forms a methyl free radical plus HCl. The methyl free radical reacts in a subsequent step with a chlorine molecule, forming methyl chloride and another Cl radical.

$$Cl\cdot + CH_4 \longrightarrow CH_3\cdot + HCl$$

$$CH_3\cdot + Cl_2 \longrightarrow CH_3Cl + Cl$$

The new Cl radical either attacks another methane molecule and repeats the above reactions, or it reacts with a methyl chloride molecule to form a chloromethyl free radical $CH_2Cl\cdot$ and HCl.

$$CH_3Cl + Cl\cdot \longrightarrow CH_2Cl\cdot + HCl$$

The chloromethyl free radical then attacks another chlorine molecule and produces dichloromethane along with a Cl radical:

$$CH_2Cl\cdot + Cl_2 \longrightarrow CH_2Cl_2 + Cl\cdot$$

This formation of Cl free radicals continues until all chlorine is consumed. Chloroform and carbon tetrachloride are formed in a similar way by reaction of $CHCl_2\cdot$ and $CCl_3\cdot$ free radicals with chlorine.

Product distribution among the chloromethanes depends primarily on the mole ratio of the reactants. For example, the yield of methyl chloride could be increased to 80% by increasing the CH_4/Cl_2 mole ratio to 10:1 at 450°C. If methylene chloride is desired, the CH_4/Cl_2 ratio is lowered and the methyl chloride recycled. Decreasing the CH_4/Cl_2 ratio generally increases poly-substitution and the chloroform and carbon tetrachloride yield. An alternative way to produce methyl chloride is the reaction of methanol with HCl. Methyl chloride could be further chlorinated to give a mixture of chloromethanes (methylene chloride, chloroform, and carbon tetrachloride).

Uses of Chloromethanes

The major use of methyl chloride is to produce silicons. Other uses include the synthesis of tetramethyl lead as a gasoline octane booster, a methylating agent in methyl cellulose production, a solvent, and a refrigerant. Methylene chloride has a wide variety of markets. One major use is a paint remover. It also is used as a degreasing solvent, a blowing agent for polyurethane foams, and a solvent for cellulose acetate. Methylene chloride also is used as a decaffeinating solvent of coffee in its supercritical state. Chloroform is mainly used to produce chlorodifluoromethane (Fluorocarbon 22) by the reaction with hydrogen fluoride:

$$CHCl_3 + 2HF \longrightarrow CHF_2Cl + 2HCl$$

This compound is used as a refrigerant and as an aerosol propellent. It also is used to synthesize tetrafluoroethylene, which is polymerized to a heat-resistant polymer (Teflon):

$$2CHClF_2 \longrightarrow C_2F_4 + 2HCl$$

Carbon tetrachloride is used to produce chlorofluorocarbons by the reaction with hydrogen fluoride using an antimony pentachloride ($SbCl_5$) catalyst:

$$CCl_4 + HF \longrightarrow CCl_3F + HCl$$

$$CCl_4 + 2HF \longrightarrow CCl_2F_2 + 2HCl$$

The formed mixture is composed of trichlorofluoromethane (Freon-11) and dichlorodifluoromethane (Freon-12). These compounds are used as aerosols and as refrigerants. Due to the depleting effect of chlorofluorocarbons (CFCs) on the ozone layer, the production of these compounds is to be reduced appreciably (Montreal Protocol).

Ozone is continuously being generated in the stratosphere by the absorption of short-wavelength ultraviolet (UV) radiation, while at the same time it is being destroyed by various chemical reactions that convert ozone back to oxygen. The rates of generation and removal affect the concentrations of ozone present. The balance between these two processes is influenced by the stratospheric concentrations of chlorine, nitrogen, and bromine which act as catalysts speeding up the removal process. Chlorofluorocarbons (CFCs) are known to be the most prominent type of ozone-destructive gases that play an important role in the greenhouse effect. Due to the very stable nature of CFCs, they are relatively unaffected by the usual pollutant destruction processes in the troposphere. When they drift to the stratosphere, CFC molecules can be broken by stronger ultraviolet radiation, freeing the reactive chlorine which is now available to destroy ozone molecules. For the case of CFC-12,

$$CCl_2F_2 + h\nu \longrightarrow Cl + CClF_2$$

The free chlorine acts as a catalyst and a single chlorine atom may break down tens of thousands of ozone molecules before it returns to the troposphere. In the troposphere, chlorine reacts with hydrogen and forms hydrogen chloride that is rained out. Since ozone absorbs biologically-damaging UV radiation before it reaches the earth's surface, its depletion increases the risks associated with UV exposure. Ultraviolet radiation and over-exposure are linked with skin cancers, cataracts, and suppression of immune system response. In 1985, this problem started to attract everyone's attention by the dramatic announcement of the discovery of a "hole" in the ozone layer over

Antarctica the size of the continental United States. Since then, the world began to acknowledge the seriousness of the problem [69].

Much research is being conducted to find alternatives to CFCs with little or no effect on the ozone layer [2]. Among these are HCFC-123 ($HCCl_2CF_3$) to replace Freon-11 and HCFC-22 ($CHClF_2$) to replace Freon-12 in such uses as air conditioning, refrigeration, aerosol, and foam. These compounds have much lower ozone depletion value compared with Freon-11, which was assigned a value of 1. Ozone depletion values of HCFC-123 and HCFC-22 relative to Freon-11 (= 1) are 0.02 and 0.055, respectively [2]. The lower values are due to the presence of hydrogen in their formulae.

D. Synthesis Gas (Steam Reforming of Natural Gas)

Synthetic gas can be produced from a variety of feedstocks. Natural gas is the preferred feedstock when it is available from gas fields (nonassociated gas) or from oil wells (associated gas). The first step in the production of synthesis gas is to treat natural gas to remove hydrogen sulfide. The purified gas is then mixed with steam and introduced to the first reactor (primary reformer). The reactor is constructed from vertical stainless steel tubes lined in a refractory furnace. The steam to natural gas ratio is 4–5 depending on natural gas composition (natural gas may contain ethane and heavier hydrocarbons) and the pressure used. A promoted nickel-type catalyst contained in the reactor tubes is used at temperature and pressure ranges of 700–800°C and 30–50 atm, respectively. The product gas from the primary reformer is a mixture of H_2, CO, CO_2, unreacted CH_4, and steam. The main reforming reactions are:

$$CH_4 \text{ (g)} + H_2O \longrightarrow CO \text{ (g)} + 3H_2 \text{ (g)} \qquad \Delta H^o = +206 \text{ kJ/mol}$$

$$\Delta H^o_{800°C} = +226 \text{ kJ/mol}$$

$$CH_4 \text{ (g)} + 2H_2O \longrightarrow CO_2 \text{ (g)} + 4H_2 \text{ (g)} \qquad \Delta H^o = +165 \text{ kJ/mol}$$

For the production of methanol, this mixture could be used by the conventional technology directly with no further treatment except adjusting the H_2/(CO + CO_2) ratio to approximately 2:1. For producing hydrogen for ammonia synthesis, however, further treatment steps are needed. First, the required amount of nitrogen for ammonia must be obtained from atmospheric air. This is done by partially oxidizing unreacted methane in the exit gas mixture from the first reactor in another reactor (secondary reforming).

The main reaction occurring in the secondary reformer is the partial oxidation of methane with a limited amount of air. The product is a mixture of hydrogen, carbon dioxide, carbon monoxide, plus nitrogen, which does not react under these conditions. The reaction is:

$$CH_4 + 0.5(O_2 + 3.76N_2) \longrightarrow CO + 2H_2 + 1.88N_2 \qquad \Delta H = -32.1 \text{ kJ/mol}$$

The reactor temperature can reach over 900°C in the secondary reformer due to the exothermic reaction heat. The second step after secondary reforming is removing carbon monoxide, which poisons the catalyst used for ammonia synthesis. This is done in three further steps, shift conversion, carbon dioxide removal, and methanation of the remaining CO and CO_2. Every stage involved in ammonia synthesis is environmentally important. Any emission or mishandling would represent not only a short-term disastrous threat to human health, but also a long-term environmental problem of grave consequences.

Shift Conversion

The product gas mixture from the secondary reformer is cooled then subjected to shift conversion. In the shift converter, carbon monoxide is reacted with steam to yield carbon dioxide and hydrogen. The reaction is exothermic and is:

$$CO + H_2O \text{ (g)} \longrightarrow CO_2 \text{ (g)} + H_2 \text{ (g)} \qquad \Delta H^{\circ} = -41 \text{ kJ/mol}$$

The feed to the shift converter contains large amounts of carbon monoxide, which should be oxidized. An iron catalyst promoted with chromium oxide is used at a temperature range of 425–500°C to enhance the oxidation. Exit gases from the shift conversion are treated to remove carbon dioxide. This may be done by absorbing carbon dioxide in a physical or chemical absorption solvent or by adsorbing it using a special type of molecular sieves. Carbon dioxide, recovered from the treatment agent as a byproduct, is mainly used with ammonia to produce urea. The product is a pure hydrogen gas containing small amounts of carbon monoxide and carbon dioxide, which are further removed by methanation.

Methanation

Catalytic methanation is the reverse of the steam reforming reaction. Hydrogen reacts with carbon monoxide and carbon dioxide, converting them to methane. Methanation reactions are exothermic and methane yield is favored at low temperatures:

$$CO \text{ (g)} + 3H_2 \longrightarrow CH_4 + H_2O \text{ (g)} \qquad \Delta H = -206 \text{ kJ/mol}$$

$$CO_2 \text{ (g)} + 4H_2 \text{ (g)} \longrightarrow CH_4 \text{ (g)} + 2H_2O \qquad \Delta H = -164.8 \text{ kJ/mol}$$

The forward reactions also are favored at higher pressures. However, the space velocity becomes high with increased pressures, and contact time becomes shorter, decreasing the yield. The actual process conditions of pressure, temperature, and space velocity are practically a compromise of several factors. Raney nickel is the preferred catalyst. Typical methanation reactor operating conditions are 200–300°C and approximately 10 atm. The product is a gas mixture of hydrogen and nitrogen having an approximate ratio of 3:1 for ammonia production.

E. Ammonia-Related Pollutants

Ammonia is the principal component in the fertilizer production. The world ammonia capacity for 1995–1996 is estimated at 123,640,000 tons of nitrogen per year [70]. Ammonia plants are high energy consumers, and selection of the feedstocks is the most important factor in determining the capital investment and production costs. The availability and cost of raw materials are factors to be taken into account when deciding on the construction of a new ammonia plant. The primary feedstocks for ammonia production include natural gas, naphtha, heavy residual oil, coke gas, and coal. Of all these feedstocks, natural gas is the raw material of choice when available because it ensures minimum investment and production costs, a plant that is easy to operate, and minimal environmental problems. Therefore, steam reforming of natural gas has become the most widespread process for ammonia production.

Figure 6.4 is a flow diagram of ammonia production starting from natural gas. The main effluents produced at each process stage have been indicated. The ammonia production process includes the following main stages: natural gas desulfurization, primary reforming, secondary reforming, high temperature-CO shift conversion, low temperature-CO shift conversion, CO_2 removal, methanation, compression, and NH_3 synthesis. The main effluents resulting from the process are the flue gas from the primary reformer, the process condensates, the carbon dioxide, and the purge gas from the ammonia synthesis loop.

This process for production of synthetic ammonia by catalytic steam reforming of natural gas is a relatively clean process and presents no unique environmental problems. To assess the environmental impacts of a modern ammonia plant on air, water, and soil, each step in the ammonia synthesis namely, desulfurization, reforming, shift conversion, carbon dioxide removal, final purification, ammonia synthesis, and refrigeration should be examined. The sources of pollutants need to be identified and matched with cost-effective solutions for minimization/elimination by using the best available pollution control measure.

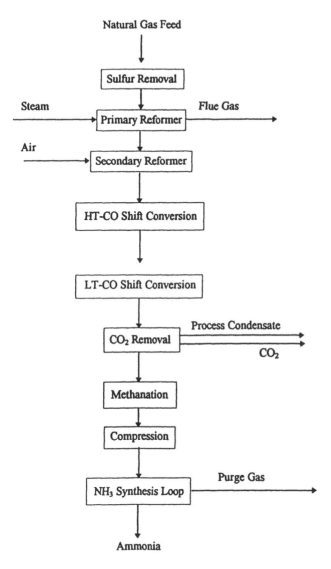

Figure 6.4 Steam reforming of natural gas. *Source*: [21].

Table 6.1 Emission Factors for a Typical 1000-MTPD Amonia Plant

Emission point	Emission	kg/metric ton[a]	lb/ton[a]
Desulfurization unit regeneration[b]	Total sulfur[c,d]	0.0096	0.019
	CO	6.9	13.8
	Nonmethane VOC[e]	3.6	7.2
Primary reformer flue gas			
Natural gas	NO_x	2.7	5.4
	SO_x	0.0024	0.0048
	CO	0.068	0.136
	Particulates	0.072	0.144
	Methane	0.0063	0.0126
	Nonmethane VOC	0.0061	0.0122
Distillate fuel oil	NO_x	2.7	5.4
	SO_x	1.3	2.6
	CO	0.12	0.24
	Particulates	0.45	0.90
	Methane	0.03	0.06
	Nonmethane VOC	0.19	0.38
CO_2 solvent regeneration	Ammonia	1.0	2.0
	CO	1.0	2.0
	CO_2	1220	2440
	Nonmethane VOC	0.52	1.04
Condensate steam stripper[f]	Ammonia	1.1	2.2
	CO_2	3.4	6.8
	Nonmethane VOC	0.6	1.2

[a] Emission factors are expressed in weight of emmisions per unit weight of ammonia product.
[b] Intermittent source from carbon beds average 10 hr once every 30 days.
[c] Worst-case assumption—that all adsorbed sulfur is emitted during regeneration.
[d] Normalized to a 24-hr emission factor.
[e] Volatile organic compounds.
[f] In new plants the high-pressure condensate stripper recycles condensate pollutants.
Source: [21]

The profitability of this process is highly dependent on energy cost and capital investment. Energy and capital cost penalties associated with pollution control systems must therefore be minimized as far as practicable. The first step is to reduce the sulfur concentration in the ammonia plant feedstock to less than 0.1 ppmv to prevent poisoning the reformer catalyst. Once desulfurized, the feedstock is partially reacted with steam in a primary reformer to produce hydrogen (H_2) and carbon monoxide (CO). To provide a source of nitrogen, air is injected in the secondary reforming stage. The CO in the reformed gas is reacted with steam in a two-stage shift converter to produce carbon dioxide (CO_2) and hydrogen. The shift-converter effluent is cooled and most of the unreacted steam is condensed and separated as process condensate.

The raw synthesis gas is then scrubbed to remove CO_2. The residual CO and CO_2 are catalytically reacted with H_2 to form methane and water vapor. The remaining trace amounts of CO_2 and water of saturation are finally removed by molecular sieves. The purified synthesis gas (containing hydrogen, nitrogen, and small amounts of methane, argon and helium) is compressed and enters the synthetic loop. The synthesis of ammonia occurs over an iron catalyst. The ammonia converter effluent is chilled indirectly with refrigerant (ammonia) to condense out the liquid ammonia product. Only a fraction of the synthesis gas feed is converted into ammonia; therefore, a large amount of unreacted gas is recycled. Inerts that tend to accumulate in the synthesis loop are purged. Hydrogen contained in the purge gas is recovered and the balance of the purge stream is sent to the fuel system. Table 6.1 lists the emission factors for a typical ammonia plant.

II. ENVIRONMENTAL ISSUES RELATED TO METHANE CONVERSION TECHNOLOGIES

The use of natural gas for manufacture of various chemicals via the syngas route or via direct methanation or via other miscellaneous methods is very important for chemical industries. However, the production of all the desired chemicals is accompanied by formation of byproducts and some end products, which when released into the environment constitute pollution hazards.

To eliminate/minimize the pollution hazard, the important step is to identify the pollutant formation at its source and try to avoid/minimize it at the conception level. The following discussion summarizes the various pollutants that accompany the natural gas conversion processes, the pollution hazards associated with them, the treatment strategies and technologies available for management and control.

A. Emission Sources and Control

Feedstock Desulfurization

Natural gas feedstocks normally contain contaminants such as sulfur, chloride, and organometallic compounds, which need to be removed to less than 0.1 ppmv to avoid poisoning the various catalysts in the ammonia plant. Sulfur compounds are mostly hydrogen sulfide (H_2S), mercaptans (R'SH), thiophenes (R'S), and carbonyl sulfide (COS).

Activated Carbon   Sulfur removal from natural gas by adsorption at ambient temperature on carbon, activated with cupric oxide, is widely used. Carbon physically adsorbs sulfur compounds to its surface and the cupric oxide reacts with hydrogen sulfide. The activated carbon is typically regenerated every 30 days by passing steam through the bed at a temperature of 230°C (450°F) for 8–10 hr while air is injected. Oxygen in the air reacts with the metal sulfide to form the metal oxide and sulfur dioxide. These reactions are:

$$CuO + H_2S ---> CuS + H_2O$$

$$2CuS + 3O_2 ---> 2CuO + 2SO_2$$

During regeneration of an activated carbon bed, SO_2, CO, hydrocarbons, and steam are vented to the atmosphere. The carbon also adsorbs heavier hydrocarbons, which are removed and vented to the atmosphere along with the sulfur compounds in the regeneration gas. Hydrocarbon emissions may be high during regeneration, reaching 3.6 gm/kg of ammonia product. Volatile organic compounds (VOCs) in the form of hydrocarbons are health hazards and their emission must be monitored and controlled.

Activated carbon has limited application since not all sulfur compounds can be removed and the ill-smelling effluent can cause discomfort to people nearby. Air pollution caused due to the organic sulfur compounds, hydrocarbons, and CO contained in the regeneration gas can be controlled by incineration. However, emission of sulfur oxides must be controlled. Alternatively, the emissions from activated carbon regeneration can be eliminated more efficiently by substituting ZnO for the activated carbon desulfurization beds. The ZnO not only eliminates emissions, but saves energy and is a more effective desulfurization technique.

Iron Oxide.   In this process, H_2S present in the feed gas is reacted with iron oxide to form iron sulfide. The iron sulfide can be regenerated either in situ or outside in air. After oxidation of the sulfur in situ, the SO_2 is released to the atmosphere with some diluent gas, such as the primary reformer flue gas. Regeneration in air releases SO_2 slowly in the atmosphere. Emission of SO_2 in the atmosphere is very strictly under regulatory control. However, this is not a very efficient method since the sulfur levels acceptable for ammonia plants cannot be achieved and a second-stage desulfurizer must be provided. Disposing of the spent iron oxide also can pose a problem due to the large volumes involved.

Zinc Oxide.   New ammonia plants use zinc oxide (ZnO) beds to desulfurize entirely or provide the final step for the feedstock and operate as per the following reaction:

$ZnO + H_2S --> ZnS + H_2O$

The ZnO catalyst can absorb up to 20 wt% sulfur and is replaced rather than regenerated. Eliminating the regeneration step, lowers the energy consumption, and minimizes the environmental impact.

Hydrotreating Organic Sulfur Compounds. Many organic sulfur compounds and carbonyl sulfide (COS) are not removed by zinc oxide and must be hydrotreated to H_2S. Hydrotreating involves passing preheated feed gas over cobalt-molybdenum (CoMo) or nickel-molybdenum (NiMo) catalysts in the presence of hydrogen to convert the sulfur to H_2S, typically by the following reactions:

$C_2H_5SH + H_2 --> C_2H_6 + H_2S$
(mercaptan)

$C_4H_4S + 4H_2 --> C_4H_{10} + H_2S$
(thiophene)

If large amounts of COS are present in the feed gas, the COS is hydrolyzed in a separate vessel by the following reaction:

$COS + H_2O --> H_2S + CO_2$

The H_2S generated by hydrotreating or hydrolysis subsequently is removed by zinc oxide or incinerated.

Sulfur Removal by Organic or Inorganic Solvents. In case the natural gas feedstock contains a high level of sulfur, typically greater than 150 ppmv, it is usually more advantageous to remove the sulfur using a solvent followed by a final cleanup with ZnO. Solvents such as monoethanolamine (MEA), diethanolamine (DEA), and hot potassium carbonate solutions have been widely used in the industry. The solution is regenerated by steam, and the released sulfur compounds (mostly H_2S) are sent to the primary reformer for incineration. A very high sulfur level in the feed gas would require elemental sulfur recovery by a Claus-type process.

Chlorides. Chlorine and other halogens also are poisons to catalysts and must be removed to a level similar to sulfur. Chlorides may enter the pipelines when the pipeline is exposed to the atmosphere, or they may be left behind after hydrostatic testing. Chlorides are removed by catalysts and solvents used for sulfur removal.

Metals. Metals or organometallic compounds such as arsenic, copper, lead, and vanadium also are poisons to the primary reforming catalysts. Metals are present in trace quantities and are usually removed during normal desulfurization processes. If present, mercury is removed by adsorption on activated carbon at ambient temperature.

Steam Reforming

Steam reforming is a catalytic process for the conversion of light hydrocarbons and steam into hydrogen and carbon oxides. Most of the side reactions are retarded by the use of excess steam. First, the hydrocarbon feed is mixed with steam and passed over catalyst at a high temperature.

The limiting reactions for methane and the CO shift reaction:

$$CH_4 + H_2O \longleftrightarrow CO + 3H_2$$

$$CO + H_2O \longleftrightarrow CO_2 + H_2$$

$$CH_4 + 2H_2O \longleftrightarrow CO_2 + 4H_2$$

The net reaction is endothermic and is driven towards equilibrium in catalyst-packed tubes at high temperatures. Next, the partially reformed gas is passed to the secondary reformer, where the final conversion of methane takes place. Nitrogen for ammonia synthesis is provided by the injection of air into the secondary reformer. The oxygen in the air provides the necessary heat through combustion to convert the partially reformed gas.

Emissions from Primary Reformer. The primary reformer of feed gas is conventional combustion emission source for oxides of nitrogen (NO_x), carbon monoxide (CO), and particulate matter (PM).

NO_x: Source, Generation, Causes, and Contributing Factors. Nitrous oxides are produced in all combustion processes using air, primarily as nitric oxide (NO) in the hottest regions of the combustion zone. Some nitrogen dioxide (NO_2) also is formed, but its concentration amounts to 5 wt% or less of the total NO_x. Nitrous oxides in the combustion of flue gas is derived from two mechanisms: thermal NO_x and fuel NO_x. Thermal NO_x is formed when the temperature of the combustion zone is high enough to cause nitrogen in the air to react with oxygen to form NO_x. Fuel NO_x is derived from the oxidation of free or bound nitrogen in the fuel [11].

There is an inverse relationship between NO_x and CO formation; this plays an important role in emission control. In complete combustion, the carbon atoms in a hydrocarbon fuel are oxidized to carbon dioxide:

$$C + O_2 \longrightarrow CO_2$$

However, real-world combustion processes do not bring the time, temperature, and turbulence together perfectly to achieve complete combustion, and some amount of CO generation is inevitable. Generally speaking, the higher the peak combustion temperature, the lower the CO generation. Unfortunately, the trend is just the reverse for NO_x generation; the higher the combustion temperature, the greater the NO_x generation. Therefore, emission control for industrial sources must be a compromise between NO_x and CO control. The very same applies to the automobile catalytic converter whose primary function is to minimize the emission of CO, NO_x, and VOCs

Candidate NO_x Control Technologies. Several potential control technologies are available for reducing the NO_x emissions from the primary reformer [5,6]. The candidate technologies are:

1. Purge gas (fuel) treatment
2. Combustion equipment design
   a. Steam injection
   b. Staged-air low-NO_x burners
   c. Staged-fuel low-NO_x burners
3. Flue gas recirculation (FGR)
4. Selective catalytic reduction (SCR)
5. Selective noncatalytic reduction (SNCR) (e.g., Ammonia injection or Thermal $DeNO_x$)
6. Urea injection

Burners. The primary reformer operates at relatively low excess combustion air levels (10% design, 5–20% in practice), and it is not practical to reduce it significantly lower. However, combustion control by burner design is a NO_x control technique that has been demonstrated successfully on utility boilers and other stationary combustion sources. Favored "low-NO_x" techniques include staging the combustion air and staging the fuel. Staged fuel is preferred over staged air as the low NO_x technique for gas fuel burners. Staged air is the generally accepted approach for liquid fuel burners. By minimizing the peak flame temperature and controlling the nitrogen-oxygen contact in the hottest zones, NO_x formation could be reduced significantly.

The inherent drawback of any technique used to control NO_x emissions by lowering flame temperature is that CO emissions will increase. Using staged-fuel burners, the CO concentration in the flue gas is expected to be in the range 50–100 ppmv, values considered acceptable with combustion equipment designed for NO_x control [6].

Candidate CO Control Technologies: Catalytic Oxidation. The CO abatement unit uses a catalyst with a platinum group metal base to react the CO at high temperature with the oxygen present in the flue gas to produce carbon dioxide. The flue gas temperature range required for satisfactory operation is not as critical as with the SCR process. Temperatures in the range of 260–

650°C (500–1200°F) are acceptable depending on the catalyst substrate. Catalyst life is expected to be 3–5 yr.

Shift Conversion

In the shift conversion step, carbon monoxide reacts with steam to form equivalent amounts of hydrogen and carbon dioxide. Upon cooling of the effluent gas, most of the unreacted steam is condensed and separated as process condensate. Modern ammonia plants utilize a two-step, in-series shifting, carried out at high and then low temperatures to increase conversion efficiency. Use of the dual-shift conversion system lowers overall plant steam requirements, and the lower CO leakage results in reduction in plant feed requirements due to more complete conversion of CO to hydrogen. Under normal operating conditions there is no emission from the shift converters.

Carbon Dioxide Removal

Carbon dioxide removal in ammonia plants is usually accomplished by organic or inorganic solvents with suitable activators and corrosion inhibitors. In a few circumstances, CO_2 is removed by pressure swing adsorption (PSA) (see Chapter 3). The removed CO_2 is sometimes vented to the atmosphere, but in many instances it is recovered for the production of urea and dry ice. Urea is the primary use of carbon dioxide and, in case of a natural gas feed, all of the CO_2 is consumed by the urea plant. This practice is especially significant since CO_2 is a proven greenhouse gas. Typically, 1.3 tons of CO_2/ton of NH_3 is produced in a natural gas-based ammonia plant. The CO_2 vented to the atmosphere usually contains water vapor, dissolved gases from the absorber (e.g., H_2, N_2, CH_4, CO, Ar), traces of hydrocarbons, and traces of solvent. Water wash trays in the top of the stripper and double condensation of the overhead help to minimize the amount of entrained solvent. The solvent reclaimer contents are neutralized with caustic before disposal. Waste may be burned in an incinerator with an afterburner and a scrubber to control NO_x emissions.

The solvents themselves may be a nuisance and pose environmental problems, if released, contaminating ground water or nearby streams and rivers, adversely affecting vegetation growth and harming fish life. Therefore, the CO_2-removal section has to be well protected against accidental spills and leaks. In case they do occur, the paved and curbed ground surface allows collection of the solvent into a sump located underground. From there, the released solvent is recovered by filtration.

Carbon dioxide removal systems can be made environmentally acceptable by properly specifying the instrumentation, water wash of the overhead vapors, efficient use of demisters, routine control of the solvent physical properties and chemical composition, and care in startup and emergency shutdown situations. Industry has not been required yet to install emission control devices on carbon dioxide vent streams. There is a growing awareness of carbon dioxide generation and emission to atmosphere as a contributing factor to potential global warming. Any attempt to reduce significantly the CO_2 vented to the atmosphere would require a major shift in the nitrogen fixation process. The most effective way to reduce CO_2 production is by using natural gas raw material instead of heavy hydrocarbons and coal. Converting the ammonia to urea essentially eliminates CO_2 emission at the plant site.

Fugitive Emission Control

Fugitive emissions result from incomplete sealing of equipment at the point of interface of process fluid with the environment. Control techniques for equipment leaks include leak detection and repair programs, and equipment installation or configuration.

Pumps and Valves. Special types of equipment or equipment configurations can be used to eliminate or capture fugitive emissions from pump seals. There are leakless pumps: pumps that are designed with no interface between the process fluid and the environment, such as diaphragm pumps and canned motor pumps. Sophisticated pump seals also can be used to capture or eliminate fugitive emissions. Dual-seal systems, with barrier fluids at pressures greater than process pressures or low-pressure systems vented to control devices, may be used in some applications. Leak detection and repair programs also can be used but their efficiency depends on factors such as frequency of monitoring, effectiveness of maintenance, action level, and the underlying tendency to leak. Fugitive emissions from pressure relief valves may be virtually eliminated through the use of rupture disks to prevent leakage through the seal. Fugitive emissions may also be added to gases collected in a flare system by piping the relief valve to a flare header. Installation of caps, plugs, or blind flanges can reduce the leakage of vapors through block valves at the open end of pipes.

Compressors. Compressors are a source of emission of the compressed gas through seals and of the lubricating and seal oils. After separation from the seal oil, the seal gases can be vented, recycled to fuel, recovered back into the process, or sent to a flare. Usually, a mechanical separator and a carbon filter are provided when the seal gases are to be recovered. Dry seals used for new commercial centrifugal compressors greatly reduce gas emissions and eliminate the potential for oil emissions. Lubricated reciprocating compressors are equipped with oil traps and carbon filters in the discharge lines to capture oil carryover. The oil is usually mixed with water and in some cases contains toxic process gases. Therefore, they must be properly degassed and decanted. The oil can then be put into containers and taken to disposal sites either for incineration or for land fill. There is an increased emphasis on recycle, and there are a number of companies that can reprocess the waste oils.

Catalyst Disposal

The manufacture of most products from natural gas feed ultimately relies on a series of catalytically enhanced chemical reactions. For example, a typical 1000–metric ton/day ammonia plant has at least eight unit operations that make use of fixed-bed catalysts, with an overall catalyst volume of approximately 310 m^3. The catalyst operations vary in useful economic life from 2 to over 10 years, depending on service. Historically, all of the catalysts were disposed of in sanitary landfills, since they are basically inert and pose no environmental or health problems. Today, with stricter regulation of the nation's landfills, under the Resource Conservation and Recovery Act (RCRA), greater attention is being given to recycling of catalyst for recovery of the main metal components. Today, numerous, cost-effective processes exist to reclaim valuable metal components from spent catalyst. Complete separation of the spent catalyst into its component parts, and subsequent reuse in the industry, leaves no environmental liability.

Startup and Shutdown Emissions

The potential for substantial emissions can result from emergency shutdowns due to equipment failure, upsets in process conditions, human error, etc. Therefore, it is critical to perform a proper risk analysis on the plant and to provide, in the design phase, the necessary preventive or control sequence in case of an emergency. Instrument interlocks, emergency backup equipment (e.g., batteries), sparing of equipment, instrument voting systems, adequate system inventory, computer control, scrubbers, and so on, are all used to avoid or minimize emissions during an emergency.

B.   Greenhouse Effects Due To Methane
Conversion Processes

Greenhouse Gases and Climate Change

The past record of climate indicates that global temperatures have varied considerably. In fact, it is the general nature of climate to be continually changing. The concern about increases in atmospheric greenhouse gas concentrations is that global temperatures may reach levels over the next century that have not been experienced on Earth in the previous 120,000 years. Scenarios for assumed future emissions of greenhouse gases evaluated by the International Panel on Climate Change (IPCC) [23] suggest that the global mean temperature could increase by as much as 3°C before the end of the next century. Current projections suggest that the radiative equivalent to a doubling of the carbon dioxide (CO_2) concentration as a result of the combined effect of all greenhouse gases will occur within the next 40–70 years. Numerical presentations of climate processes, referred to as climate models, suggest that such a radiative forcing could lead to an increase in global average temperature of 1.5–4.5°C, once the oceans warmed and a new equilibrium climate was reestablished [29,32]. Such a rapid change would require adaptation unprecedented within human history.

The biospheric production of relatively small amounts of trace gases such as carbon dioxide (CO_2), methane (CH_4), and nitrous oxide (N_2O) is of special interest, as they trap infrared radiation. These greenhouse gases and other biogenic trace gases, such as carbon monoxide, nitrogen oxides (NO_x), and a range of volatile organic compounds play a crucially important role in atmospheric chemistry by affecting tropospheric concentrations of ozone [46,69].

Since the industrial revolution, carbon dioxide concentrations [65] have increased by 25 %. Figure 6.5 shows statistical information regarding the change in atmospheric CO_2 concentration. Figure 6.6 shows the main greenhouse gases and their anthropogenic sources [65]. Figure 6.7 shows the contribution from each of the human-made greenhouse gases to the changes in radioactive forcing from 1980 to 1990 [65]. Typical sources or key processes leading to emissions of the greenhouse gases are discussed next.

Fossil fuel burning is one of the key processes involving emissions of greenhouse gases. In this regard, combustion of coal and heavier hydrocarbons generates more CO_2 than combustion of natural gas. However, it should be noted that methane itself is a greenhouse gas. Industrial chlorofluorocarbons also are important sources for greenhouse gas emissions. The emission of hydrocarbons, CO, and NO_x from biomass burning during the dry season contributes to the greenhouse effect. Increased rice production also is linked to significant increase in CH_4 concentration. Increased N_2O emission is linked with increased application of N-fertilizers and tropical deforestation activities. It has been estimated that carbon dioxide will account for about half of the future temperature increases, while methane, nitrous oxide, ozone, and CFCs will be responsible for the rest (Figure 6.7). Therefore, minimized emission of CO_2 as well as the reutilization of process-produced CO_2, definitely will be an attractive thing to do [2,69].

The Earth's Energy Balance

As a means to understand the greenhouse effect, it is useful to examine the annual and global average radiative energy balance of the combined Earth-atmosphere system. The sun emits most of its energy between 0.2 and 0.4 µm, primarily in the ultraviolet (UV), visible, and near-infrared (IR) wavelength regions. A very small fraction of this energy is intercepted by the Earth

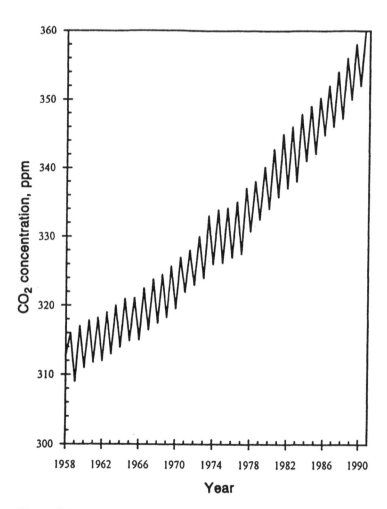

Figure 6.5 Growth in concentration of atmospheric CO_2 as measured at Mauna Loa, Hawaii. *Source*: [70].

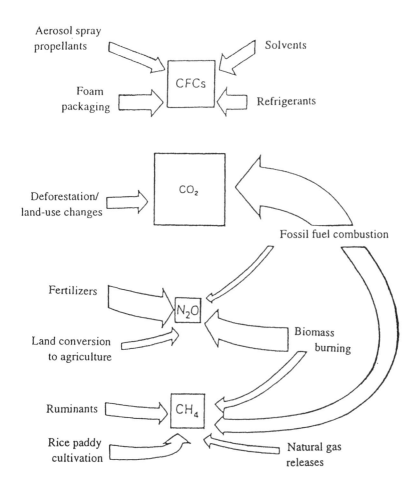

Figure 6.6   Greenhouse gases and their anthropogenic sources. *Source*: [65].

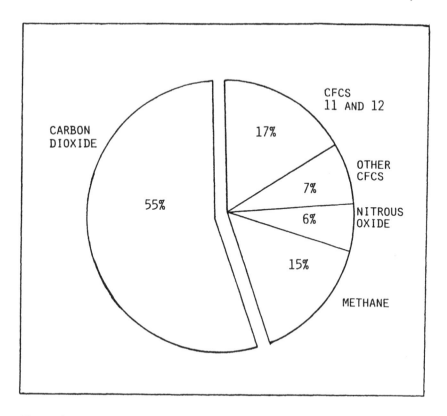

Figure 6.7 Contribution to human-made greenhouse gases to changes in radiative forcing from 1980 to 1990. *Source*: [65].

as it orbits the Sun. The atmosphere absorbs approximately 23% of the incoming solar radiation, principally by ozone (O_3) in the UV and visible ranges, and by water vapor in the near-IR. The Earth re-emits the energy it absorbs back to space in order to maintain an energy balance. Satellite measurements indicate that the incoming and outgoing radiation of the atmosphere are in balance [42].

Since the Earth is much colder than the Sun, the bulk of this emission takes place at longer wavelengths than those for incoming solar radiation. Most of this radiation is emitted in the wavelength range of 4–100 μm, which is the region generally referred to as longwave or IR radiation. Although water vapor, carbon dioxide (CO_2), and other greenhouse gases are relatively inefficient absorbers of solar radiation, these gases are strong absorbers of IR radiation. Clouds also play an important role in determining the energy balance. The greenhouse gases and clouds re-emit the absorbed longwave radiation based on their local atmospheric temperature, which tends to be cooler than the Earth's surface temperature. Some of this radiation reaches space. Some of the radiation, however, is emitted downwards, leading to a net trapping of longwave radiation and a warming of the surface. As the concentrations of greenhouse gases increase, this net trapping of IR radiation is enhanced.

Direct Radiative Influence

The contribution of a gas to the greenhouse effect depends on the wavelength at which the gas absorbs IR radiation, the concentration of the gas, the strength of the absorption per molecule (line strength), and whether or not other gases absorb at the same [37,38]. Gases absorb and emit radiation at wavelengths that correspond to transitions between discrete energy levels. Absorption at IR (greenhouse) wavelengths occurs for triatomic or larger molecules where vibrational and rotational energy transitions occur at appropriate wavelengths. Although each transition is associated with a discrete wavelength, the interval over which absorption occurs is "broadened" either by addition or removal of energy due to molecular collision (pressure broadening) or by the Doppler frequency shift that results from the random velocities of molecules (Doppler broadening). If absorption is strong, there may be complete absorption (saturation) around the central wavelength of the spectral line.

Climate Response

It is relatively easy to determine the direct radiative forcing effect on surface temperature due to the increases in emissions of greenhouse gases because the radiative properties of the greenhouse gases are reasonably well known [28,32,37,38]. The indirect effects resulting from atmospheric chemical processes and from aerosol formation are much more difficult to quantify. There also are many uncertainties remaining about the extent of the processes acting to amplify (through positive feedbacks) or reduce (through negative feedbacks) the expected warming from greenhouse gases. Some of the important feedbacks that have been identified are described next.

H_2O *Greenhouse Feedback.*   As the lower atmosphere (the troposphere) warms, it can hold more water vapor. The enhanced water vapor traps more IR radiation and amplifies the greenhouse effect. Ramanathan [36] indicates that, based on studies with one-dimensional climate models, this feedback amplifies the air temperature by a factor of about 1.5 and the surface warming by a factor of about 3. The IPCC [23] determined a surface temperature amplification factor of 1.6 for water vapor feedback.

Ice-Albedo Feedback.   Warming induced by the global greenhouse effect melts sea ice and snow cover. Whether it is ocean or land, the underlying surface is much darker (i.e., it has a lower albedo) than the ice or snow. Therefore, it absorbs more solar radiation, thus amplifying the initial warming. Ice-albedo feedback amplifies the global warming by 10–20%, with larger effects near sea-ice margins and in polar oceans [36].

Cloud Feedback.   Cloud feedback mechanisms are extremely complex and are still poorly understood. Changes in cloud type, amount, altitude, and water content can all affect the extent of the climatic feedback.

Ocean-Atmosphere Interactions.   The oceans influence the climate in two fundamentally important ways. First, because of the importance of the water-vapor greenhouse feedback, the air and land temperature responses are affected by the warming of the ocean surface. If the oceans do not respond to the greenhouse heating, the H_2O feedback would be turned off because increased evaporation from the warmer ocean is the primary source of increasing atmospheric water vapor. Second, oceans can sequester the radiative heating into the deeper layers which, because of their enormous heat capacity, can significantly delay the overall global warming effect over land surfaces.

Significance of Climate Change

The effect of changing composition of the atmosphere on climate is of significance to human and natural systems. The U.S. Department of Energy [71] identified 10 distinct categories: energy systems, agriculture, water resources, forestry, air quality, fisheries, sea level and coastal zone, infrastructure and human habitat, human health, and unmanaged terrestrial ecosystems. Potential consequences that could occur in four important areas during the course of the period to 2030 are summarized next.

Sea Level Rise.   Sea levels are expected to rise as a response to global warming, although the rate and timing remain uncertain. Most of the contribution to sea level rise is expected to derive from thermal expansion of the oceans and the increased melting of mountain glaciers. The prospect of such an increase in the rate of sea level rise is of concern to low-lying coasts. The most severe effects of the sea level rise may result from extreme events such as storm surges, the incidence of which may also be affected by climate change.

Effect on Human Health.   Direct effects on human health of the emitted greenhouse gases are believed to be small. Other links between climate change and human health have not been well established, although modest effects could result. For example, temperature increases could stress human health such as heat stress, but in cold areas they could reduce stress. Some diseases and pests could be more prevalent. However, good predictions of effects on health cannot be made without good predictive data on changes in local temperatures, humidities, and levels of precipitation. Stratospheric O_3 depletion could affect health because of the corresponding increase in UV radiation. In the next 50 years, the increases in emissions of greenhouse gases to the atmosphere are estimated to result in global warming by 0.3°C per decade [23].

Agriculture and Food Supplies. Increased atmospheric CO_2 could potentially enhance growth of some food crops as well as increase water use efficiency. However, changing climatic patterns could require changes in cropping patterns and consequently in infrastructures and costs, perhaps bringing benefits to some regions while negatively impacting others [28]. Rapid changes in climate or severe climate change could make adaptation more difficult.

Ecosystems. Ecosystem structure and species distribution are particularly sensitive to the rate of change of climate [23]. Ecosystems will respond to local changes in temperature, precipitation, and soil moisture, and also to the occurrence of extreme events. Increasing concentrations of CO_2 increase plant growth. However, rapid changes in climate threaten a reduction in biodiversity. Some existing species of plants and animals might be unable to adapt, being insufficiently mobile to migrate at the rate required for survival. Many uncertainties still remain about the adaptability of natural ecosystems to climate change.

III. REGULATION AND POLLUTION CONTROL

The coming decades will undoubtedly witness increased industrialization. As in the developed countries, the production of industrial byproducts and wastes will increase in the less industrialized countries. Based on the experience of the past few decades, increased emphasis will be placed on preventing and limiting environmental pollution. Having encountered the effects of environmental pollution earlier, the developed nations have been in the forefront in establishing pollution control laws and regulations covering proposed facilities, existing facilities, and decommissioned facilities.

We live in a new environmental era. The number and complexity of environmental regulations have grown dramatically in the past two decades and will continue through the turn of the century. In the United States, the first environmental regulation affecting the quality of air came with the Federal Air Pollution Control Act of 1955, then with the Clean Air Act of 1963. The subsequent ones include: the Motor Vehicle Air Pollution Control Act of 1965, the Air Quality Act of 1967, and the Clean Air Act Amendments of 1970.

The Clean Air Act Amendments of 1970 along with the National Environmental Policy Act, which created the Environmental Protection Agency (EPA), put some strong measures into air pollution control enforcement. The U.S. Environmental Protection Agency is responsible for implementation of the statutes of this law. The first air pollutants covered by this law of concern to the fuels industry were SO_x and NO_x. Since the passage of the Clean Air Act Amendments of 1970, a series of additional U.S. Federal environmental laws affecting various industries have been passed:

1. Federal Air Pollution Control Act (Clean Air Act), 1970
2. Federal Water Pollution Control Act (Clean Water Act), 1972
3. Safe Drinking Water Act, 1974
4. Energy Supply and Environmental Coordination Act (ESECA), 1974
5. Toxic Substances Control Act, 1976
6. Resource Conservation and Recovery Act (RCRA), 1977
7. Clean Air Act Amendments of 1977
8. Comprehensive Environmental Response, Compensation, and Liability Act (CERCLA), 1980
9. Pollution Prevention Act, 1990
10. Clean Air Act Amendments of 1990

All of these laws are technology based except the Safe Drinking Water Act, which is health based. The technology-based environmental laws set the limits for pollutant discharges on the basis of the best practicable control technology available at the time of enactment of the law. The health-based law sets limit on the concentration of specific chemicals in the water on the potential health hazard for which the water is used.

A. Federal Air Pollution Control Act

The Federal Air Pollution Control Act also is known as the Clean Air Act. This law, which was derived from the original Clean Air Act of 1955 and the Air Quality Act of 1967, was enacted in 1970, with amendments in 1977 and 1990. With this act, the U.S. Environmental Protection Agency (EPA) was required to establish national ambient air quality standards and was empowered to control hazardous air pollution emissions. Amendments to the act put into place a standardized basis for rules relating to new source performance standards and hazardous air pollution standards. The Clean Air Act regulates emissions of hazardous air pollutants. Title III of the Clean Air Act lists 189 pollutants, among which are included particulate matter, NO_x, HNO_3, HF, SO_2, H_2SO_4, radionucleides (including radon), several organics, etc., which are of concern to fuel (conversion) industries. Individual states in the United

States must implement this act using the federal standards as minimum, and can set their own emission standards for any or all species.

The Clean Air Act and its amendments contain over 800 pages, and additional detailed regulations will continue to be written and implemented for individual chemicals. Similarly, specific rules and regulations for individual states will evolve with time using the Clean Air Act as the minimum standard. In this act, toxins are considered to be at the "major source" threshold (subject to regulation) when the emission rate is 10 tons/yr or higher for one of the listed pollutants or at 25 tons/yr or higher for any combination of pollutants. Regulation of the emissions will occur automatically for new facilities through the state permit systems. Existing facilities have three years to reach compliance after the date that relevant air toxic regulations become effective. To avoid the requirements for obtaining operating permits, plants that are not major sources must document that they are "nonmajor sources" of air pollution. Existing facilities can be prepared for the new permitting process by verifying compliance with existing permits on a periodic basis, collecting the data and the information needed for the new permits, or reducing emissions to levels below the major source threshold [54,56,59].

B. Federal Water Pollution Control Act

The Federal Water Pollution Control Act of 1972 and its amendments, the Clean Water Act of 1977 and the Water Quality Act of 1987, focused on improving water quality standards in the United States [54]. These requirements are based on health standards intended for various water use categories, which are:

1. Water for full-contact recreational use
2. Water for support of fish and wildlife
3. Public water supply
4. Water for agricultural and industrial use

Both direct and indirect discharges of water are covered by this act. Point sources regulated in the fuel industries include water treatment plants, process waste streams, liquid discharges from cooling towers, and sewage discharges. Discharge of any pollutant into U.S. waters from point sources is prohibited by this act unless the discharge is authorized in a permit from the National Pollution Discharge Elimination System (NPDES). The NPDES permit specifies required monitoring, control and reporting, and allowable concentrations of pollutants in the discharge [54].

The Clean Water Act allows individual states to operate the NPDES permit system instead of the federal EPA. Again the individual state requirements must meet the federal requirements as a minimum. At the present time, about 80% of the states issue NPDES permits, and the EPA issues permits for the remaining states. In addition to the industrial water discharges, the Clean Water Act covers storm water discharges from industrial areas where material handling equipment or activities, raw materials, intermediate products, final products, waste materials, byproducts, or industrial machinery are exposed to storm water.

Effluent guidelines and standards exist in the EPA NPDES Guidance Manual [72] for a large number of industries. A permit is required under the Clean Water Act for discharge of process wastewaters from new sources and from new dischargers. In the permit application, toxic pollutants and hazardous substances must be identified that are present in the process wastewater discharge. Examples of the chemicals that may be found in fuels-processing water discharges that must be identified and quantified include nitrogen dioxide, sulfuric acid, and polycyclic aromatic hydrocarbons (PAH), To meet the standards for a permit, information is required on the quantities, treatment facilities, and processing for the listed chemicals discharged in the process wastewater.

C. Safe Drinking Water Act

The Safe Drinking Water Act (SDWA) of 1974 had health-based standards for setting the maximum contaminant levels in water used for the public water supply. The act contains requirements for monitoring, reporting, and public notification. The standards in this act also are used as reference points for the protection and remediation of water resources under several other EPA programs, such as the Clean Water Act, and as standards for state programs. Although the SDWA is oriented primarily toward public water supply systems, industry must take proper precautions to prevent contamination of groundwater and surface waters that serve as a source of public drinking water.

Potential liability can result from not having an effective environmental management system in the process industries. Severe penalties and very heavy liability can result from many different contaminant sources. A summary of some of the major sources of potential water contamination from the process industries follows:

1. Continuous or periodic emissions, spills and leaks
2. Wastes and byproducts deposited in landfills
3. Wastes in surface containment facilities, such as ponds, pits, and lagoons
4. Large accidental releases of intermediates, products, or waste products
5. Open storage of product or waste products
6. Improper waste management by another party of off-site waste material from a production facility

Later amendments to the SDWA listed additional contaminants with standards on the maximum contaminant level in drinking water. The chemicals listed as contaminants include organics, inorganics, metals, herbicides, and pesticides. Some of the chemicals encountered in the fuel industry on the SDWA contaminant list are nitrates, fluorides, sulfates, zinc, cadmium, radon, and materials causing turbidity in water.

D. Toxic Substances Control Act

The Toxic Substances Control Act (TSCA) of 1976 covers organic and inorganic substances, including combinations that have been determined to be a risk to human health or toxic in the environment. An initial inventory of chemical substances was made by requiring industry to report all chemical substances produced or imported in 1977. This act also controls newly manufactured chemicals. The act requires a premanufacture notification of any new chemical or the reporting of any proposed significant new use of the chemical. Import certifications are required for chemicals and mixtures under this act. The inventory of chemical substances was updated in 1986 and 1990 and will continue being updated at 4–year intervals. The TSCA requires special reporting and record-keeping requirements for specific chemical compounds. Chemical compounds that require special reporting and record keeping are listed by the EPA in the publications identified as 40–CFR-704, GPO#055–000–00254–1, and GPO#055–000–00361–1.

E. Resource Conservation and Recovery Act

The Resource Conservation and Recovery Act (RCRA) of 1976 was enacted to establish a federal waste management system for hazardous waste. It also authorized a program for regulation of hazardous waste that included inspection and enforcement authority. The RCRA has since been amended several times. Wastes covered by the RCRA now include waste types categorized as hazardous wastes, municipal solid wastes, special large-volume wastes, and nonhazardous industrial solid wastes [55]. Recent emphasis of the RCRA waste management philosophy has been to favor pollution prevention, recycling, and waste treatment. Criminal penalties exist for noncompliance with the RCRA. Individual states are required to compile and maintain an inventory of active and inactive waste sites within the states. The RCRA is one of the major congressional acts that has resulted in much detailed regulation, penalty provision, and high liability provisions for contaminated sites.

The RCRA definition of "hazardous waste" has been most frequently used in the field. It defines "a solid waste, or combination of solid wastes, which because of its quantity, concentration, or physical, chemical, or infectious characteristics may (1) cause, or significantly contribute to, an increase in mortality or an increase in serious irreversible or incapacitating reversible illness or, (2) pose a substantial present or potential hazard to human health or the environment when improperly treated, stored, transported, or disposed of or otherwise managed."

F. Comprehensive Environmental Response, Compensation, and Liability Act

The Comprehensive Environmental Response, Compensation, and Liability Act (CERCLA) enacted in 1980 had as its goal responding to and cleaning up sites from past hazardous waste activities and responding to current waste releases. The CERCLA also is known as the "Superfund" act since part of this act includes regulations, penalty provisions, and liability for contaminated-site cleanup. The CERCLA specifies that responsible parties may be liable for government response costs or any other necessary costs incurred in the event of a hazardous substance release or threatened release from a facility. The definition of a hazardous substance regulated by the CERCLA is listed under the CAA, CWA, TSCA, or RCRA acts described earlier. In case of an accident or emergency, release of a hazardous substance above the reportable quantities level requires reporting the release to the National Response Center. Reportable quantities are specified in 40–CFR Part 302 of the CERCLA. For example, the accidental collapse of a gypsum pond wall could release fluids to a river. The chemical composition and quantity of fluid released would be covered under the reportable quantities provisions. In reporting an emergency release, the following information is required:

1. Location of release
2. Description of material released
3. Quantity of material released
4. Time and date of release

As part of the CERCLA, the Superfund Amendments and Reauthorization Act (SARA) of 1987, companies were required to report on the fate of more than 300 chemicals. This information is reported publicly and the numbers are totaled for individual chemicals on a nationwide basis. This Toxics Release Inventory (TRI) has been used as a basis for measuring progress in reducing pollution. The EPA has helped to ensure the accuracy of TRI reporting through thousands of plant site visits and fines totaling millions of dollars. Since 1987, total releases of these 300+ chemicals have been reported. In the 25 chemicals of highest tonnage release, chemicals encountered in the fuels industry include ammonia, ammonium nitrate, ammonium sulfate, nitric acid, phosphoric acid, and sulfuric acid. Based on data available for 1990, a significant reduction of 44% in the release rate of the top 25 chemicals was achieved in the period 1987–1990. This reduction is even more significant when considering the 11% increase in total production of these chemicals in the same period. One effect

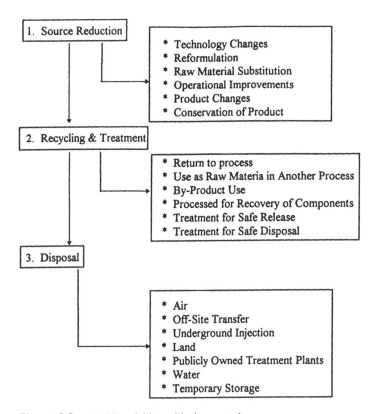

Figure 6.8   Multilevel hierarchical approach.

of the 1987 toxic chemical release requirements has been a strong focus by industry on prevention of pollution by various methods of reducing wastes produced in chemical facilities. These methods of pollution prevention include process changes, equipment changes, chemical substitutions, and product changes.

G. Pollution Prevention Act

The Pollution Prevention Act (PPA) of 1990 was enacted to orient pollution control more in the direction of pollution prevention and in reducing pollution at the source compared with previous legislation, which focused on waste that had already been produced. The U.S. EPA delineated a multilevel hierarchical approach to waste management in the PPA. The objective of this hierarchy is to give priority first to source reduction, second to recycling and treatment, and last to waste disposal. This management hierarchy is shown in Figure 6.8.

The pollution philosophy has been promoted by the EPA by starting an Office of Pollution Prevention. This office has now been merged with the EPA Toxic Substances Office. In this philosophy, the best waste management system is believed to be that which avoids the generation of waste initially rather than one that reacts to waste production through recycling, waste treatment, and disposal. In many cases, the prevention of pollution can be more profitable than other methods of waste management. Economic incentives for pollution prevention include higher product production efficiency, better resource conservation, less legal liability, lower cost of waste product management, and less stringent regulatory levels. Pollution prevention results will become more and more subject to measurement on both the national scale and on an individual company basis with the more recent waste reporting requirements of the EPA Toxic Release Inventory and the Clean Air Act.

REFERENCES

1. Adams, C. E., Jr., Removing nitrogen from wastewater. *Environ. Sci. Technol.*, 7(8): 696–701 (1973).
2. Al-Najjar, I. M., *CFC's Symposium: Phase out Chlorofluorocarbons*, Chambers of Commerce and Industry, Dammam, Saudi Arabia. No. 24, 1992, pp. 398–441.
3. Anderson, E., Lead cut gives alcohols crack at gasoline blend market. *C&EN*, April 8: 17–18 (1985).
4. Asplund, G. and Grimvall, A. ES&T features: Organohalogens in nature. *Environ. Sci. Technol.*, 25(8): 1346–1350 (1991).
5. Bartok, W., Crawford, A. R., and Skopp, A. Control of NO emissions from stationary sources. *Chem. Eng. Prog.*, 67(2): 64–76 (1971).
6. Bentley, K. M. and Jelinek, S. F., NO_x control technology for boilers fired with natural gas or oil, *Tappi J.* 72: 4 (1989).
7. Browning, J. E., Ash—The usable waste. *Chem. Eng.*, 80: 68–70 (1973).
8. Calvert, J. G., Hydrocarbon involvement in photochemical smog formation in los angeles atmosphere. *Env. Sci. & Technol.*, 10(3): 256–262 (1976).
9. Calvert, J. G., Test of the theory of ozone generation in los angeles atmosphere. *Env. Sci. & Technol.*, 10(3): 248–255 (1976).
10. Chakravarti, D. and Dranoff, J. S., Photochlorination of methane in a two-zone photoreactor. *AIChE J.*, 30(6): 986–988 (1984).
11. Chameides, W. L. and Stedman, D. H., Ozone formation from NO_x in "clean air". *Env. Sci. & Technol.*, 10(2): 150–153 (1976).
12. *Compilation of Air Pollutant Emission Factors*, AP-42, 4th Ed., Vol. 1, U.S. Environmental Protection Agency, Research Triangle Park, NC, Table 5.2–1, p. 1.4–3, 1985.
13. Czuppon, T. A. and Knez, S. A., Emerging technologies for the ammonia industry, *Fertilizer Focus*, Aug.: 11–15 (1991).

14. Czuppon, T. A., Knez, S. A., and Rovner, J. M., Ammonia, in *Kirk-Othmer Encyclopedia of Chemical Technology*, 4th Ed., (Kroschwitz, J. I. and Howe-Grant, M., Eds.), Wiley, New York, Vol. 2, p. 665, 1992.

15. Ecklund, E. E. and Mills, G. A., Alternative fuels: Progress and prospects, Part 1. *Chemtech*, Sept.: 549–556, (1989).

16. Fellows, W. D., Application of the thermal deNOx process to glass melting furnaces, *1989 Fall International Symposium*, American Flame Research Committee, International Flame Research Foundation, 1989.

17. Finlayson-Pitts, B. and Pitts, J. N., Jr., Volatile organic compounds: Ozone formation, alternative fuels and toxics. *Chemistry & Industry*, Oct. 18: 796–800 (1993).

18. Fox, J. M., III, Chen, T. P., and Degen, B. D., An evaluation of direct methane conversion processes. *Chem. Engr. Progr.*, April: 42–50 (1990).

19. Hangebrauck, R. P., von Lehmden, D. J., and Meeker, J. E. Emissions of poly-nuclear hydrocarbons and other pollutants from heat generation and incineration processes. *J. Air Pollut. Control Assn.*, 14(7): 267–278 (1964).

20. Hatch, L. F. and Matar, S., *Petrochemicals from Methane. From Hydrocarbons to Petrochemicals*, Gulf Publishing Co., Houston, 1981, p. 49.

21. Hodge, C. A. and Popovici, N. N. (Eds.), *Pollution Control in Fertilizer Production*, Marcel Dekker, Inc., New York, 1994, p. 37.

22. Hung, S. L. and Pfefferle, L. D., ES&T research: Methyl chloride and methylene chloride incineration in a catalytically stabilized thermal combustor. *Environ. Sci. Technol.*, 23(9): 1085–1091 (1989).

23. IPCC (Intergovernmental Panel on Climate Change). *Climate Change: The IPCC Scientific Assessment*, (Houghton, J. T., Jenkins, G. J., and Ephraums, J. J., Eds.), Cambridge University Press, Cambridge, MA, 1990.

24. Jorgensen, S. E. Recovery of ammonia from industrial wastewaters. *Water Res.*, 9(12): 1187–1191 (1975).

25. Katz, D. L. et al., Evaluation of coal conversion processes to provide clean fuels. Part II. Final report. EPRI 206–0–0. PB-234 203, Electric Power Research Institute, Palo Alto, CA, 1974.

26. Koon, J. H. and Kaufman, W. J., Ammonia removal from municipal wastewaters by ion exchange. *J. Water Pollut. Control Fed.*, 47(3): 448–465 (1975).

27. LeBlanc, J. R., Technical aspects of reducing energy consumption in new and existing ammonia plants, *Chem. Econ. and Engin. Rev.*, 18(5): 22–26 (May 1986).

28. MacCracken, M. C., *Greenhouse Warming: What Do We Know?*, UCRL-99998, Lawrence Livermore National Laboratory, Livermore, CA, 1988.

29. MacCracken, M. C. and Luther, F. M. (Eds.). *Projecting the Climatic Effects of Increasing Carbon Dioxide*, DOE/ER-0237, U.S. Department of Energy, Washington, D.C.; available from the National Technical Information Service, U.S. Department of Commerce, Springfield, VA, 1985.

30. Mayland, B. J. and Heinze, R. C. Continuous catalytic absorption for NOx emission control. *Chem. Eng. Prog.*, 69(5): 75–76 (1973).

31. Mercer, B. W. et al., Ammonia removal from secondary effluents by selective ion exchange. *J. Water Pollut. Control Fed.*, 42(2): R95–R107, 1970.

32. *Changing Climate: Report of the Carbon Dioxide Assessment Committee*, National Academy of Science Press, Washington, D.C., 1983.

33. *Nitrogen Oxides Emissions and Controls*, EPA/600/7–88/015, U.S. Environmental Protection Agency, Research Triangle Park, NC, 1988.

34. Offen, G. R. et al., Stationary combustion NO_x control, *JAPCA*, 37: 864 (1987).

35. Prezelj, M., Pool octanes via oxygenates: Gasoline pool octane-volume contributions and process reviews are given for methanol, ethanol, isopropanol, tert-butanol, sec-butanol, MTBE, and TAME. *Hydrocarbon Processing*, Sept.: 68–70, (1987).

36. Ramanathan, V. The greenhouse theory of climate change: A test by inadvertent global experiment, *Science*, 240: 293–299 (1988).

37. Ramanathan, V. et al., Climate-chemical interactions and effects of changing atmospheric trace gases, *Rev. Geophys.*, 25: 1441–1482 (1987).

38. Ramanathan, V. et al., Trace gas trends and their potential role in climate change, *J. Geophys. Res.*, 90: 5547–5566 (1985).

39. Rawlings, G. D. and Reznik, R. B., *Source Assessment: Synthetic Ammonia Production*, EPA-600/2–77–107, NTIS PB276718/AS, U.S. Environmental Protection Agency, Research Triangle Park, NC, 1977.

40. *Recovery of Spent Catalysts*, Division of Petroleum Chemistry Symposia, American Chemical Society, Kansas City, MO, Vol. 27(3), Sept. 12–17, 1982.

41. Romero, C. J. et al., *Treatment of Ammonia Plant Process Condensate*, EPA-600/2–77–200, NTIS PB273069/AS, U.S. Environmental Protection Agency, Research Triangle Park, NC, 1977.

42. Shine, K. P., The Greenhouse Effect, in *Ozone Depletion: Health and Environmental Consequences*, (Jones, R. R. and Wigley, T., Eds.), John Wiley & Sons, New York, 1989.

43. Smith, E. P., Considerations in determining O_2 and CO control for combustion efficiency and quality, *Industrial Heating*, 54: 47 (1987).

44. *Source Category Survey: Ammonia Manufacturing Industry*, EPA-450/3–80–014, U.S. Environmental Protection Agency, Research Triangle Park, NC, 1980.

45. Stasiuk, W. N., Hetling, L. J., and Shuster, W. W., Nitrogen removal by catalyst-aided breakpoint chlorination. *J. Water Pollut. Control Fed.*, 46(8): 1974–1983 (1974).

46. Stern, A. C. et al., *Fundamentals of Air Pollution*, 2nd Ed., Academic Press, New York, 1984, pp. 86–88.

47. Stevenson, R. M., *Introduction to the Chemical Process Industries*, Reinhold Publishing Corporation, New York, 1966, p. 293.

48. Twigg, M. V., *Catalyst Handbook*, 2nd ed., Wolfe Publishing, Prescott, AZ, 1989, pp. 182–190.

49. Valenti, M., *Alternative Fuels: Paving the Way to Energy Independence. Mech. Engin.*, 113: 42–46 (1991).

50. Waibel, R. T. et al. *Advanced burner technology for stringent NO_x regulations*, API Midyear Refining Meeting, Orlando, FL, May 1990, American Petroleum Institute, New York.

51. BNA Editorial Staff, *U.S. Environmental Laws*, Bureau of National Affairs, Inc., Washington, D.C., (1986).

52. Canter, L. and Know, R., *Ground Water Pollution Control*, Lewis Publishers, Chelsea, MI, 1986.

53. Deutsch, S. and Tarlock, A. (Eds.), *Land Use and Environmental Law Reviewed 1991*, Clark Boardman, Callaghan, NY, 1991.

54. *Environmental Compliance Manual: A guide to Pollution Control Regulations*, J. Keller & Associates, Washington, D.C., 1991.

55. *EPA Informational Briefing on RCRA Reauthorization*, May 1991, U.S. Environmental Protection Agency, Washington, D.C., 1991.

56. EPA Publication 20z-1011, U.S. Environmental Protection Agency, Washington, D.C., September, 1990.

57. EPA Publication 9200.5–008H, U.S. Environmental Protection Agency, Washington, D.C., September, 1990.

58. *GII Environmental Statutes* (various editions 1982–1987), Government Institutes, Inc., Rockville, MD, 1987.

59. Hall, R., Environmental management, the viewpoint of the regulator at the national level, *Phosphorus in the Environment Workshop*, Tampa, FL, March 25–27, 1992.

60. Jessup, D., *Guide to State Environmental Programs*, The Bureau of National Affairs, Washington, D.C., 1988.

61. Lowry, G. and Lowry, R., *Lowry's Handbook of Right-To-Know and Emergency Planning*, Lewis Publishers, Chelsea, MI, 1988.

62. *NAP Ground Water Quality Protection; State and Local Strategies*, National Academy of Science Press, Washington, D.C., 1986.

63. C&EN editorial staff, Record year for environmental fines, *C&EN*, 70(46): 12 (Nov. 16, 1992).

64. Rona, D., *Environmental Permits: United States*, Van Nostrand Reinhold, New York, 1988.

65. Rosswall, T., Greenhouse Gases and Global Change: International Collaboration, *Environ. Sci. Technol.*, 25(4): 567–572 (1991).

66. Selmi, D. and Manaster, K., *State Environmental Law*, Clark Boardman, Callaghan, NY, 1991.

67. Thayer, A., Pollution reduction, *C&EN*, 70(46): 22–51 (1992).

68. Wright, J., *Managing Hazardous Wastes: A Programmatic Approach*, The Council of State Governments, Lexington, KY, (1986).

69. Wentz, C., *Hazardous Waste Management*, McGraw Hill, New York, (1990).

70. *World Nitrogen Survey*, World Bank Technical Paper 174, Washington, D.C., 1992, p. 37.

71. Roback, E., U.S. Department of Energy Risk Assessment Methodology, National Institute of Standards and Technology, Gaithersburg, MD, 1990.

72. EPA Guidance Manual for Facilities that Generate, Treat, Store, and Dispose of Hazardous Waste, Government Institute, Rockville, MD, 1994.

Index